Stereolithography

Paulo Jorge Bártolo
Editor

Stereolithography

Materials, Processes and Applications

Springer

Editor
Paulo Jorge Bártolo
Centre for Rapid and Sustainable Product
Polytechnic Institute of Leiria
Leiria, Portugal
pbartolo@estg.iplei.pt

ISBN 978-0-387-92903-3 e-ISBN 978-0-387-92904-0
DOI 10.1007/978-0-387-92904-0
Springer New York Dordrecht Heidelberg London

Library of Congress Control Number: 2011921342

Printed on acid-free paper

Springer is part of Springer Science+Business Media (www.springer.com)

Preface

Stereolithography is one of the most popular Additive Manufacturing processes, involving the solidification of a liquid photosensitive polymer by a laser beam scanned across its surface.

The use of stereolithography is still in its infancy. Much of the technology and operating procedures are based on empirical correlations and work experience, and very little is known about the physical and chemical changes occurring in the material due to light irradiation. A good understanding of the curing process is an important factor to improve the precision and quality of the models, as well as to develop well-adapted polymeric systems.

This book provides the current state of the art on stereolithographic processes covering aspects related to the most recent advances in the field, in terms of fabrication processes (two-photon polymerisation, micro-stereolithography, infrared stereolithography and stereo-thermal-lithography), materials (novel resins, hydrogels for medical applications and highly reinforced resins with ceramics and metals), computer simulation and applications.

I am deeply grateful to all the contributing authors.

Leiria, Portugal — Paulo Jorge Bártolo

Contents

1 **Stereolithographic Processes** 1
Paulo Jorge Bártolo

2 **History of Stereolithographic Processes** 37
Paulo Jorge Bártolo and Ian Gibson

3 **Infrared Laser Stereolithography** 57
André Luiz Jardini Munhoz and Rubens Maciel Filho

4 **Microstereolithography** 81
Arnaud Bertsch and Philippe Renaud

5 **Polymeric Materials for Rapid Manufacturing** 113
Fred J. Davis and Geoffrey R. Mitchell

6 **Materials for Stereolithography** 141
Serge Corbel, Olivier Dufaud, and Thibault Roques-Carmes

7 **Polymerizable Hydrogels for Rapid Prototyping: Chemistry, Photolithography, and Mechanical Properties** 161
Jurgen Stampfl and Robert Liska

8 **Simulation Methods for Stereolithography** 183
Benay Sager and David W. Rosen

9 **Theoretical and Modeling Aspects of Curing Reactions** 209
Paulo Jorge Bártolo

10 **Injection Molding Applications** 243
Russell Harris

11 Photonic and Biomedical Applications of the Two-Photon Polymerization Technique 257
Aleksandr Ovsianikov, Maria Farsari, and Boris N. Chichkov

12 Practical Use of Hydrogels in Stereolithography for Tissue Engineering Applications 299
Karina Arcaute, Brenda K. Mann, and Ryan B. Wicker

Index 333

Contributors

Karina Arcaute
W.M. Keck Center for 3D Innovation, University of Texas at El Paso,
El Paso, TX, USA

Paulo Jorge Bártolo
Centre for Rapid and Sustainable Product,
Polytechnic Institute of Leiria, Leiria, Portugal
pbartolo@estg.iplei.pt

Arnaud Bertsch
Ecole Polytechnique Fédérale de Lausanne, Lausanne, Switzerland
arnaud.bertsch@epfl.ch

Boris N. Chichkov
Laser Zentrum Hannover, Hannover, Germany
and
Institute of Electronic Structures & Laser (IESL), Heraklion, Greece
b.chichkov@lzh.de

Serge Corbel
Laboratoire Réactions et Génie des Procédés, UPR 3349 CNRS,
Nancy-Université, 1, rue Grandville, 20451, 54001 Nancy Cedex, France
serge.corbel@ensic.inpl-nancy.fr

Fred J. Davis
Department of Chemistry, The University of Reading, Whiteknights, Reading,
RG6 6AF, UK

Olivier Dufaud
Département de Chimie Physique des Réactions, UMR 7630 CNRS-INPL,
Nancy-Université, 1, rue Grandville, 20451, 54001 Nancy Cedex, France

Maria Farsari
Institute of Electronic Structures & Laser (IESL), Heraklion, Greece

Rubens Maciel Filho
School of Chemical Engineering, State University of Campinas, Campinas, São Paulo, Brazil

Ian Gibson
Centre for Rapid and Sustainable Product Development, Polytechnic Institute of Leiria, Leiria, Portugal
and
Department of Mechanical Engineering, National University of Singapore, Kent Ridge, Singapore

Russell Harris
Loughborough University, Leicestershire, UK
R.A.Harris@lboro.ac.uk

André Luiz Jardini Munhoz
School of Chemical Engineering, State University of Campinas, Campinas, São Paulo, Brazil
jardini@feq.unicamp.br

Robert Liska
Vienna University of Technology, Vienna, Austria

Brenda K. Mann
Department of Bioengineering, University of Utah, Salt Lake City, UT, USA

Geoffrey R. Mitchell
Centre for Rapid and Sustainable Product Development, Polytechnic Institute of Leiria, Leiria, Portugal
geoffrey.mitchell@ipleiria.pt

Aleksandr Ovsianikov
Laser Zentrum Hannover, Hannover, Germany

Philippe Renaud
Ecole Polytechnique Fédérale de Lausanne, Lausanne, Switzerland

Thibault Roques-Carmes
Département de Chimie Physique des Réactions, UMR 7630 CNRS-INPL, Nancy-Université, 1, rue Grandville, 20451, 54001 Nancy Cedex, France

David W. Rosen
The Woodruff School of Mechanical Engineering, Georgia Institute of Technology, Atlanta, GA 30332-0405, USA
david.rosen@me.gatech.edu

Benay Sager
The Woodruff School of Mechanical Engineering,
Georgia Institute of Technology, Atlanta, GA 30332-0405, USA

Jurgen Stampfl
Vienna University of Technology, Vienna, Austria
jstampfl@pop.tuwien.ac.at

Ryan B. Wicker
W.M. Keck Center for 3D Innovation, University of Texas at El Paso,
El Paso, TX, USA
rwicker@utep.edu

Chapter 1
Stereolithographic Processes

Paulo Jorge Bártolo

1.1 Introduction

Over the last few decades many changes in technology, particularly in the areas of computers and information management, have been introduced into the manufacturing sector. In modern engineering processes, designers, engineers, and technicians commonly use various computer-assisted technologies to evaluate their products at each stage in the product development cycle [1–4]. These computer-assisted technologies comprise:

- Computer-aided design (CAD)
- Computer-aided manufacturing (CAM)
- Computer-aided engineering (CAE)

CAD is the technology concerned with the use of computer systems to assist in the creation and modification of a design [5–8]. Geometric modeling systems have advanced greatly in the last decade so that current CAD systems are now capable of handling sophisticated geometry and assembly of part models [9]. Moreover, computers now have huge memory capacity, high processing speed, and user-friendly CAD systems have been developed. Following these trends several examples of computer-assisted design software and three-dimensional (3D) animation software have been developed. These systems can be used for the normal activity of two-dimensional (2D) drawing or to build and manipulate the corresponding 3D representation. With these systems it is possible to generate photo-realistic images and render models according to elaborate shading algorithms [10–12].

Recently, the concept of "nD modeling" or "nD CAD" has been developed due to an increased interest of both the academia and industry [13]. nD CAD corresponds to the expansion of 3D CAD modeling with other design attributes such as time/scheduling, cost, energy, maintainability, etc., in an attempt to improve the

P.J. Bártolo (✉)
Centre for Rapid and Sustainable Product, Polytechnic Institute of Leiria, Leiria, Portugal
e-mail: pbartolo@estg.iplei.pt

P.J. Bártolo (ed.), *Stereolithography: Materials, Processes and Applications*,
DOI 10.1007/978-0-387-92904-0_1, © Springer Science+Business Media, LLC 2011

efficiency and effectiveness of the product design process [13–17]. This will help to improve the decision making process and the product design performance. Another important advancement in the CAD field is the concept of Haptic CAD. Haptic CAD requires a haptic device, which typically has a handle connected to sensors and actuators, and a computational system that interfaces with the device [18–23]. Haptic interface can provide force and/or torque feedback to the users [18, 19]. From a computational point of view, haptic interaction relates to all aspects of touch and body movement and the application of these senses to computer interaction. Therefore, haptic interfaces have the potential to increase the quality and capability of human–computer interaction by exploiting our sense of touch and ability to skilfully manipulate objects. These interfaces enable direct physical interactions with computer models providing a useful and intuitive augmentation to visual displays. For efficient and high-quality design of mechanical products, Kimura and Yamane [24] proposed a prototype system integrating appropriate CAD systems, simulation software tools and haptic devices (Fig. 1.1).

CAM is the technology concerned with the use of computer systems to plan, manage, and control manufacturing operations [6, 25, 26]. CAM systems have become possible through the development of computers, electronically controllable, accurate, versatile, and productive machine tools, and through the development of sophisticated software technologies. Through the use of CAM systems it is possible to generate numerical control (NC) instructions to control a machine based on geometric information from a CAD database [6, 25].

CAE is a technology concerned with the use of computer systems to analyze CAD geometry, allowing simulations of how the product will behave [6, 27]. Therefore, through a CAE system, the CAD geometry can be redefined and optimized if necessary. CAE tools are available for a wide range of analyses, such as: stress and strain evaluation, heat transfer analyses, magnetic field distribution, fluid dynamics, vibration and kinematics analyses, etc., [28–32]. Finite element analysis tools usually solve each of these types of problems, by transforming the physical problem being studied into a simplified model made up of interconnected elements.

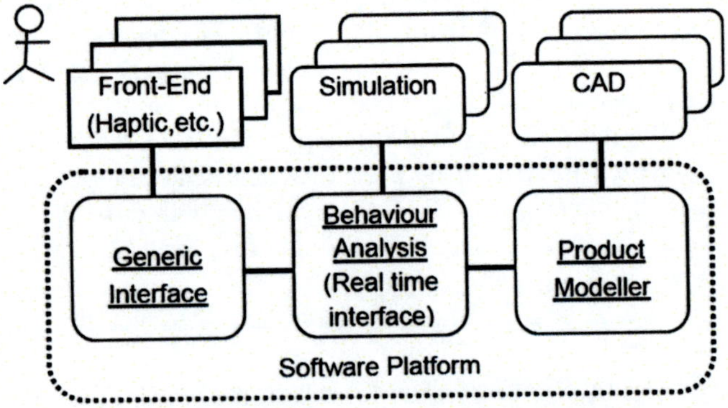

Fig. 1.1 Human interface design platform [24]

Besides these existing technologies, very sophisticated techniques have recently been developed for generating 3D objects. These techniques have been integrated into a new technology known as Additive Manufacturing processes, comprising different techniques using different materials as indicated in Figs. 1.2 and 1.3

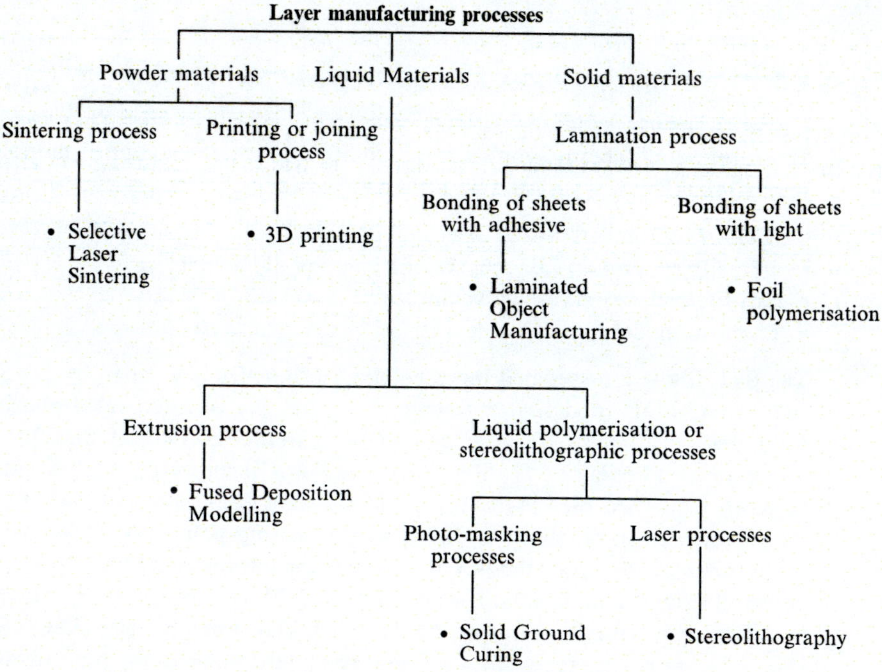

Fig. 1.2 Classification of additive manufacturing technologies [33]

solid name	The STL filename
facet normal n_i, n_j, n_k	The unit normal vector of facet
outer loop	
vertex v_{x1}, v_{y1}, v_{z1}	Coordinates of vertex 1
vertex v_{x2}, v_{y2}, v_{z2}	Coordinates of vertex 2
vertex v_{x3}, v_{y3}, v_{z3}	Coordinates of vertex 3
endloop	
endfacet	
endsolid name	

Fig. 1.3 The structure of an ASCII STL file format

[34–37]. However, all of these different techniques use the same principle, i.e. the transformation of a geometric CAD model into a physical model produced layer by layer. The main advantage of these processes is that they build a part, even a complex part, in one step without requiring planning of process sequences or specific equipment for handling materials [34–37].

1.2 Information Flow in Additive Manufacturing

The information chain common to all additive processes comprise the following steps [34, 38]:

- Solid model generation
- STL file generation (3D meshes of triangular elements)
- SLI file generation (model slicing)
- Physical model production

The STL file is a tessellated representation, originally developed by 3D Systems Inc., where a 3D model is represented by a number of three sided planar facets (triangles), each facet defining part of the external surface of the object [39]. There are two kinds of STL files: binary and ASCII files [10, 38]. The difference between these two files is the format of the data definition. A binary file stores the topological information in 32-bit single floating-point format defined by the IEEE standard 754, while the ASCII file stores the information as ASCII strings with keyword strings as indicators. The size of the ASCII STL format is larger than that of binary format, but is human readable [10, 38]. The structure of an ASCII STL file (Fig. 1.3), starts with the word *solid* followed by the name of the file and ends with the word *endsolid* [10, 38]. Between these two words, the triangles are defined through the specification of the facet normal and the vertices' coordinates.

The generation of the STL representation follows two important rules [40] shown in Fig. 1.4:

- Facet orientation rule: the facets define the surface of the 3D object. The orientation of the facet involves the definition of the vertices of each triangle in a counterclockwise order

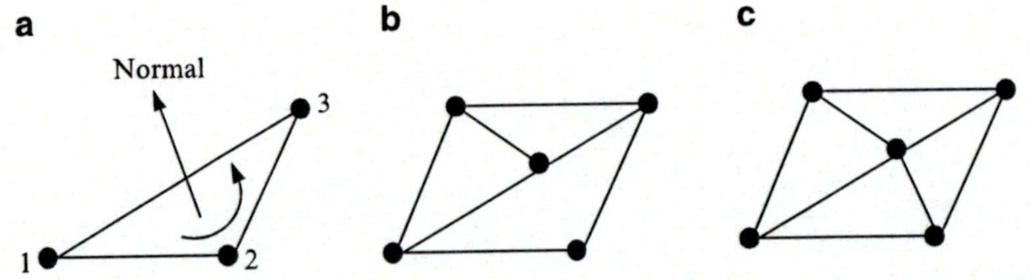

Fig. 1.4 **(a)** Orientation of a triangular facet in a STL file. **(b)** Violation of vertex-to-vertex rule. **(c)** Correct triangulation [10]

- Adjacency rule: each triangular facet must share two vertices with each of its adjacent triangles. For every facet edge, there must be one other facet sharing the same edge. Since the vertices of a facet are ordered, the direction on one facet's edge is exactly opposite to that of another facet sharing the same edge. This condition is known as the Möbius rule

To assure that the adjacency rule is obeyed it is necessary to verify the following consistency rules [10]:

- The number of triangles must be even
- The number of edges must be a multiple of three
- The number of triangles must be equal to two thirds of the number of edges
- The number of vertices is given by

$$V = 0.5 \times T + 2 \tag{1.1}$$

where, V is the number of vertices and T is the number of triangles.

To assure that the set of STL triangles comprises a closed volume, it is necessary to verify the Euler's rule for solids. According to this rule, the definition of a proper solid implies that [41]:

$$T - E + V - H = 2 \times (B - P) \tag{1.2}$$

with E being the number of edges, H the number of face holes, B the number of separate disjoint bodies, and P is the number of passages, i.e. holes through the entire body.

For a triangular polyhedron satisfying the adjacency rule, (1.2) is reduced to:

$$-1/2\,T + V = 2 \times (B - P) \tag{1.3}$$

The disadvantages of the use of a tessellated representation as a standard in Additive Manufacturing are the following ones:

- It is a first-order approximation of the original CAD model
- High degree of redundancy
- The user needs to input acceptable chordal tolerance (Fig. 1.5). Increasing the number of triangles smoothens the surface, but also leads to much larger data files

Slicing refers to intersecting a CAD model with a plane in order to determine 2D contours (Fig. 1.6). The slicing can either be uniform, where the layer thickness is kept constant, or adaptive, where the layer thickness changes based on the surface geometry of the CAD model [42, 43]. Adaptive slicing involves slicing with varying layer thickness. In this strategy, surfaces of high curvature are sliced with thinner layer thickness and surface of low curvature are sliced with thicker layer thickness. Adaptive slicing yields better surface quality, as the staircase effect decreases and the variations in the cusp height across the layers is minimized [43].

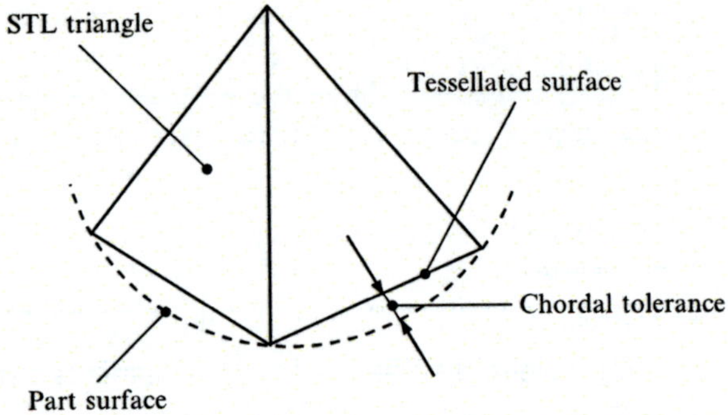

Fig. 1.5 Chordal error [10]

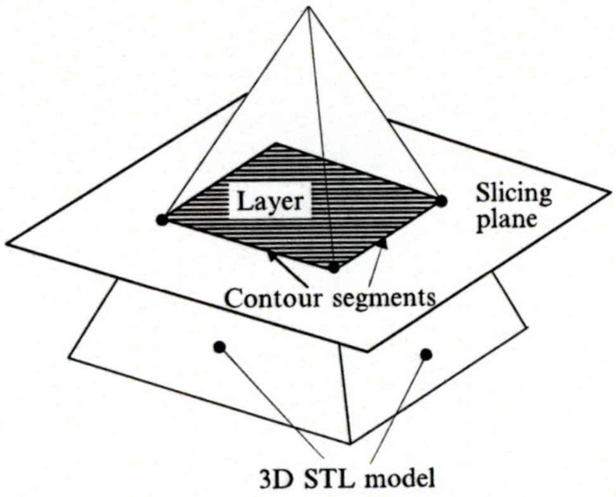

Fig. 1.6 Definition of a layer through the intersection between the 3D STL model and a slicing plane [10]

The main advantage of adaptive slicing is that it gives the user explicit control over the surface quality. Most adaptive slicing routines use a single cusp height for the whole model. This implies that all surfaces have the same surface finish requirement (Fig. 1.7).

1.3 Stereolithography

Stereolithography is one of the most important Additive Manufacturing technologies currently available. This technology involves the curing or solidification of a liquid photosensitive polymer through the use of an irradiation light source, which supplies the energy that is needed to induce a chemical reaction (curing reaction), bonding large numbers of small molecules, and forming a highly cross-linked polymer. Figure 1.8 identifies different stereolithographic ways to start the curing reaction.

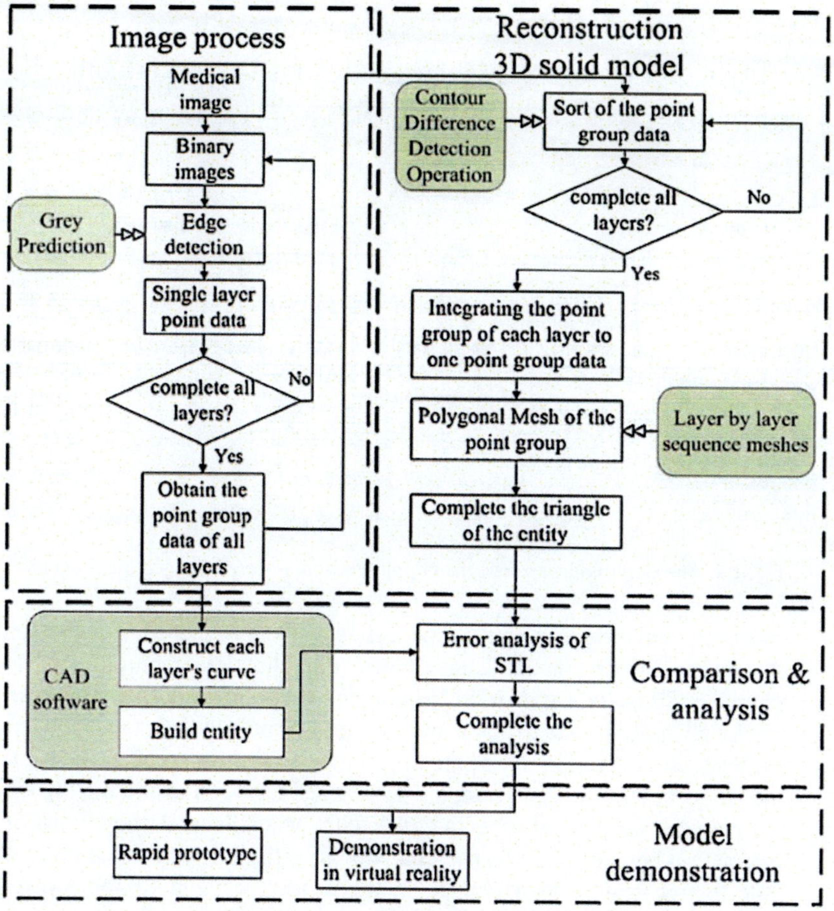

Fig. 1.7 Information flow to produce biomedical models using Additive Manufacturing [44]

1.3.1 Light-Polymer Interaction

Stereolithographic processes transform a multifunctional prepolymer into a cross-linked polymer through a chain reaction initiated by reactive species (free radicals or ions) generated by light exposure. Since most monomers or prepolymers do not produce initiating species upon irradiation, it is necessary to introduce low molecular weight organic molecules (initiators) that will start polymerisation, through photophysical and photochemical processes [45].

The first step in the light-polymer interaction is a photophysical process, which does not involve chemical changes, but only set electrons into motion [46]. When light is absorbed by any chromophore (functional groups which contain electrons originating from π and n orbitals), the photon must interact with an atom or group

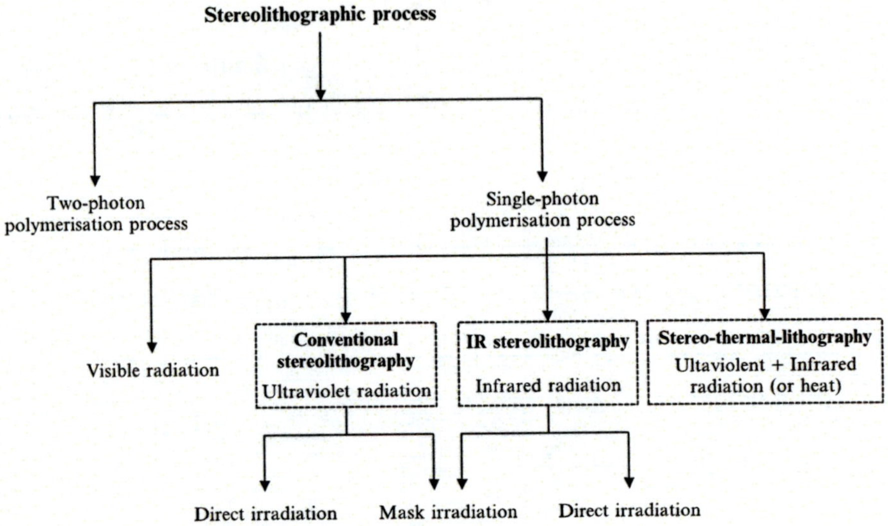

Fig. 1.8 Different modes to start the curing reaction in stereolithographic processes

of atoms and promote transitions between the ground state and excited states. According to the Franck–Condon principle the time required for absorption a photon and resultant transition to an excited state, is very short (10^{-5} s) [47].

For absorption to take place, two conditions must be satisfied. First, the energy E of the photon, given by $h\nu$ where h is Planck's constant (6.62×10^{-34} J s) and ν is the frequency of the radiation, must be at least as high as the difference in energy ΔE between the ground state and the excited state. Second, there must be a specific interaction between the electric component of the incident radiation and the molecule, which results in a change in the dipole moment of the molecule during the transition [47]. Light absorption is given by $\varepsilon\beta L$, where ε is the molar absorptivity or extinction coefficient, β is the concentration of the absorption species (photoinitiators), and L is the light-path length.

As the incident photon exchanges its energy with the molecule, a valence electron is promoted from the highest occupied molecular orbital to an unoccupied molecular orbital, with the formation of an excited singlet state molecule [45, 46, 48]. However, this excited molecule is a short living species (less than 10^{-8} s) that disappears by various competitive processes, dissipating the excited energy [45, 46, 48]. The absorption of light by a molecule and the subsequent evolution of its excited states are illustrated in Fig. 1.9 through the Jablonski energy diagram [34, 45, 49]. Two processes can be identified [34, 45, 49]:

- Photophysical processes:
  - Radiative
  - Nonradiative
- Photochemical processes.

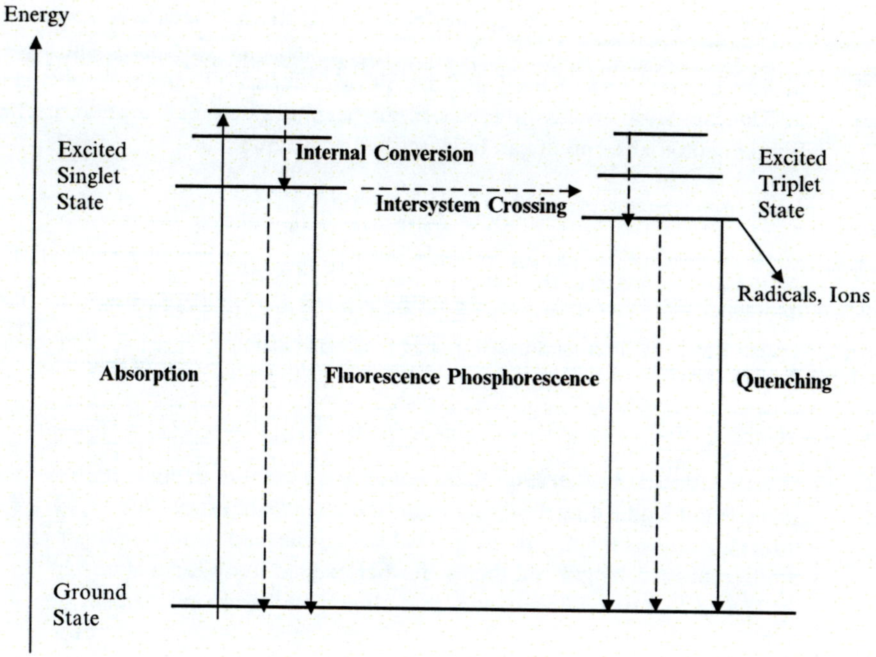

Fig. 1.9 The Jablonski energy diagram [34, 45, 49]

Radiative mechanisms involve the absorption of a photon by a molecule in its ground state (S_0) and the emission of a photon from an electronically excited state by either fluorescence, $S_1 \rightarrow S_0 + h\nu'$ (de-excitation of an excited state with the same spin multiplicity as the ground sate) or phosphorescence, $T_1 \rightarrow S_0 + h\nu''$ (de-excitation of an excited state with different spin multiplicity as the ground sate) [34, 45, 49]. Nonradiative processes include internal conversion (IC) and the intersystem crossing (ISC), which occurs among states of different spin multiplicity creating excited triplet states (T_1) [34, 45, 49].

The de-excitation of an excited state can also involve energy transfer processes to another molecule

$$S_1 + Q_s \rightarrow S_0 + Q_s^* \tag{1.4}$$

or

$$T_1 + Q_t \rightarrow S_0 + Q_t^* \tag{1.5}$$

where, Q_s and Q_t are quenchers of the excited singlet state and excited triplet state, respectively [50].

The photochemical process is a transformation of the starting molecule, through cleavage processes, electron transfer reactions, hydrogen abstraction, etc. Due to the short life of the excited singlet state molecules many photochemical reactions occur only via excited triplet states which are longer lived species (greater than 10^{-6} s) [45–50]. Singlet states or triplet states are electronic states where the molecule possesses all paired electrons or two unpaired electrons, respectively [50].

The excitation process involves single or multi-photon absorption mechanisms. Single photon absorption can be described as follows:

$$S_0 \xrightarrow{h\nu_1} S_1 \xrightarrow{\text{ISC}} T_1 \tag{1.6}$$

A bi-photonic system, the simplest example of multi-photon absorption, involves both sequential or a simultaneous processes. There are three fundamental types of sequential bi-photonic excitation. The first process is:

$$S_0 \xrightarrow{h\nu_1} S_{v/r} \xrightarrow{h\nu_2} S_1 \tag{1.7}$$

where, a molecule, in ground state S_0, absorbs a photon of relatively low energy $h\nu_1$ in order to reach a higher vibrational/rotational excited state. After that, the excited molecule returns quickly to the ground state or instead absorbs another photon with a different wavelength and then is further excited to a singlet state.

The second type of sequential two-photon excitation mechanism is:

$$S_0 \xrightarrow{h\nu_1} S_1 \xrightarrow{h\nu_2} S_n \tag{1.8}$$

According to this process a molecule, in ground state S_0, absorbs a photon, of higher energy than in previous case, in order to reach an excited singlet state. At this point the excited molecule is raised to a higher excited singlet state (S_n) by the second photon.

Finally the third type of excitation is described by:

$$S_0 \xrightarrow{h\nu_1} S_1 \xrightarrow{\text{ISC}} T_1 \xrightarrow{h\nu_2} T_n \tag{1.9}$$

In this case, a molecule absorbs a photon with an appropriate wavelength and forms an excited singlet state, which then undergoes intersystem crossing (ISC) to form the triplet state. This triplet state soon decays into the ground state if it is not further stimulated. However, if it is exposed to radiation of appropriate energy it is capable of forming higher excited triplet state which is highly reactive.

1.3.2 The Curing Mechanism

The curing (solidification) reaction of stereolithographic resins is an exothermic polymerisation process characterized by chemical cross-linking reactions that create an infusible, insoluble, and highly cross-linked 3D network [51]. This reaction is initiated by supplying an appropriate form of energy that depends on the adopted stereolithographic strategy. A generic representation of the cure process is indicated in Fig. 1.10.

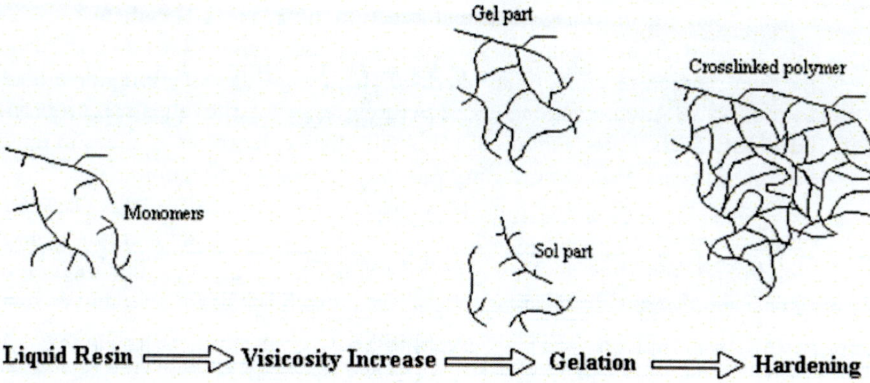

Fig. 1.10 The cure mechanism

Two main events occur during the curing reaction, independent of the stereolithographic strategy used to start the polymerisation process [52–60]:

- Gelation, i.e. liquid-to-rubber transition
- Vitrification, i.e. liquid or rubber-to-glass transition.

Gelation, described by Gilham [52] as a "chemical explosion," is a nonreversible event and corresponds to the incipient formation of an infinite molecular network (molecular gelation), which is associated with a dramatic increase of viscosity (macroscopic gelation) [52–54, 60]. At this point, the system will no longer flow and two phases coexist: a *gel phase* and a *sol phase* [57, 58]. The gel phase is the gelled part, insoluble in nondegrading solvents while the sol phase, which remains soluble, can be extracted with solvents [56–58]. Molecular gelation is defined as the incipient formation of branched molecules of infinite molecular weight, and occurs at a particular fractional conversion for each system [53–56, 58], which depends on the functionality, reactivity, and stoichiometry of the reactive species [54, 61, 62].

Macroscopic gelation is associated with a significant increase in viscosity and a corresponding decrease in processability [52, 54–58, 61]. The abrupt transformation, from a viscous liquid to an elastic gel or rubber, corresponds to the gel point and defines the onset of gelation [54]. As the reaction further progresses, the amount of sol phase presented in the system decreases and the polymer becomes more and more cross-linked [62]. Molecular mobility also decreases due to the restrictions imposed by the cross-linked structure [63].

Vitrification is a gradual, thermo-reversible process that corresponds to the formation of a glassy solid material due to an increase in both the cross-linking density and the molecular weight of the polymer being cured [52, 54, 64]. From vitrification, the rate of reaction will undergo a significant decrease and the reaction becomes very slow as it becomes controlled by the diffusion of the reactive species [65, 66]. This way, vitrification marks the change, from a reaction that is predominantly kinetically controlled into a reaction that is diffusion-controlled [65, 67]. In this regime, the time it takes for the reactive groups to find each other is much longer than the intrinsic chemical reaction time [63]. The diffusion-controlled effect, apart from causing a slow-down of the conversion reaction, determines the final degree of conversion obtained [65, 66]. This effect may also lead to the formation of nonhomogeneous structures, altering the physical properties and durability of the cured polymer [34]. Additionally, it creates changes in the specific volume of a material contributing this way to stress generation [68].

During the curing reaction, it is observed an increase in the glass transition temperature of the material, which is a consequence of the increase in molecular weight and cross-linking density [34]. Normally, it is expected that increased extent of cure will increase molecular weight and cross-linking, which removes free chain-ends and restrains polymer chain motions [69]. Therefore, the specific volume, which is the sum of free volume and occupied volume, is expected to decrease with increased extent of cure [69]. Consequently, it might be expected that the density and modulus of the material will increase as the conversion increases [70].

The kinetics of the curing process is strongly influenced by light intensity, temperature, and resin composition (e.g., concentration of initiator). Many authors [34, 71–73] have pointed out that photoinitiated cure reactions are characterized by high initiation rates. Consequently, during these reactions, the system cannot be in volume equilibrium because volume shrinkage is much slower than the rate of chemical reaction. This effect generates a temporary excess of free-volume, which increases the mobility of the reactive species in the system, inducing higher rates and conversions than the ones obtained with in volume equilibrium. Therefore, the higher the light intensity, temperature or initiator concentration, the higher the rate of gel formation (and the free-volume excess) that leads to higher conversions [34].

1.4 Conventional Stereolithography

1.4.1 Processes

Conventional stereolithography is a single-photon photo-fabrication process that builds shapes using ultraviolet light to selectively solidify photosensitive resins. There are two basic approaches (Fig. 1.11):

- Direct or laser writing
- Mask-based writing

These two approaches can also be classified into two types, free-surface and constrain-surface [74].

The direct or laser writing approach consists of a computer, a vat containing a photosensitive polymer, a moveable platform on which the model is built, a laser to irradiate and cure the polymer, and most currently a dynamic mirror system to direct the laser beam. The computer uses the sliced model information to

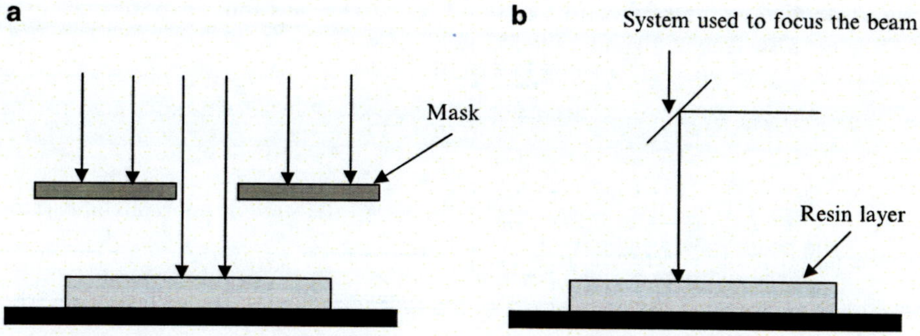

Fig. 1.11 Writing methods in stereolithographic processes. (a) Mask-based method. (b) Direct focused beam-writing method [34]

control the mirrors, which direct the laser beam over the polymer surface "writing" the cross-section of one slice of the model, by polymerisation of a set of elementary volumes called *voxels*. After drawing a layer, the platform dips into the polymer vat, leaving a thin film from which the next layer will be formed. The next layer is drawn after a wait period to recoat the surface of the previous layer.

The purpose of the recoating process is to cover the model with resin once it has been moved down by one layer thickness. However, the recoating process may give rise to undulations on the resin surface and surface bubbles that can be trapped inside the model during the curing process. The definition of the waiting time after the dipping/sweeping recoating process is an important parameter because too short time reduces the accuracy, causing poor bonding between layers, while too long waiting time increases the building process [75]. Moreover, surface tension often holds the liquid resin layer into a convex shape that requires significant time to reach equilibrium. This phenomenon is particularly important if the liquid resin has high viscosity. The time to reach equilibrium can be decreased, but not eliminated, by using low viscosity resins and/or by using a moving blade across the coated area. Finally, the building process is repeated until all layers have been drawn to form the 3D model.

As the UV radiation hits the surface of the liquid resin, it is dispersed and absorbed. This effect limits the penetration of the radiation and thus the depth of curing. This depth, known as the cure depth, is defined as follows:

$$C_{\mathrm{d}} = D_{\mathrm{p}} \ln\left[\frac{E_{\max}}{E_{\mathrm{c}}}\right] \tag{1.10}$$

where, C_{d} is the maximum cure depth, D_{d} is the penetration depth at which the beam intensity is reduced to $1/e$ of its surface value, $E_{\max}$ is the maximum exposure energy on the resin surface and E_{c} is critical energy required for the transition of the resin from the liquid phase to the solid phase. This way, the cure depth is a function of exposure and consequently, may be controlled by varying the radiation level applied to the surface of the liquid resin. The maximum exposure energy on the resin surface is given by the following expression [6, 7]:

$$E_{\max} = \sqrt{\frac{2}{\pi}}\frac{P}{w_0 v_s} \tag{1.11}$$

where, P is the laser power within the vat of the stereolithography apparatus, w_0 is the beam radius and v_s is the scanning speed.

The energy at a given position x on the resin surface is given by:

$$E(x) = E_{\max} \exp\left[\frac{-x^2}{w_0^2}\right] \tag{1.12}$$

At any point within the resin thickness, the energy is defined as:

$$E(x, y, z) = E(x, y, 0) \exp\left[\frac{-z}{D_P}\right] \tag{1.13}$$

In order to build a layer, the laser traces the shape of its outline and crosshatches in-between, according to the selected hatch style. The result is a model with a honeycomb interior and solid skins on top and bottom, encapsulating liquid resin. The hatch or build style is the process of solidifying the cross-section. The distance between parallel vectors used to hatch each layer defines the hatch spacing. If the hatch spacing is very small, the solidifying vectors will overlap, minimizing the amount of under-cured resin in the layer, so minimizing the need of postcure. Large hatch spacing allows the liquid polymer to be trapped inside the model. The polymerised line width is described as:

$$L_w = B\sqrt{\frac{C_d}{2D_p}} \tag{1.14}$$

where, B is the laser spot size.

Depending on the size of the layers and the orientation of the surfaces of the model regarding the laser beam, more or less plane and smooth surfaces can be obtained. Stair stepping effects are noticeable if the surface is not perpendicular to the laser beam and the layer thickness is higher. Several techniques and computational tools were proposed to reduce or eliminate the stair stepping effect either based on adaptive slicing or optimal building orientation and to optimize processing parameters [76–80]. Holzer and Fadel [81] proposed the use of a slicing algorithm based on a slope-slicing strategy and the modification of the stereolithography machine by increasing the number of degrees of freedom of the working platform.

Once produced 3D models must carry out a finishing process performed through a variety of operations such as glass bead blasting, sanding, milling, drilling, tapping, polishing, painting, etc.

Once the model is finished, it undergoes a series of post processes that make it safe to handle. The model is cleaned by removing the excess of resin, rinsing with water, and finally rinsing with ethyl alcohol, before removing the supports. The insufficient extent of cure obtained during the building process means that it is necessary to complete the cure in an ultraviolet chamber during the so-called postcure operation. Important consequences of insufficient polymerisation are warping due to relaxation, diffusion, and evaporation of low molecular weight components, and postcure shrinkage due to density changes associated with the postcure of the liquid resin trapped in the lattice structure.

Stereolithographic parts have limited functionality due to low strength and stiffness, poor creep performance, and environmental instability. However, their functionality can be enhanced by applying a layer of copper or nickel or their corresponding alloys [82, 83].

Shrinkage, which can cause internal stresses in the model, is another important problem associated with stereolithography and is a logical consequence of forming large molecules from small ones during the polymerisation process, which results in an increase in density [84]. Besides, the curing process of each layer produces flexure of the layers previously solidified, causing the phenomenon of curl distortion. Moreover, when the object is removed from the platform, additional distortions can occur owing to the liberation of internal forces that have been developed during the building process between the model and the platform.

Mask-based writing systems or surface curing stereolithographic processes build models by shining a flood lamp through a mask, which lets light pass through it. The exposure energy will start the curing process on the exposure area forming each cross-section of the 3D physical object. The first mask-based process commercially available was the Solid Ground Curing (SGC) developed by Pomerantz at Cubital [85]. The process included the following steps: photo-polymer spread, photo-mask generation, UV exposure, vacuum off unreacted resin, was spread, and milling. A mask generator produces a negative image of each cross-section on a glass plate. After irradiation, noncured resin is removed by vacuum cleaner and wax is spread to fill the gaps. Finally, the solidified wax is machined flat to provide the support for the next layer. SGC is no longer being sold.

Mask-based systems generally require the generation of a lot of masks with precise mask alignments. One solution for this problem is the use of a liquid crystal display (LCD), a Digital Micromirror Device (DMD) or a digital processing projection system as a flexible mask. The DMD technology presents several advantages regarding the LCD as indicated in Table 1.1.

1.4.2 Resin Compositions

Resin compositions used in conventional stereolithography comprise the following components: photoinitiators, polymerizable oligomers or prepolymers, a reactive diluent and additives.

Table 1.1 Comparison between LCD and DMD technologies

	LCD	DMD
UV compatibility	No	Yes
Modulation efficiency	12.5%	88%
Light transmission	21%	71%
Optical fill factor	64%	85%
Pitch size	26 × 24 μm	14–17 μm
Pixel size	33 × 33 μm	13–16.2 μm
Contrast	100:1	350:1
Switching speed	20 ms	20 μs

1.4.2.1 Photo initiators

The choice of the photoinitiator is of prime importance in stereolithography, since it directly governs the rate of cure. A suitable photoinitiator system must have high initiation efficiency, good solubility in the prepolymer, and storage stability. Additionally, the photoproducts produced by these initiators should not be colored or toxic or induce some degradation of the polymer upon aging.

The photoinitiators currently used can be classified into two major categories: radical or cationic initiators. There are also some initiators (iodonium and sulfonium salts, etc.) that are able to initiate curing reactions through both radical and cationic processes [76]. Photoinitiators are usually added in quantities from 0.5 to 12 wt% enabling cure depths between 0.1 and 2.5 mm.

Radical Initiators

A radical photoinitiator is used to absorb the incident radiation and then undergo processes leading to radical formation. These radical initiators must present a high absorption in the emission range of the laser. In addition, the excited states formed must both have a short lifetime to avoid quenching by oxygen or the prepolymer, splitting into reactive radicals species with the highest possible quantum yield.

There are two main types of radical initiators commonly used in stereolithographic applications [86–88]:

- Initiators that promote radical polymerisation by photo-cleavage (type I) such as, benzoin ether derivatives, benzyl ketal, hydroxyalkylphenones, α-amino ketones, acylphosphine oxides, etc.
- Initiators that promote radical polymerisation by hydrogen abstraction (type II) such as, benzophenone, xanthones, thioxanthones, aromatic diketones, phenylglyoxalates, camphorquinone, etc.

Most continuous-wave lasers have their strongest emission lines in the visible range, so there is a growing demand for photopolymers sensitive to above 400 nm radiations. However, the initiation efficiency of many systems that were tested so far was found to be inferior to that of UV initiators, mainly owing to the lower energy of visible photons [89].

Atmospheric oxygen quenches excited triplet states reducing the quantum yields of the initiating radicals [34, 86, 90]. This is particularly critical with type II photoinitiators [76].

Cationic Initiators

Cationic photopolymerizations use acids possessing anions of very low nucleophilicity. Cationic initiators are usually onium salts, such as iodonium and sulfonium salts, that produce reactive species upon UV irradiation [91, 92]. The reactivity of the salts is very much dependent upon the counterion with the reactivity increasing in the series $BF_4^- < PF_6^- < AsF_6^- < SbF_6^-$ [93]. Most of the onium salts exhibit absorbance below 350 nm; however, it is possible to increase the light yield of these salts using sensitizers like polycyclic aromatic hydrocarbons or aromatic ketones. Other cationic photoinitiators are metallocene salts that can be used in combination with oxidizing agents.

1.4.2.2 The Prepolymer

Initially, the most common stereolithographic polymeric resins were free radical systems based on acrylate and methacrylate monomers. These monomers cure rapidly and are easily modified at the ester functionality, allowing obtaining materials with a variety of properties. However, acrylate systems are relatively volatile, have an unpleasant odor and present potential hazards. Additionally, free radical curing reactions are inhibited by oxygen [94].

Epoxides (less reactive) and vinyl ethers (very reactive) are the most important systems that polymerise through a cationic mechanism [95–97]. These systems have better mechanical properties than acrylates exhibiting low shrinkage, but are usually less reactive. Contrary to free radical reactions, cationic species are not scavenged by molecular oxygen, they exhibit high active center concentrations, and the reaction proceeds spontaneously long after irradiation has ceased (dark polymerisation), increasing the green strength even at room temperature (Fig. 1.12) [92, 98–100]. Figure 1.13 shows monomers that are polymerizable with cationic photoinitiators. Table 1.2 presents a comparison between free radical and cationic polymeric systems.

Hybrid systems have also been proposed, comprising both acrylate compounds and epoxy or vinyl ether compounds forming an interpenetrating network.

1.5 Infrared Stereolithography

Infrared stereolithography is a novel stereolithographic process that uses IR radiation to cure thermosensitive resins [101–105]. Contrary to the conventional stereolithography, which is a UV-initiated polymerisation process, IR stereolithography is a thermal-initiated process.

Fig. 1.12 Laser-initiated cationic curing reactions. Dark polymerisation is represented by – [98]

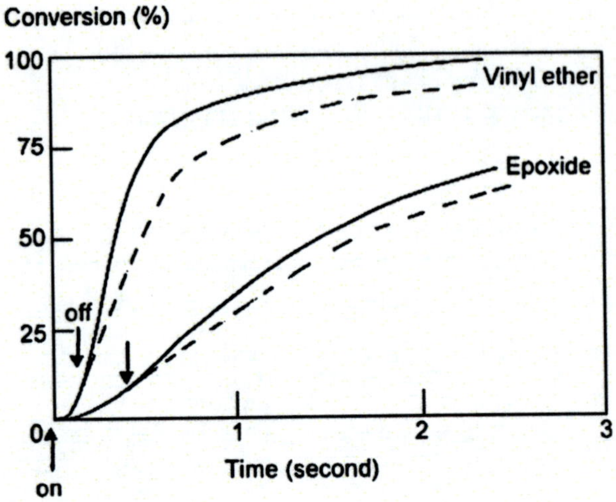

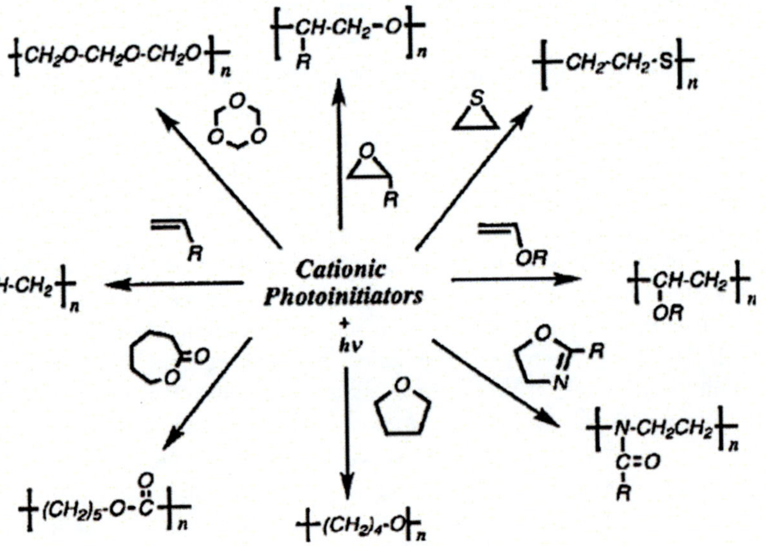

Fig. 1.13 Polymerisable monomers with cationic photoinitiators

Table 1.2 Comparison between radicalar and cationic systems

	Free-radical (acrylate)	Cationic (epoxy)
Cost	Low	High
Mechanical properties	Low	High
Shrinkage	High	Low
Reactivity	High	Low
Inhibition	Oxygen	Water

This process uses a CO_2 laser emitting IR radiation at $\lambda = 10.6\ \mu m$, a special scan head to focus the laser beam and a proper movable working table electronically controlled. The interaction between the laser beam and the thermosensitive material is controlled by the pulse energy of the laser, the dwell time, scan speed, laser beam width, and by other optical and thermal parameters of the resin [101–105].

IR stereolithography is commonly utilized with epoxy systems [101]. However, the use of unsaturated polyester resins mixed with peroxide initiators can also be used. Traditionally, the resin used in IR stereolithography is a diglycidyl ether of bisphenol A containing silica. The cross-linking of epoxy resins can be carried out either through the epoxy groups or the hydroxy groups. Amines are used as curing agents to cross-linking epoxy resins either by a catalytic mechanism or by bridging across epoxy molecules. Primary and secondary amines act as reactive hardeners whilst the tertiary amines are catalytic. The diethylenetriamine commonly used in IR stereolithography is a highly reactive primary aliphatic amine with five active hydrogen atoms available for cross-linking. The curing reaction with this type of amine is highly exothermic. Less reactive curing agents such as dimethylaminopropylamine and diethylaminopropylamine can also be used.

Silica plays an important role in the curing process, strongly determining the dimensional accuracy of the obtained models [101, 102]. When small amounts of silica are used, the cure process is not localized, as heat diffuses to areas outside the irradiated volume. If the amount of silica is high, the curing is not complete or may not occur as the silica absorbs most of the laser energy. The type of silica is also quite relevant [105].

1.6 Stereo-Thermal-Lithography

To produce multimaterial functionally graded components, researchers from the Centre for Rapid and Sustainable Product Development of the Polytechnic Institute of Leiria (Portugal) are developing a new stereolithographic fabrication process named stereo-thermal-lithographic process [33, 34]. This process uses ultraviolet radiation and thermal energy (produced by IR radiation) to initiate the polymerization reaction in a medium containing both photo- and thermal-initiators (Fig. 1.14) [33, 34, 106, 107]. The concentrations of both initiators are carefully selected and the reaction only starts when there is a particular combination of UV radiation and thermal energy [33, 34]. This way, the amount of each initiator must be low to inhibit the start of the polymerization by only one of these two effects. However, at the point where the two effects intersect each other, the amount of radicals generated is sufficiently high to initiate the polymerization process. Temperature is used to produce both radicals through the fragmentation of thermal-initiators and simultaneously increase the initiation and reaction rate of the photoinitiated curing reaction. As a result, the extent of cure is increased and no post cure will be needed. This is an important advantage of this process as to produce high quality models it is usually necessary to ensure a fairly high degree of gel formation. Moreover, if a

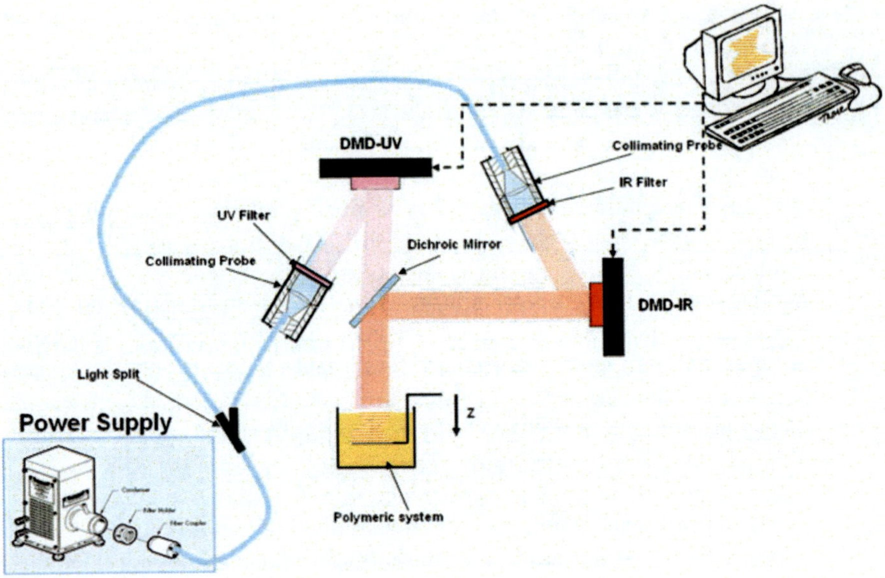

Fig. 1.14 The stereo-thermal-lithographic process: irradiation process

small amount of solid material is formed, a large number of unreacted molecules are still present, so it is possible that an excessive heat generation will occur during the post cure phase, leading to the distortion and warping of the built model.

The main advantages of STLG over conventional stereolithography are as follows [33, 34]:

- The generation of radicals is more efficient
- Small concentrations of the two types of initiator are used, enabling the radiation to penetrate deeper into the polymer
- The combination of UV radiation and temperature increases the reaction rate and hence the fractional conversion values
- The curing reaction is more localized, improving the accuracy of the produced models
- The system has more tunability

Four subsystems can be considered. Subsystem A uses ultraviolet radiation to solidify a liquid resin that contains a certain amount of photoinitiator. This subsystem corresponds to an approach similar to conventional stereolithography. Subsystem B uses thermal energy produced by infrared radiation to solidify a liquid resin that contains a certain amount of thermal initiator. Subsystem C uses both heat produced using infrared radiation and ultraviolet radiation to solidify a liquid resin that contains a certain amount of photoinitiator. Subsystem D uses both heat produced using infrared radiation and ultraviolet radiation to solidify a liquid resin containing a certain amount of thermal initiator and photoinitiator.

In addition to these key advantages, the system also contains a rotating multi vat that enables the fabrication of multi-material structures (Fig. 1.15). This represents

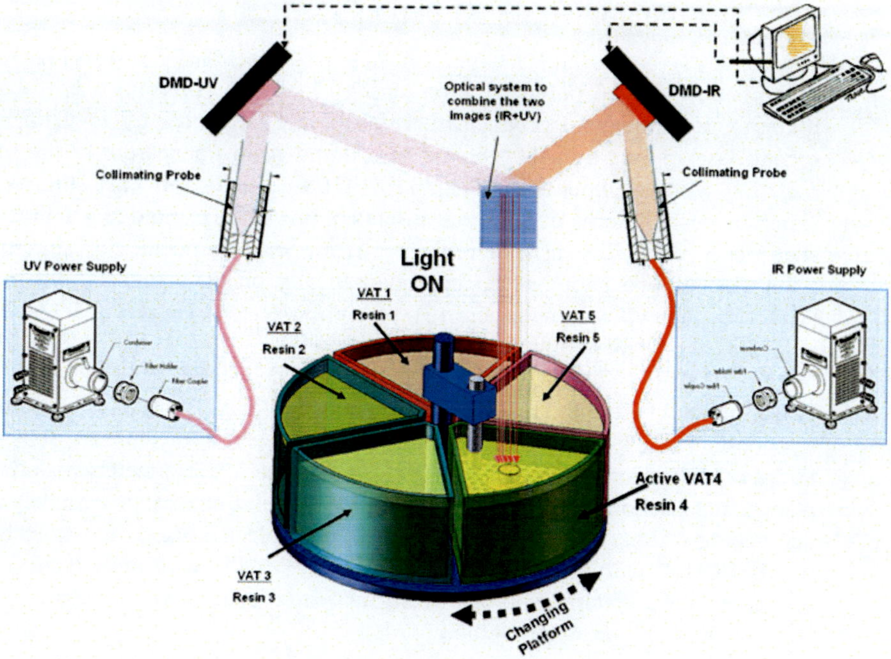

Fig. 1.15 The micro stereo-thermal-lithographic process: multi-vat system

an important advancement in this field. STLG is being developed to produce multi-material microscopic engineering prototypes through nanostructures for exploitation in waveguiding and photonic crystals, multi-material functional graded scaffolds for tissue engineering, other biomedical components, and micro functional metallic or ceramic parts.

The initiation process used in the stereo-thermal-lithographic process can be described as follows [33, 34]:

- Path 1 (thermal energy effects):

$$\text{Initiator} \xrightarrow{\text{heat}} \text{Radicals} \rightarrow \text{Radical recombination and/or inhibition}$$

- Path 2 (UV radiation effects):

$$\text{Initiator} \xrightarrow{\text{UV radiation}} \text{Radicals} \rightarrow \text{Radical recombination and/or inhibition}$$

- Intersection of path 1 and path 2 (thermal energy + UV radiation effects):

$$\left.\begin{array}{r} \text{Initiator} \xrightarrow{\text{heat}} \text{Radicals} \\ \text{Initiator} \xrightarrow{\text{UV radiation}} \text{Radicals} \end{array}\right\} \begin{array}{c} \text{Sufficient numbers of radicals} \\ \text{to start the polymerisation} \end{array}$$

1.7 Two-Photon-Initiated Polymerisation

Two-photon-initiated polymerisation curing processes represents a useful stereolithographic strategy to produce micro/nanoscale structures by focusing femtosecond laser pulses into the volume of a liquid resin transparent to the infrared radiation without photomasks (Fig. 1.16) [108–117]. In this case, the molecule simultaneously absorbs two photons instead of one being excited to a higher singlet state. The probability of electronic excitation of a molecule by simultaneous absorption of two photons depends quadratically on the incident light intensity [109, 116]. This allows a submicron 3D resolution, on top of enabling 3D fabrication at greater depth and an ultrafast fabrication.

Lim et al. [114, 115] developed the so-called nano-stereolithography (NSL) based on the two-photo polymerisation process (Fig. 1.17). The system uses a mode-locked Ti:sapphire laser beam, with wavelength of 780 nm, repletion of 80 MHz and pulse width less than 100 fs. The beam is scanned across the focal plane using a set of two Galvano mirrors with a resolution of approximately 2.5 nm per step, and along the vertical axis using a piezoelectric stage. The laser beam is tightly focused with an objective lens (NA 1.25 × 100, with immersion oil) on a photopolymeric system. Microparts are fabricated using a voxel matrix scanning method or a contour offset method [114, 115].

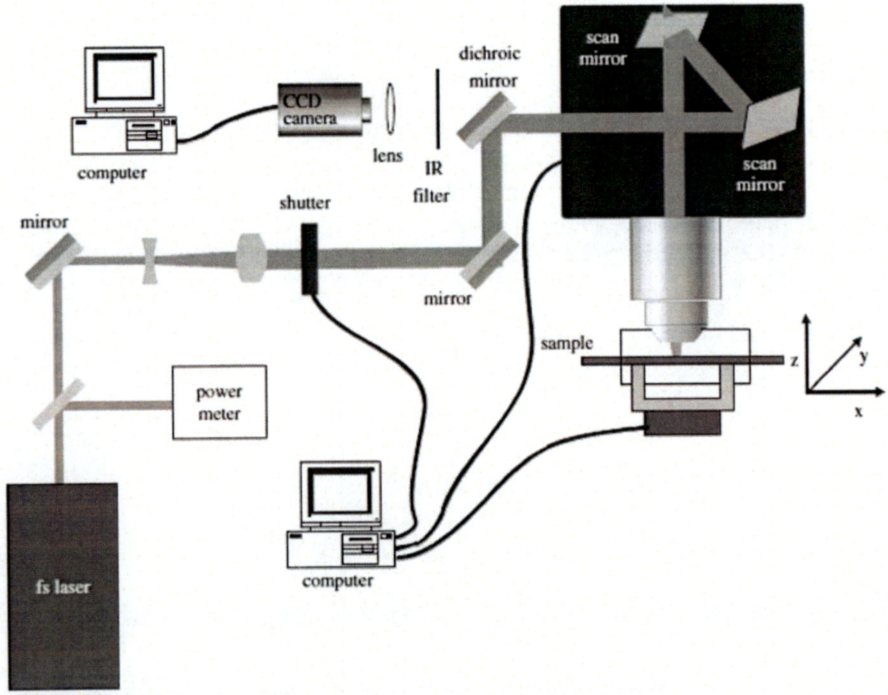

Fig. 1.16 Typical setup for multi-photon polymerization [108]

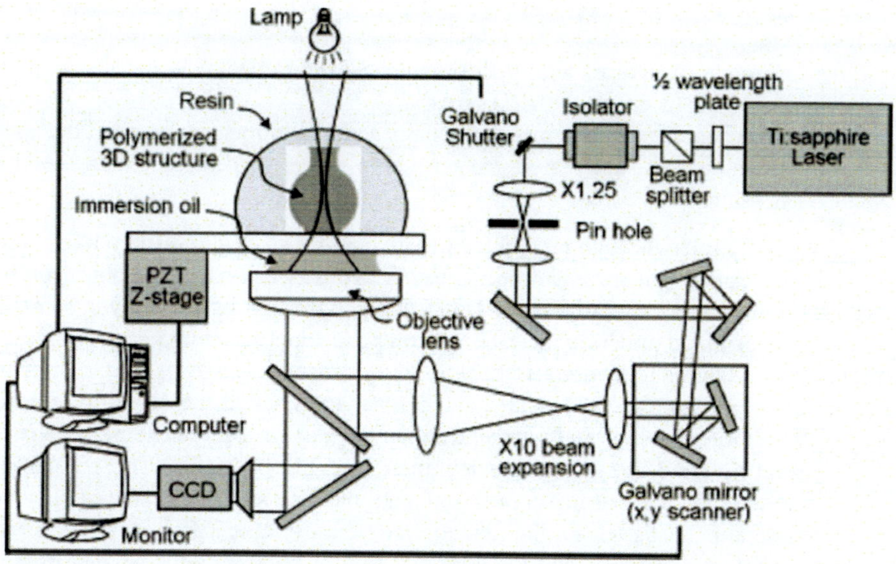

Fig. 1.17 The NLS system [114]

Although two-photon polymerisation using resins incorporating conventional initiators has been reported, such polymeric systems exhibit low photosensitivity as conventional photoinitiators suffer from small two-photon absorption cross-sections in the near-IR wavelength where femtosecond lasers are typically used [110, 116, 118]. Commonly used materials are multi functional inorganic–organic hybrid polymers whose properties can be tuned from those that are characteristic for organic polymers to those that are similar to inorganic glasses [113, 119, 120].

1.8 Advanced Materials for Novel Applications

1.8.1 Liquid Crystals

Liquid crystalline (LC) polymers are materials that exhibit a degree of order intermediate between those of amorphous liquids and crystalline solids, corresponding to an important class of stereolithographic materials, with high stiffness and high thermal stability [121, 122]. Usually, LC polymers contain flexible and rigid (mesogenic) fragments shaped as rods (calamitic LCs) and disks (discotic LCs), connected to each other by different ways [123, 124]. Main-chain liquid crystalline polymers are formed when the mesogens are part of the polymer backbone, while side-chain liquid crystalline polymers are generated when mesogens are connected as side chains to the polymer backbone [121–126]. The rigid mesogenic segments produce polymer networks with high glass transition temperature and liquid crystalline order.

Three main types of liquid crystals have been identified [122–125]:

- Nematic, structures with one-dimensional order
- Smectic, structures with 2D order
- Cholesteric, twisted nematic structures with one-dimensional order, but additional organization usually arise from a chiral center or centers within the chemical structure

The LC molecules can be aligned before cure by an external force like a magnetic field, resulting in an anisotropic cross-linked network when the photopolymerization "locks-in" the alignment [122]. The alignment of each layer of resin in a predetermined direction produces a "composit-like" structure with optimized directional and planar properties [122].

Several authors [127, 128] have extensively studied the effect of temperature, photoinitiator type and concentration, alignment of the nematic phase, and operating conditions on the cure kinetics of LC polymers. The results indicate that the polymerisation conversion and reaction rate increase as the reaction temperature rises within the nematic temperature range. Polymerization rate also increases by incrementing the concentration of photoinitiator present in the resin. Optimization of resin-cure characteristics can lead to higher conversions and to higher polymerisation rates. Based on these research works, more advanced resins with higher glass transitions ($T_g > 300°C$) are being developed. These materials with high temperature, mechanical stiffness, and strength can be applied to a great variety of areas as high temperature structural materials.

1.8.2 Ceramics and Metals

Stereolithography of ceramic and metallic materials consists of a UV curable ceramic suspension prepared with a prepolymer that will acts as the binder material, a photoinitiator, ceramic or metallic powder and additives [90, 129–145]. Upon polymerisation, the polymer bonds the ceramic/metallic particles conferring the necessary cohesion to the obtained ceramic or metallic matrix. This matrix structure is then subjected to binder removal through an appropriate thermal treatment and sintering, that ensures the final properties of the model (Fig. 1.18).

Particle size and light scattering effects are the most important mechanism determining the spatial resolution (Fig. 1.19). The curing depth is given by [130, 136]:

$$C_{\mathrm{d}} \approx \left(\frac{d}{Q}\right)\left(\frac{1}{\Phi}\right)\ln\left(\frac{E}{E_{\mathrm{c}}}\right) \tag{1.15}$$

with

$$Q = \left(\frac{\Delta n}{n_0}\right)^2\left(\frac{d}{\lambda}\right)^2 \tag{1.16}$$

Fig. 1.18 Turbine wheel produced by stereolithography of poly(methylsilsesquioxane)/alumina [145]

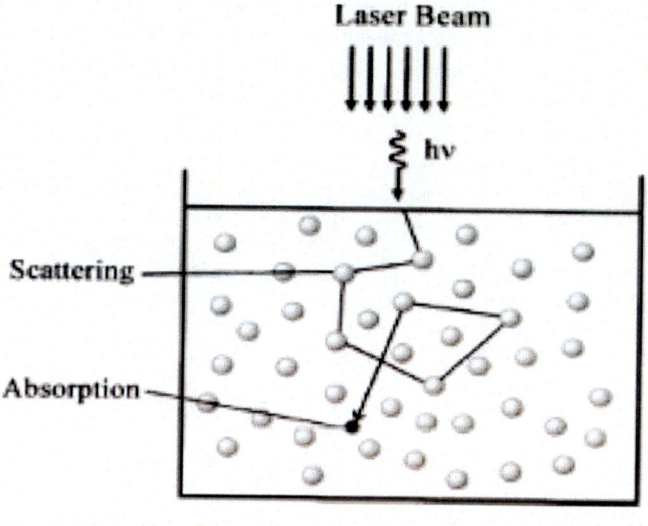

Fig. 1.19 Schematic effect of laser scattering due to solid particles dispersed in resin [135]

where, d is the mean particle size of the powder, Φ is the volume fraction of the powder in suspension, n_0 is the refractive index of the polymeric solution, Δn is the refractive index difference between the powder material and the polymeric material, and λ is the radiation wavelength. For ceramic powders, Sun [137] found that the light scattering effects are strongest when the size of the powder approaches the laser wavelength and when the refractive index contrast between the particles and the prepolymer is high.

Typical ceramic suspensions used in stereolithographic applications are:

- Piezoelectric ceramics such as lead zirconate titanate (PZT) powders, with an average size of 5 μm, a density of 7.6 g/cm^2, and a refractive index of 2.5, mixed with an acrylate prepolymer such as 1,6-hexanediol diacrylate (HDDA) or trimethylolpropane triacrylate (TMPTA) [92]. Epoxy and epoxy-acrylate prepolymers such as 3,4-epoxycyclohexylmethyl-3,4-epoxycyclohexylcarboxylate trimethylolpropane triacrylate are also used [138]
- Inert ceramics as alumina powder with a mean diameter of 0.5 μm and a refractive index of 1.7, mixed with acrylate prepolymers such HDDA [133, 139–141]
- Biocompatible ceramics such as hydroxyapatite [142].

A dispersant is used to increase the ceramic fraction in a low viscosity and homogeneous suspension [132]. This is due to the fact that an increase in the dispersant concentration slightly decreases the viscosity of the prepolymer as shown in Fig. 1.20. These results are due to the electrostatic and stearic effects between the dispersant and the prepolymer.

Metallic stereolithography has been explored using polymeric systems consisting of two different types of resins, an isophthalic unsaturated polyester resin and an epoxy resin, two types of photoinitiators: 2,2-Dimethoxy-1,2-diphenylethan-1-one (radical initiator) and 4-methylphenyl[4-(2-methylpropyl)phenyl]-hexafluorophosphate (cationic initiator), and tungsten carbide (WC) and cobalt (Co) with different powder sizes [143, 144]. UP resins were diluted with styrene to reduce the viscosity of the reinforced hybrid polymeric systems. Experimental results show that this dilution has no significant effect over the curing kinetics.

The reinforcement of the resins with metallic powder increases the viscosity. However, for the same level of reinforcement, the viscosity is higher for powders

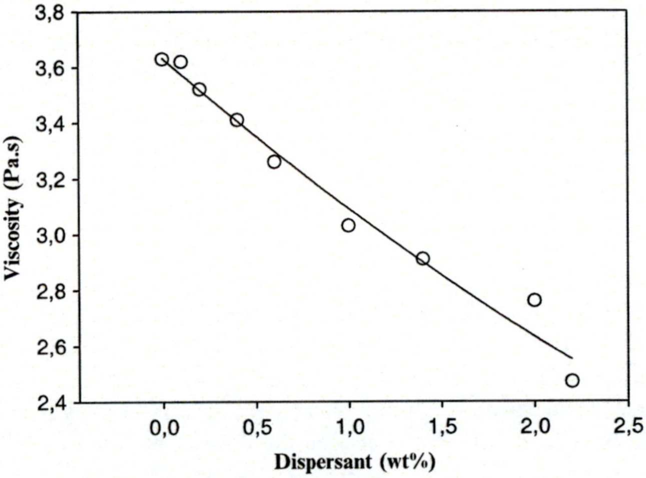

Fig. 1.20 The effect of the dispersant concentration on the viscosity of a diacrylate prepolymer containing 0.5 wt% of photoinitiator [133]

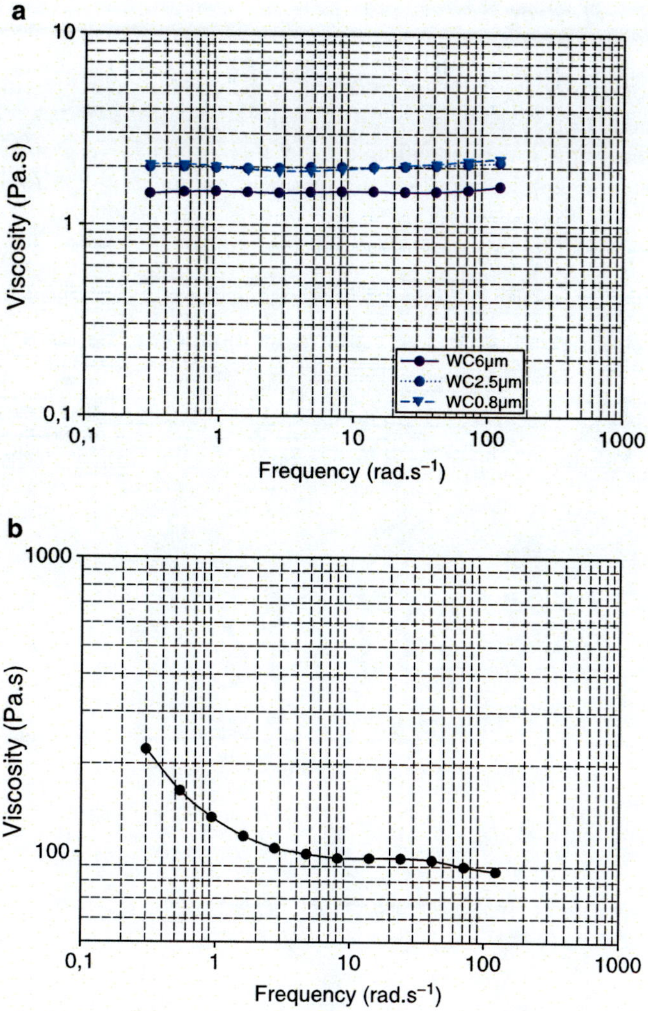

Fig. 1.21 The variation of viscosity as a function of frequency for (**a**) hybrid resins (60 wt% of unsaturated polyester and 40 wt% of epoxy resins) reinforced with 60 wt% of tungsten carbide powders of different particle dimensions; (**b**) hybrid resins (60 wt% of unsaturated polyester and 40 wt% of epoxy resins) reinforced with 60 wt% of cobalt powder particles of 1.5 μm [144]

with lower particle sizes as shown in Fig. 1.21a [143, 144, 146]. The use of powder particles of low density is also associated with an increase in viscosity. For cobalt powder particles it is possible to observe from Fig. 1.21b that the viscosity significantly decreases with frequency until a value near log 5 rad s^{-1}(shear-thinning behavior), from which the viscosity remains almost constant [143, 144].

For resins reinforced with 60 wt% of metallic powder, the results show that for low power sizes, the packing effect is dominant contributing to both the decrease in light penetration and the overall fractional conversion [143, 144, 146].

1.8.3 Hydrogels

Hydrogels are cross-linked hydrophilic polymers that represent a novel class of polymers to be used in stereolithographic applications [147–153]. These materials are receiving much attention due to their potential use in a wide variety of biomedical applications, including tissue engineering scaffolds, drug delivery,

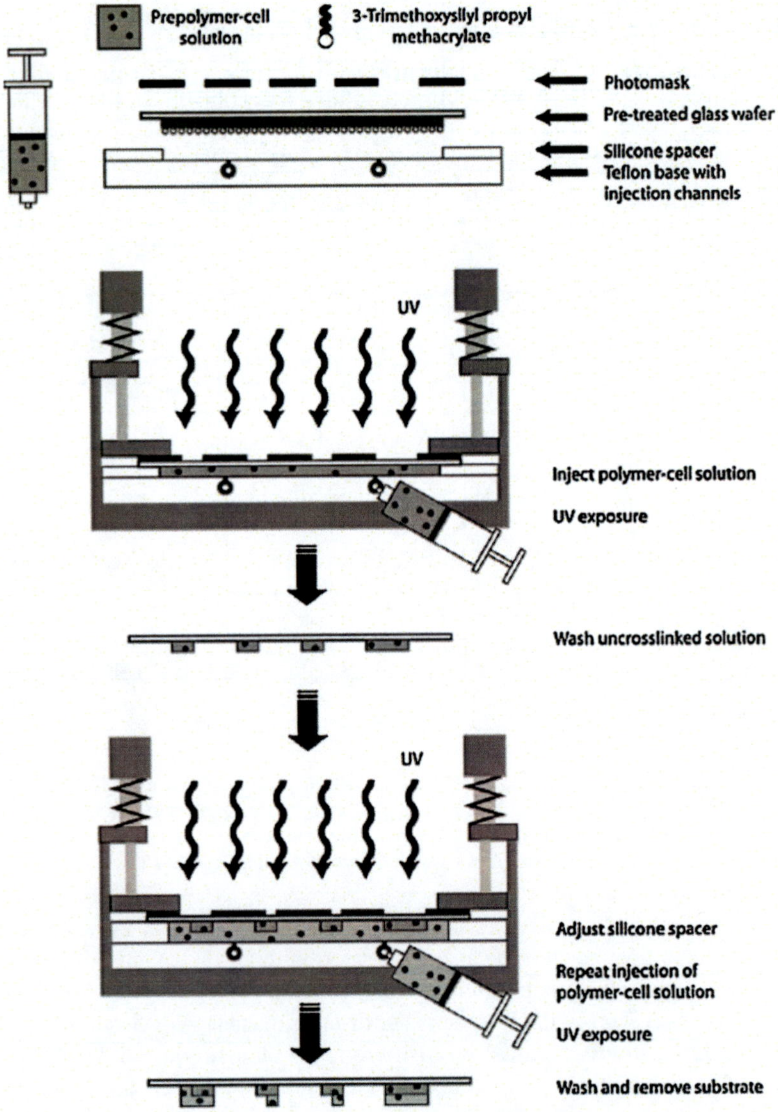

Fig. 1.22 Process for the formation of hydrogel microstructures containing living cells [158]

contact lenses, corneal implants, and wound dressing [147–153]. They usually exhibit excellent biocompatibility and high permeability for oxygen nutrients. The mechanical properties of many hydrogels can be tailored to match those of soft tissues, so these materials can be attractive scaffolds for soft tissue development or regeneration either by acting as tissue barriers or local drug delivery systems, or as cell-carrier materials for tissue replacement [147–154].

Some types of hydrogels can be photopolymerized using UV radiation and appropriate photoinitiators. The UV polymerisation of hydrogels occurs at sufficiently mild conditions, like low light intensity, short irradiation time, physiological temperature, and low organic solvent levels, enabling the reaction to be carried out in the presence of cells [155].

Hydrogels used in stereolithographic applications comprise natural polymers, synthetic polymers, or combination of both natural and synthetic polymers [147–157]. Natural polymeric hydrogels include polymers like hyaluronic acid, chitosan, collagen, gelatine, dextran, etc. Among synthetic polymers are those based on poly(propylene fumarate) (PPF), poly(2-hydroxyethyl methacrylate) (poly(HEMA)), hyaluronic acid-based materials, polyvinyl alcohol derivatives, etc.

Mask-based writing system can be used to pattern hydrogel structures with high resolution. Liu and Bhatia [158] reported a method, where multiple steps of micropatterned photopolymerisation processes can be coupled to produce 3D cell matrix structures with micro-scale resolution (Fig. 1.22).

References

1. F.W. Liou, *Rapid prototyping and engineering applications – a toolbox for prototype development*, CRC Press, Boca Raton, 2008
2. E.B. Magrab, S.K. Gupta, F.P. McCluskey, P.A. Sandborn, *Integrated product and process design and development*, CRC Press, Boca Raton, 2010
3. G. Pahl, W. Beitz, *Engineering design – a systematic approach*, Springer, London, 1996
4. I. Gibson, Rapid prototyping: from product development to medicine and beyond, *Virtual and Physical Prototyping*, 1, 31-42, 2006
5. A. Thakur, A. G. Banerjee, S.K. Gupta, A survey of CAD model simplification techniques for physics-based simulation applications, *Computer-Aided Design*, 41, 65–80, 2009
6. K. Lee, *Principles of CAD/CAM/CAE systems*, Addison-Wesley, Reading, Massachusetts, 1999
7. F.E.H. Tay, A. Roy, CyberCAD: a collaborative approach in 3D-CAD technology in a multimedia-supported environment, *Computers in Industry*, 52, 127–145, 2003
8. S.K. Ong, Y. Shen, A mixed reality environment for collaborative product design and product development, *CIRP Annals – Manufacturing Technology*, 58(1), 139–142, 2009
9. T.R. Langerak, Parameter reconstruction of freeform shapes for improved product modelling, *Tools and Methods of Competitive Engineering*, Vol.1, Edited by I. Horváth and Z. Rusák, Delft University of Technology, 2008
10. N.M. Alves, P.J. Bartolo, Integrated tools for virtual and physical automatic construction, *Automation in Construction*, 15, 257–271, 2006
11. E. Izquierdo, J.-R. Ohm, Image-based rendering and 3D modelling: a complex framework, *Signal Processing: Image Communication*, 15, 817–858, 2000

12. F. Verbiest, G. Willems, L. Van Gool, Image-based rendering for photo-realistic visualization, *Virtual and Physical Prototyping*, 1, 19–30, 2006
13. S. Rowlinson, D. Yates, nDCAD: a virtual change agent for professions and procurement systems?, *Construction Management and Economics*, 21, 849–857, 2003
14. A. Mahalingam, R. Kashyap, C. Mahajan, An evaluation of the applicability of 4D CAD on construction projects, *Automation in Construction*, 19, 148–159, 2010
15. Z. Ma, Q. Shen, J. Z., Application of 4D for dynamic site layout and management of construction projects, *Automation in Construction*, 14, 369–381, 2005
16. K.W. Chau, M. Anson, J.P. Zhang, Four-dimensional visualization of construction scheduling and site utilization, *Journal of Construction Engineering and Management*, 130, 598–606, 2004
17. D. Heesom, L. Mahdjoubi, Trends of 4D CAD applications for construction planning, *Construction Management and Economics*, 22, 171–182, 2004
18. Y. Ren, S.K. Lai-Yuen, Y.S. Lee, Virtual prototyping and manufacturing planning by using tri-dexel models and haptic force feedback, *Virtual and Physical Prototyping*, 1, 3–18, 2006
19. N.M.F. Alves, P.J. Bártolo, Virtual modelling through human vision sense, *International Journal of Interactive Design and Manufacturing*, 1, 195–207, 2007
20. S. Ha, L. Kim, S. Park, C. Jun, H. Rho, Virtual prototyping enhanced by a haptic interface, *CIRP Annals – Manufacturing Technology*, 58(1), 135–138, 2009
21. M. Bordegoni, U. Cugini, M. Covarrubias, Design of a visualization system integrated with haptic interfaces, *Tools and methods of competitive engineering*, Vol.1, Edited by I. Horváth and Z. Rusák, Delft University of Technology, 2008
22. I. Gibson, Z. Gao, Virtual prototyping – similarities and differences, *Advanced Research in Virtual and Rapid Prototyping, Proceedings of the 1st International Conference on Advanced Research in Virtual and Rapid Prototyping*, Edited by P.J. Bartolo et al, Polytechnic Institute of Leiria, 2003
23. W. Zhu, Y.S. Lee, Haptic sculpting and machining planning with 5-DOF haptic interface for virtual prototyping and manufacturing, *Advanced Research in Virtual and Rapid Prototyping, Proceedings of the 1st International Conference on Advanced Research in Virtual and Rapid Prototyping*, Edited by P.J. Bartolo et al, Polytechnic Institute of Leiria, 2003
24. F. Kimura, N. Yamane, Haptic environment for designing human interface of virtual mechanical products, *CIRP Annals – Manufacturing Technology*, 55(1), 127–130, 2008
25. P.J. Jones, *CAD/CAM: features, applications and management*, MacMillan Press, London, 1992
26. G. Molcho, R. Schneor, Y. Zipori, P. Kowalsi, B. Denkena, M. Shiptalni, Computer aided manufacturability analysis closing the CAD-CAM knowledge GAP, *Proceedings of the 9th Biennial ASME Conference on Engineering Systems Design and Analysis*, 2008
27. R. Stark, H. Hayka, D. Langenberg, New potentials for virtual product creation by utilizing grid technology, *CIRP Annals – Manufacturing Technology*, 58, 143–146, 2009
28. O.C. Zienkiewicz, *The finite element method*, McGraw-Hill, London, 1977
29. R.D. Cook, D.S. Malkus, M.E. Plesha, *Concepts and applications of finite element analysis*, Wiley, New York, 1989
30. S. Ranganath, C. Guo, P. Hegde, A finite element modeling approach to predicting white layer formation in nickel supperalloys, *CIRP Annals – Manufacturing Technology*, 58(1), 77–80, 2009
31. A.D. Prete, T. Primo, A. Elia, CAE tools as valid oportunity to improve quality control systems performances for metal formed components, *Proceedings of the 9th Biennial ASME Conference on Engineering Systems Design and Analysis*, 2008
32. A. Saxena, B. Sahay, *Computer aided engineering design*, Springer, Dordrecht, 2009
33. P.J. Bartolo, G. Mitchell, Stereo-thermal-lithography: a new principle for rapid prototyping, *Rapid Prototyping Journal*, 9, 150–156, 2003
34. P.J. Bartolo, *Optical approaches for macroscopic and microscopic engineering*, PhD Thesis, University of Reading, UK, 2001

35. G.N. Levy, Digital layer manufacturing chances and challenges, *Proceedings of the 9th Biennial ASME Conference on Engineering Systems Design and Analysis*, 2008
36. J.-P. Kruth, M.C. Leu, T. Nakagawa, Progress in additive manufacturing and rapid prototyping, *CIRP Annals – Manufacturing Technology*, 47(2), 525–540, 1998
37. G. N. Levy, R. Schindel, J.-P. Kruth, Rapid manufacturing and rapid tooling with layer manufacturing (LM) technologies, state of the art and future perspectives, *CIRP Annals – Manufacturing Technology*, 52(2), 589–609, 2003
38. C.C. Kai and L.H. Fai, *Rapid prototyping: principles and applications in manufacturing*, Wiley, Chichester, 1997
39. K.S. Lee, S.H. Kim, Non-uniform deformation of an STL model satisfying error criteria, *Computer-Aided Design*, 42, 239–247, 2010
40. Y.H. Chen, C.T. Ng, Y.Z. Wang, Data reduction in integrated reverse engineering and rapid prototyping, *International Journal of Computer Integrated Manufacturing*, 12, 97–103, 1999
41. M.K. Agoston, *Algebraic topology*, Marcel Dekker, New York, 1976
42. S.H. Choi, K.T. Kwok, A tolerant slicing algorithm for layered manufacturing, *Rapid Prototyping Journal*, 8, 161–179, 2002
43. Y.F. Zhang, Y.S. Wong, H.T. Loh, An adaptive slicing approach to modelling cloud data for rapid prototyping, *Journal of Materials Processing Technology*, 140, 105–109, 2003
44. C.S. Wang, W.H.A. Wang, M.C. Lin, STL rapid prototyping bio-CAD model for CT medical image segmentation, *Computers in Industry*, 61, 187–197, 2010
45. J.P. Fouassier, *Photoinitiation, photopolymerization, and photocuring: fundamentals and applications*, Hanser, New York, 1995
46. A. Gilbert and J. Baggot, *Essentials of molecular photoch*emistry, Blackwell Science Publishers, London, 1991
47. J. Guillet, *Polymer photophysics and photochemistry*, Cambridge University Press, Cambridge, 1985
48. H. Ushiki and K. Horie, Influence of molecular structure on polymer photophysical and photochemical properties, *Handbook of polymer science and technology*, Vol. 4, Edited by N. P. Cheremisinoff, Dekker, New York, 1989
49. C. Decker, Effect of UV radiation on polymers, *Handbook of polymer science and technology*, Vol. 3, Edited by N.P. Cheremisinoff, Dekker, New York, 1989
50. H.B. Olayan, H.S. Hamid, E.D. Owen, Photochemical and thermal crosslinking of polymers, *JMS Review Macromolecular Chemistry and Physics* C36(4), 671–719, 1996
51. E. Selli and I.R. Bellobono, Photopolymerization of multifunctional monomers: kinetic aspects, in *Radiation curing in polymer science and technology*, Vol. III: *Polymerisation mechanisms*, Edited by J. P. Fouassier and J. F. Rabek, Elsevier Science Publishers, London, 1993
52. J.K. Gillham, Award address formation and properties of network polymeric materials, *Polymer Engineering and Science*, 19, 676–682, 1979
53. J.K. Gillham, Characterization of thermosetting materials by torsional braid analysis, *Polymer Engineering and Science*, 16, 353, 1976
54. R.B. Prime, Thermosets in *Thermal characterization of polymeric materials*, Vol. 2, Edited by A. Turi, Academic Press, London, 1997
55. S. Gan, J.K. Gillham and R.B. Prime, A methodology for characterizing reactive coatings: time-temperature-transformation (TTT) analysis of the competition between cure, evaporation, and thermal degradation for an epoxy-phenolic system, *Journal of Applied Polymer Science*, 15, 803–816,1989
56. F.W. Billmeyer, *Textbook of polymer science*, Wiley, New York, 1984
57. A. Ravve, *Organic chemistry of macromolecules*, Marcel Dekker, New York, 1967
58. P.J. Flory, *Principles of polymer chemistry*, Cornell University Press, Ithaca, 1967
59. J. Lange, N. Altmann, C.T. Kelly, P.J. Halley, Understanding vitrification during cure of epoxy resins using dynamic scanning calorimetry and rheological techniques, *Polymer*, 40, 5949–5955, 2000

60. J.C. Dominguez, M.V. Alonso, M. Oliet, F. Rodriguez, Chemorheological study of the curing kinetics of a phenolic resol resin gelled, *European Polymer Journal*, 46, 50–57, 2010
61. X. Ramis and J.M. Salla, Time-temperature transformation (TTT) cure diagram of an unsaturated polyester resin, *Journal of Polymer Science Part B: Polymer Physics*, 30, 371–388, 1997
62. S. Lunak, J. Vladyka and K. Dušek, Effect of diffusion in the glass transition region on critical conversion at the gel point during curing of epoxy resins, *Polymer*, 19, 931–933, 1978
63. A. Hale, Thermosets, *Handbook of Thermal Analysis and Calorimetry*, Edited by S.Z.D. Cheng, Elsevier, London, 2002
64. S. Montserrat, F. Roman, P. Colomer, Vitrification and dielectric relaxation during the isothermal curing of an epoxy-amine resin, *Polymer*, 44, 101–114, 2003
65. W.X. Zukas, Torsional braid analysis of the aromatic amine cure of epoxy resins, *Journal of Applied Polymer Science*, 53, 429–440, 1994
66. G. Wisanrakkit and J.K. Gillham, Continuous heating transformation (CHT) cure diagram of an aromatic amine/epoxy system at constant heating rates, *Journal of Applied Polymer Science*, 42, 2453–2463, 1991
67. N. Fang, C. Sun, X. Zhang, Diffusion-limited photopolymerization in scanning micro-stereolithography, *Applied Physics A*, 79, 1839–1842, 2004
68. B.A. Osinski, Alpha-T-T and T-T-alpha-T diagrams as a new element in comprehensive modeling of thermoset processing, *Polymer*, 34, 752–758, 1993
69. P. Pang and J.K. Gillham, Anomalous behavior of cured epoxy resins: density at room temperature versus time and temperature of cure, *Journal of Applied Polymer Science*, 37, 1969–1991, 1989
70. J.D. Ferry, Viscoelastic properties of polymers, Wiley, New York, 1980
71. S.L. Simon and J.K. Gillham, Conversion-temperature-property diagram for a liquid dicyanate ester/high-Tg polycyanurate thermosetting system, *Journal of Applied Polymer Science*, 51, 1741–1752, 1994
72. J.P. Fouassier, *Photoinitiation, photopolymerization, and photocuring-fundamentals and applications*, Hanser, Munich, 1996
73. C. Decker, New developments in UV-curable acrylic monomers, in *Radiation curing in polymer science and technology*, Vol. III: *Polymerisation mechanisms*, Edited by J. P. Fouassier and J. F. Rabek, Elsevier Science Publishers, London, 1993
74. Y.M. Huang, S. Kuriyama, C.P. Jiang, Fundamental study and theoretical analysis in a constrained-surface stereolithography system, *International Journal of Advanced Manufacturing Technology*, 24, 361–369, 2004
75. D.T. Pham, C. Ji, A study of recoating in stereolithography, *Proceedings of the Institution of Mechanichal Engineers, Part C – Journal of Mechanical Engineering Science*, 217, 105–117, 2003
76. D.T. Pham, S.S. Dimov, R.S. Gault, Part orientation in stereolithography, *International Journal of Advanced Manufacturing Technology*, 15, 674–682, 1999
77. P. Lan, S. Chou, L. Chen, D. Gemmill, Determination of fabrication orientations for rapid prototyping with stereolithography apparatus, *Computer-Aided Design*, 29, 53–62, 1997
78. H.C. Kim, S.H. Lee, Reduction of post-processing for stereolithography systems by fabrication-direction optimization, *Computer-Aided Design*, 37, 711–725, 2005
79. J. Giannatsis, V. Dedoussis, Decision support tool for selecting fabrication parameters in stereolithography, *International Journal of Advanced Manufacturing Technology*, 33, 706–718, 2007
80. K. Chockalingam, N. Jawahar, U. Chandrasekar, K.N. Ramanathan, Establishment of process model for part strength in stereolithography, *Journal of Materials Processing Technology*, 208, 348–365, 2008
81. F. Holzer, G. Fadel, Design of a 3-degrees of freedom platform for the stereolithography apparatus, *Rapid Prototyping Journal*, 8, 100–115, 2002

82. B. Luan, X.Y. Liu, J. Nagata, W.J. Cheong, Residual stress analysis – an important consideration for coating of stereolithographic polymers, *Surface and Coatings Technology*, 192, 323–330, 2005
83. Z. Zhou, D. Li, Z. Zhang, Rapid fabrication of metal-coated composite stereolithography parts, *Proceedings of the Institution of Mechanichal Engineers, Part C – Journal of Process Mechanical Engineering*, 221, 1431–1440, 2007
84. D.C. Watts, A.S. Marouf, A.M. Al-Hindi, Photo-polymerization shrinkage-stress kinetics in resin composites: methods development, Dental Materials, 19, 1–11, 2003
85. I. Pomerantz, S. Gilad, Y. Dollberg, B. Ben-Ezra, Y. Sheinman, G. Barequet, M. Katz, *Three dimensional modeling apparatus*, US Pat. 5519816, 1996
86. E. Andrzejewska, Photopolymerization kinetics of multifunctional monomers, *Progress in Polymer Science*, 26, 605–665, 2001
87. D. Colombani, Chain-growth control in free radical polymerization, *Progress in Polymer Science*, 22, 1649–1720, 1997
88. T. Corrales, F. Catalina, C. Peinado, N.S. Allen, Free radical macrophotoinitiators: an overview on recent advances, *Journal of Photochemistry and Photobiology A: Chemistry*, 159, 103–114, 2003
89. C. Decker and B. Elzaouk, Laser-induced crosslinking polymerization of acrylic photoresists, *Journal of Applied Polymer Science*, 65, 833–844, 1997
90. O. Dufaud, S. Corbel, Oxygen diffusion in ceramic suspensions for stereolithography, *Chemical Engineering Journal*, 92, 55–62, 2003
91. J.V. Crivello, K. Dietliker, *Photoinitiators for free radical, cationic and anionic photopolymerization*, Wiley, New York, 1998
92. J.V. Crivello, The discovery and development of onium salt cationic photoinitiators, *Journal of Polymer Science: Part A: Polymer Chemistry*, 37, 4241–4254, 1999
93. V. Sipani, A.B. Scranton, Dark-cure studies of cationic photopolymerizations of epoxides: characterization of the active center lifetime and kinetic rate constants, *Journal of Polymer Science: Part A: Polymer Chemistry*, 41, 2064–2072, 2003
94. M. Shirai, K. Mitsukura, H. Okamura, Chain propagation in UV curing of di(meth)acrylates, *Chemistry of Materials*, 20, 1971–1976, 2008
95. C. Decker, C. Bianchi, D. Decker, F. Morel, Photoinitiated polymerization of vinyl ether-based systems, *Progress in Organic Coatings*, 42, 253–266, 2001
96. C. R. Chatwin, M. Farsari, S. Huang, M.I. Heywood, R.C.D. Young, P.M. Birch, F. Claret-Tournier, J.D. Richardson, Characterisation of epoxy resins for microstereolithographic rapid prototyping, *International Journal of Advanced Manufacturing Technology*, 15, 281–286, 1999
97. M. Bjpai, V. Shukla, A. Kumar, Film performance and UV curing of epoxy acrylate resins, *Progress in Organic Coatings*, 44, 271–278, 2002
98. C. Decker, High-speed curing by laser irradiation, *Nuclear Instruments and Methods in Physics Research B*, 151, 22–28, 1999
99. F. Boey, S.K. Rath, A.K. Ng, M.J.M. Abadie, Cationic UV cure kinetics for multifunctional epoxies, *Journal of Applied Polymer Science*, 86, 518–525, 2002
100. C.E. Carcione, A. Greco, A. Maffezzoli, Photopolymerization kinetics of an epoxy based resin for stereolithography, *Journal of Thermal Analysis and Calorimetry*, 72, 687–693, 2003
101. A.L. Jardini, R.M. Filho, M.A.F. Scarparo, S.R. Andrade, L.F.M. Moura, Infrared laser stereolithography: prototype construction using special combination of compounds and laser parameters in localised curing process, *International Journal of Materials and Product Technology*, 21, 241–254, 2004
102. A.L.M. Jardini, R.F. Maciel, M.A.F. Scarparo, S.R. Andrade, L.F.M. Moura, Advances in stereolithography: a new experimental technique in the production of a three-dimensional plastic model with an infrared laser, *Journal of Applied Polymer Science*, 92, 2387–2394, 2004

103. M.A.F. Scarparo, A. Kiel, Z. Zhiyao, C.A. Ferrari, Q.J. Chen, J.H. Miller and S.D. Allen, Study of resin based materials using CO_2 laser stereolithography, *Polymer*, 38, 2175–2181, 1997
104. M.L. Barros, M.A.F. Scarparo, A. Kiel, E. Gerck, J.J. Hurtak, Stereolithography with thermosensitive resins using CO_2 laser, *Journal of Applied Polymer Science*, 54, 1575–1578, 1994
105. A.L. Jardini, R.Maciel, M.A. Scarparo, S.R. Andrade, L.F. Moura, The development in infrared stereolithography using thermosensitive polymers, *Advanced Research in Virtual and Rapid Prototyping, Proceedings of the 1st International Conference on Advanced Research in Virtual and Rapid Prototyping*, Edited by P.J. Bartolo et al, Polytechnic Institute of Leiria, 2003
106. P.J. Bartolo, G. Mitchell, Advanced photo-fabrication system for thermosetting materials, *Proceedings of the PPS-19, Polymer Processing Society*, Melburn, Australia, 2003
107. P.J. Bartolo, G. Mitchell, A new photo-fabrication system through the use of multiple light effects within organic polymers, *POLYCHAR-10 World Forum on Polymer Applications and Theory*, Denton, USA, 2002
108. E. Stratakis, A. Ranella, M. Farsari, C. Fotakis, Laser-based micro/nanoengineering for biological applications, *Progress in Quantum Electronics*, 33, 127–163, 2009
109. Y.X. Yan, X.T. Tao, Y.H. Sun, G.B. Xu, C.K. Wang, J.X. Yang, X. Zhao, Y.Z. Wu, Y. Ren, M.H. Jiang, Two new asymmetrical two-photon photopolymerization initiators: synthesis, characterization and nonlinear optical properties, *Optical Materials*, 27, 1787–1792, 2005
110. K.J. Schafer, J.H. Hales, M. Balu, K.D. Belfield, E.W. van Stryland, D.J. Hagan, Two-photon absorption cross-sections of common photo-initiators, *Journal of Photochemistry and Photobiology A: Chemistry*, 162, 497–502, 2004
111. S. Kawata, H.B. Sun, Two-photon photopolymerization as a tool for making micro-devices, *Applied Surface Science*, 208–209, 153–158, 2003
112. M. Miwa, S. Juodkazis, T. Kawakami, S. Matsuo, H. Misawa, Femtosecond two-photon stereo-lithography, *Applied Physics A*, 73, 561–566, 2001
113. U. Stute, J. Serbin, C. Kulik, B.N. Chichkov, Three-dimensional nanostructures fabricated by two-photon polymerization of hybrid polymers, *Advanced Research in Virtual and Rapid Prototyping, Proceedings of the 1st International Conference on Advanced Research in Virtual and Rapid Prototyping*, Edited by P.J. Bartolo et al, Polytechnic Institute of Leiria, 2003
114. K.S. Lee, D.Y. Yang, S.H. Park, R.H. Kim, Recent developments in the use of two-photon polymerization in precise 2D and 3D microfabrications, *Polymers for Advanced Technologies*, 17, 72–82, 2006
115. T.W. Lim, S.H. Park, D.Y. Yang, Contour offset algorithm for precise patterning in two-photon polymerization, *Microelectronic Engineering*, 77, 382–388, 2005
116. S. Wu, J. Serbin, M. Gu, Two-photon polymerisation for three-dimensional micro-fabrication, *Journal of Photochemistry and Photobiology A: Chemistry*, 181, 1–11, 2006
117. M. Zhou, H.F. Yang, J.J. Kong, F. Yan, L. Cai, Study on the microfabrication technique by femtosecond laser two-photon photopolymerization, *Journal of Materials Processing Technology*, 200, 158–162, 2008
118. K.D. Belfield, K.J. Schafer, Y.U. Liu, X.B. Ren, E.W Van Stryland, Multiphoton-absorbing organic materials for microfabrication, emerging optical applications and non-destructive three-dimensional imaging, *Journal of Physical Organic Chemistry*, 13, 837–849, 2000
119. S. Schlie, A. Ngezahayo, A. Ovsianikov, T. Fabian, H.A. Kolb, H. Haferkamp, B.N. Chichkov, Three-dimensional cell growth on structures fabricated from ORMOCER by two-photon polymerization technique, *Journal of Biomaterials Applications*, 22, 275–287, 2007
120. A. Doraiswamy, C. Jin, R.J. Narayan, P. Mageswaran, P. Mente, R. Modi, R. Auyeung, D.B. Chrisey, A. Ovsianikov, B. Chichkov, Two photon induced polymerization of organic-inorganic hybrid biomaterials for microstructured medical devices, *Acta Biomaterialia*, 2, 267–275, 2006

121. D. Klosterman, R. Chartoff, T. Tong, M. Galaska, Electron-beam curing of a novel liquid crystal thermoset resin, *Thermochimica Acta*, 396, 199–210, 2003
122. J.S. Ullett, T. Benson-Tolle, J.W. Schultz, R.P. Chartoff, Thermal-expansion and fracture toughness properties of parts made from liquid crystal stereolithography resins, *Materials and Design*, 20, 91–97, 1999
123. V. Shibaev, A. Bobrovsky and N. Boiko, Photoactive liquid crystalline polymer systems with light-controllable structure and optical properties, *Progress in Polymer Science*, 28, 729–836, 2003
124. D.J. Broer, G.N. Mol, J.A.M.M. van Haaren, J. Lub, Photo-induced diffusion in polymerizing chiral-nematic media, *Advanced Materials*, 11, 573–578, 1999
125. S. Kurihara, A. Sakamoto, T. Nonaka, Liquid-crystalline polymer networks: effect of cross-linking on the stability of macroscopic molecular orientation, *Macromolecules*, 32, 3150–3153, 1999
126. P. Oswald, P. Pieranski, *Nematic and cholesteric liquid crystals: concepts and physical properties illustrated by experiments*, CRC Press, Boca Raton, 2005
127. J.S. Ullett, J.W. Schultz, R.P. Chartoff, Novel liquid crystal resins for stereolithography, *Rapid Prototyping Journal*, 6, 8–17, 2000
128. D.J. Broer, G.N. Mol, G. Challa, In-situ photopolymerization of oriented liquid-crystalline acrylates, *Makromoleculare Chemie*, 192, 59–74,1991
129. O. Dufaud, H. Le Gall, S. Corbel, Application of stereolithography to chemical engineering – from macro to micro, *Chemical Engineering Research and Design*, 83(A2), 133–138, 2005
130. X. Zhang, X.N. Jiang, C. Sun, Micro-stereolithography of polymeric and ceramic microstructures, *Sensors and Actuators*, 77, 149–156, 1999
131. M. Wozniak, T. Graule, Y. Hazan, D. Kata, J. Lis, Highly loaded UV curable nanosilica dispersions for rapid prototyping applications, *Journal of the European Ceramic Society*, 29, 2259–2265, 2009
132. Y. Hazan, J. Heinecke, A. Weber, T. Graule, High solids loading ceramic colloidal dispersions in UV curable media via comb-polyelectrolyte surfactants, *Journal of Colloid and Interface Science*, 337, 66–74, 2009
133. C. Hinczewski, S. Corbel, T. Charier, Ceramic suspensions suitable for stereolithography, *Journal of the European Ceramic Society*, 18, 583–590, 1998
134. A. Bertsch, S. Jiguet, P. Renaud, Microfabrication of ceramic components by microstereolithography, *Journal of Micromechanics and Microengineering*, 14, 197–203, 2004
135. C. Sun, X. Zhang, Experimental and numerical investigations on microstereolithography of ceramics, *Journal of Applied Physics*, 92, 4796–4802, 2002
136. M. Fersini, F. Montagna, A. Licciulli, A. Maffezzoli, Rapid prototyping of amorphous silica through laser stereolithography, *Virtual and Rapid Manufacturing – Advanced Research in Virtual and Rapid Prototyping*, Edited by P.J. Bartolo *et al*, Taylor&Francis, London, 2008
137. C. Sun and X. Zhang, The influences of the material properties on ceramic microstereolithography, *Sensors and Actuators A*, 101, 364–370, 2002
138. O. Dufaud, P. Marchal and S. Corbel, Rheological properties of PZT suspensions for stereolithography, *Journal of the European Ceramic Society*, 22, 2081–2092, 2002
139. C. Hinczewski, S. Corbel, T. Chartier, Stereolithography for the fabrication of ceramic three-dimensional parts, Rapid Prototyping Journal, 4, 104–111, 1998
140. C. Provin and S. Monneret, Complex ceramic-polymer composite microparts made by microstereolithography, *IEEE Transactions on Electronics Packaging Manufacturing*, 25, 59–63, 2002
141. F. Doreau, C. Chaput, T. Chartier, Stereolithography for manufacturing ceramic parts, *Advanced Engineering Materials*, 2, 493–496, 2000
142. R.A. Levy, T.G. Chu, J.W. Holloran, S.E. Feinberg, S. Hollister, CT-generated porous hydroxyapatite orbital floor prosthesis as a prototype bioimplant, *American Journal of Neuroradiology*, 18, 1522–1525, 1997

143. J. Gaspar, P.J. Bartolo, F.M. Duarte, Cure and rheological analysis of reinforced resins for stereolithography, *Materials Science Forum*, 587–588, 563–567, 2008
144. P.J. Bartolo, J. Gaspar, Metal filled resin for stereolithography metal part, *CIRP Annals – Manufacturing Technology*, 57, 235–238, 2008
145. R. Sindelar, P. Buhler, F. Niebling, A. Otto, P. Greil, Solid freeform fabrication of ceramic parts from filler loaded preceramic polymers, *Proceedings of the Solid Free Form Fabrication Conference*, Austin, USA, 2002
146. J. Gaspar, P.J. Bartolo, Metallic stereolithography, *Proceedings of the 9th Biennial ASME Conference on Engineering Systems Design and Analysis*, Haifa, Israel, 2008
147. F.P.W. Melchels, J. Feijen, D.W. Grijpma, A review on stereolithography and its applications in biomedical engineering, *Biomaterials*, 6121–6130, 2010
148. K. Arcaute, B. Mann, R. Wicker, Stereolithography of spatially controlled multi-material bioactive poly(ethylene glycol) scaffolds, *Acta Biomaterialia*, 6, 1047–1054, 2010
149. C. Heller, M. Schwentenwein, G. Russmueller, F. Varga, J. Stampfl, R. Liska, Vinyl esters: low cytotoxicity monomers for the fabrication of biocompatible 3D scaffolds by lithography based additive manufacturing, *Journal of Polymer Science: Part A: Polymer Chemistry*, 47, 6941–6954, 2009
150. M. Schuster, C. Turecek, G. Weigel, R. Saf, J. Stampfl, F. Varga, R. Liska, Gelatin-based photopolymers for bone replacement materials, *Journal of Polymer Science: Part A: Polymer Chemistry*, 47, 7078–7089, 2009
151. S. Lu, K.S. Anseth, Photopolymerization of multilaminated poly(HEMA) hydrogels for controlled release, *Journal of Controlled Release*, 57, 291–300, 1999
152. J.W. Lee, G.S. Ahn, D.S. Kim, D.-W. Cho, Development of nano- and microscale composite 3D scaffolds using PPF/DEF-HA and micro-stereolithography, *Microelectronic Engineering*, 86, 1465–1467, 2009
153. M. Schuster, C. Turecek, F. Varga, H. Lichtenegger, J. Stampfl, R. Liska, 3D-shaping of biodegradable photopolymers for hard tissue replacement, *Applied Surface Science*, 254, 1131–1134, 2007
154. P.J. Bartolo, C.K. Chua, H.A. Almeida, S.M. Chou, A.S.C. Lim, Biomanufacturing for tissue engineering: present and future trends, *Virtual and Physical Prototyping*, 4, 203–216, 2009
155. N.E. Fedorovich, M.H. Oudshroon, D. Geemen, W.E. Hennink, J. Alblas, W.J.A. Dhert, The effect of photopolymerization on stem cells embedded in hydrogels, *Biomaterials*, 30, 344–353, 2009
156. J. Jansen, F.P.W. Melchels, D.W. Grijpma, J. Feijen, Fumaric acid monoethyl ester-functionalized poly(D,L-lactide)/N-vinyl-2-pyrrolidone resins for the preparation of tissue engineering scaffolds by stereolithography, *Biomacomolecules*, 10, 214-220, 2009
157. K.W. Lee, S.F. Wang, B.C. Fox, E.I. Ritman, M.J. Yaszemski, L.C. Lu, Poly(propylene fumarate) bone tissue engineering scaffold fabrication using stereolithography: effects of resin formulations and laser parameters, *Biomacromolecules*, 8, 1077–1084, 2007
158. V.A. Liu and S.N. Bhatia, Three-dimensional patterning of hydrogels containing living cells, *Biomedical Microdevices*, 4, 257–266, 2002

Chapter 2
History of Stereolithographic Processes

Paulo Jorge Bártolo and Ian Gibson

2.1 Introduction

There are a number of processes that can realize three-dimensional (3D) shapes such as those stored in the memory of a computer. An example is the use of holographic techniques [1], but these require many complex calculations to obtain the hologram and there is insufficient accuracy and clarity. A manual or a conventional mechanical process can also make a physical model, but such models require long fabricating times, high cost and excessive labor. To solve these kinds of problems, a new group of techniques called additive manufacturing (AM) technologies have been developed over the last 10 numbers [2–14].

AM is a collection of processes in which physical objects are quickly created directly from computer generated models. The basic concept of rapid prototyping is where 3D structures are formed by laminating thin layers according to two-dimensional (2D) slice data, obtained from a 3D model created on a CAD/CAM system [2–15].

Stereolithography is one of the most popular AM process. It usually involves the curing or solidification of a liquid photosensitive polymer by a laser beam scanned across its surface. The laser supplies energy that induces a chemical reaction, bonding large number of small molecules and forming a highly cross-linked polymer [16].

2.2 The Importance of a Prototype

In today's highly competitive marketplace with short life cycles of products, developing a new product to meet consumers' needs in a shorter lead time is very important for an enterprise. Facing this environment, the strategy of developing a product is transformed from "product-push" type to "market-pull". Thus, to

P.J. Bártolo (✉)
Centre for Rapid and Sustainable Product, Polytechnic Institute
of Leiria, Leiria, Portugal
e-mail: pbartolo@estg.iplei.pt

P.J. Bártolo (ed.), *Stereolithography: Materials, Processes and Applications*,
DOI 10.1007/978-0-387-92904-0_2, © Springer Science+Business Media, LLC 2011

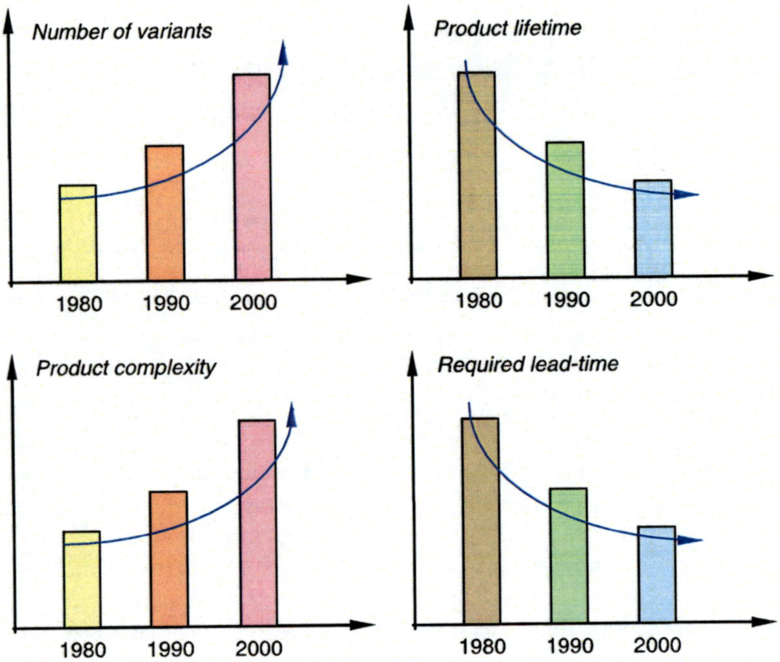

Fig. 2.1 Changes in the manufacturing industry during the last 30 years [17]

improve competitiveness, a product should not only satisfy consumers' physical requirements, but also should satisfy their needs, increasing the product complexity and reducing its lifetime [16, 17] as shown in Fig. 2.1. Besides, market segmentation has resulted in demand from individualistic consumers, which has led to the concept of "niche markets" increasing product choice [18]. Moreover, companies must meet customer expectations in terms of improved quality and lower cost of products. These new strategies adopted by modern companies lead to a tremendous change in their internal flexibility. As a consequence, the current industrial trend is moving from mass production, i.e. high volume and small range of products for manufacturing, to small volume and a wide range of products [18].

As a consequence of the increasing globalization, companies are now facing competition from low cost and newly industrialized countries putting market prices under pressure [16, 19]. Besides, technology development is increasing rapidly and ecological factors also became important sources of pressure [19]. International markets are therefore highly volatile and competition is brutal, imposing new demands on the innovative ability of companies. Moreover, it becomes increasingly important to rapidly develop new and successful products, requiring changes on how a product is developed. Thus, different groups in a company must cooperate more closely towards a common goal. This must be clear to everyone involved, and if cooperation is to be effective, it is essential to avoid communication problems [3].

According to Krouwel [20], the product development process encompasses five different phases: the information phase, concept phase, engineering phase, tooling phase, and production phase. The information phase encompasses market research, analysis of patents and competing products, etc. The concept phase corresponds to the design-modeling step and is generally the phase where a computer model is created (virtual or soft prototyping phase) [3]. In the engineering phase, engineers study the product in order to find the best and most simple technological solution in order to implement the initial concept. During this phase, a prototype of the product is usually made and tested. Only after the engineering phase is completed the tools, preparation for manufacturing (tooling phase) and the production phase will start. These first three phases represent almost 50% of the product development cycle and among them the engineering phase can represent 25–40% [20]. These phases are performed in a sequence, which means that any serious error detected in the engineering phase implies a new concept phase and repetition of the process. Therefore, for most products the majority of development time occurs in the concept and engineering validation phases, and changes to a design become more costly as they approach the production phase [3, 20]. Moreover, with more complex products, the probability of errors increases dramatically. Thus, it is important to identify any inconsistencies or problems early on. The possibility of creating a computer model of the product to be manufactured, and at the same time using that model to create a prototype, aids in this process by helping to ensure that the product which is going to be produced is exactly the way the product designers, engineers, and customers want it. Through the prototypes, product designers and engineers can get feedback on design information for optimization as well as for further manufacturing processing, reducing errors from incorrect interpretation of the design [3]. Moreover, fabrication of a prototype of the product in the concept phase provides the possibility of starting the engineering phase almost in parallel, reducing significantly the product development cycle. This way, prototypes can be important communication tools as well as useful tools for testing the concept to see if it performs as required or needs improvement, and for esthetic assessment, minimizing time-consuming discussions and evaluations [2, 3].

AM as a group of processes for the rapid production of models also provides the necessary support for the adaptation of simultaneous or concurrent engineering. Simultaneous engineering (SE) is a strategy of bringing all the teams in a company to participate together at an early stage in the design process. SE methodology requires everyone in the company to perform their tasks in parallel, in contrast with traditional manufacturing processes where the product idea moves sequentially through the company (see Fig. 2.2).

AM also enables effective implementation of reverse engineering (RE), permitting the redesign of an existing product. Through RE, the shape of an existing product is digitized, creating the correspondent surface model, which can then be manipulated (re-design process), and finally the model of the new product can be produced using AM processes [22–25].

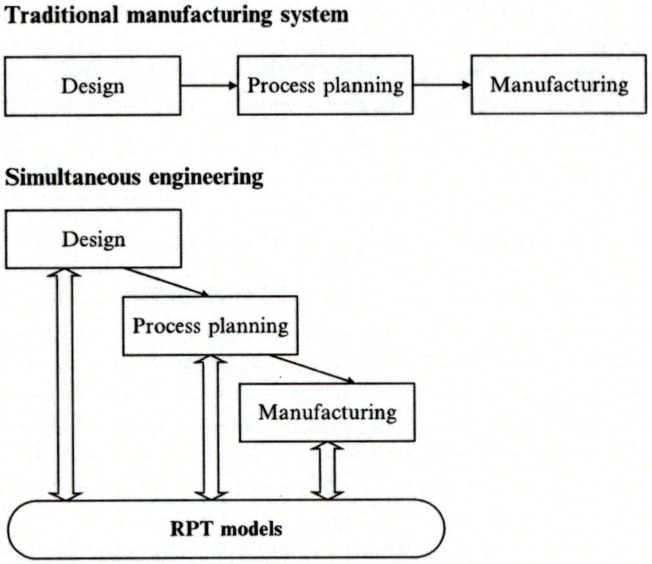

Fig. 2.2 Comparison between the traditional manufacturing system and the new manufacturing approach through AM technologies [21]

2.3 Techniques to Produce Prototypes

The traditional way to make a physical model or prototype, besides hand-made wood or clay models, is to use numerically controlled machines as CNC (computer numerical control) milling, electric-discharge machining, turning, and grinding machines [26, 27].

In such processes, the object is revealed by cutting away material from a starting block and therefore these processes are called *subtractive methods* [3, 26–28]. While conventional machine tools are usually effective in producing the desired object, they are deficient in many respects. First, a large amount of waste material for disposal is produced. Further, such methods usually require expensive object-specific tooling, the setting up of machining protocols, and generation and programming of 3D tool paths which all require much time and a great deal of human judgement and expertise. The cost and time to set up and run machine-specific tooling, along with the initial costs for tooling, make conventional manufacturing processes both time and cost intensive for small productions like models or prototypes [28].

The final difficulty associated with such processes is the impossibility of making special object configurations [3, 26, 27]. Effectively, these conventional methods are usually best suited for producing symmetrical objects and objects where only the exterior is machined. However, when a desired object has an unusual shape or specific internal features, the machining becomes more difficult and quite often the object must be divided into segments for production [3, 26, 27]. In many cases, a particular object configuration is not possible because of the limitations imposed upon the tool.

Other important classes of conventional manufacturing processes are the so-called *formative methods* [3], e.g. casting, injection molding, compressive molding, etc. Through these processes the material is forced into the desired shape using molds, in which the material is made to harden and solidify. However, these processes are still often highly expensive, time consuming, and require a broad range of expertise.

Recently, AM emerged as a step forward in the product cycle, reducing lead times for new products, as well as improving design manufacturing and tooling costs [2–14]. In AM, a single automated system can be used to produce models directly from engineering designs. Such systems are limited only by the size of the model and not by its complexity [2, 3].

AM technologies are *additive methods* [2–14] because they build objects layer by layer, and as a consequence they are also generally known as Layered Manufacturing Techniques [8, 9]. AM processes are similar processes to 2D printing and plotting technologies using both vector-based and raster-based imaging techniques. The various AM processes include laser sintering, lamination, extrusion, ink-jet printing, and photolithographic systems [2–14, 16]. AM technologies have been mainly used for [2–5, 13, 16, 29–38]:

- Physical verification of a previously defined CAD model
- Form, fit, and function testing
- Creating models without regard to draft angles, parting lines, etc
- Concept presentations and design reviews
- Direct tooling as well as masters for rapid tooling, using conversion technologies such as investment casting and silicone, epoxy and spray metal molds
- Reducing time-to-market
- Creating anatomical models constructed from computer-aided tomography data for surgical planning, prosthesis design, scaffolds for tissue engineering and dental implants
- Producing relief models for geographical applications
- Creating 3D portraits (three-dimensional photography) using data produced by 3D shape digitizing technology.

2.4 Stereolithographic Processes

Photolithographic systems build shapes using light to selectively solidify photosensitive resins. There are two basic approaches:

- *Laser lithography*
- *Photo-mask*

The laser lithography (or Stereolithography) approach is currently one of the most used AM technologies. Models are defined by scanning a laser beam over a photopolymer surface. Photo-mask systems build models by shining a flood lamp through a mask, which lets light through it and is a method commonly employed in microlithography.

2.4.1 *History and Development of the Photolithographic Systems*

Lithography is the art of reproduction of graphic objects and comprises different techniques, such as photographic reproduction, photosculpture, xerography and microlithography. Modern photolithographic AM systems harness the principle of computer generated graphics combined with photosensitive materials to produce 3D objects.

Photosensitive materials have been known at least since the time of the ancient Egyptians and probably long before them. The alchemists of the Middle Ages and Renaissance knew about the phenomena of blackening silver salts by light exposure. However, they did not realize that this phenomenon was due solely to the Sun's light and not to its heat. In fact, they argued that all changes produced in bodies exposed to sunlight were due to heat and not to light [39].

In 1775, Schultz discovered that a silver-containing precipitate used to produce phosphorous, turned purple when illuminated by sunlight, whilst the portion turned away from the light remained white. After that, he divided the mixture into two lots, one of which he kept in the dark, exposing the other to sunlight, with a thin cord tied round the bottle, and again a change in the precipitate exposed to the sunlight was observed. He repeated the experiment by covering the bottle with paper from which he had cut out words and entire sentences, this way "writing" the words and sentences in the solution [39].

Another remarkable achievement was due to Niépce (1822), when he made his first successful and permanent copy of an engraving of Pope Pius VII [39]. He dissolved bitumen of Judea in oil of lavender, and spread a thin layer on a glass plate on which he superimposed an engraving of Pope Pius VII made transparent by oiling. After exposure to light, the bitumen under the white parts of the engraving became hard, whilst that under the dark lines remained soluble [34, 39].

2.4.1.1 Origins of Modern Stereolithography

The first significant work associated with modern photolithographic AM systems only emerged during the 1970s [8, 10]. In 1971, Swainson [40] presented a patent for a system where two intersecting beams of radiation produce a phase change in

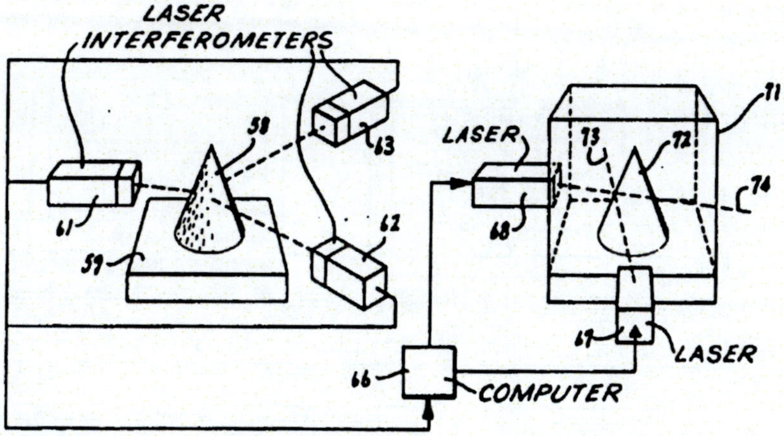

Fig. 2.3 Photochemical machining process [40]

a material to build 3D objects. The essential features of this process, named photochemical machining [41], are illustrated in Fig. 2.3. The object through this process can be formed by either photochemically cross-linking or degrading a polymer [42–44]. However, the major problem of this process was due to the photonic absorption by the photopolymeric system used, which occurs somewhere along the paths of each laser, initiating polymerisations in spots that differ from the planned ones [41]. In the 1980s, the idea was abandoned due to funding problems, without achieving optimum working parameters, adequate materials, and good accuracy of final models [34].

Kodama [45] described an automatic method for fabricating 3D models in layered stepped stages using a photosensitive polymer. Light capable of curing the polymer was directed onto the surface, and the desired shape of a layer was created by using an appropriate mask (Fig. 2.4a, b) or an optical fiber manipulated by an X–Y plotter (Fig. 2.4c).

Herbert [46] described the design of two sets of apparatus for producing replicas of solid objects, in a layer-by-layer way, using a photosensitive polymer. The purpose of the first one (Fig. 2.5) was only for the construction of solids of revolution, made by rotating a layer of polymer and focusing a spot of light on the layer. The second apparatus constructed solid objects of any desired cross-section (Fig. 2.6).

Hull conceived the idea of modern stereolithography [47–49]. According to the principles of stereolithography (Fig. 2.7), a 3D object is formed layer by layer in a stepwise fashion out of a material capable of solidification upon exposure to ultraviolet (UV) radiation [47–49]. Moreover, the non-transformed layers typically adhere to the previously formed layers through the natural adhesive properties of the photosensitive polymer upon solidification. Almost in parallel, André, who prepared different patent applications [50, 51] conducted similar work in France.

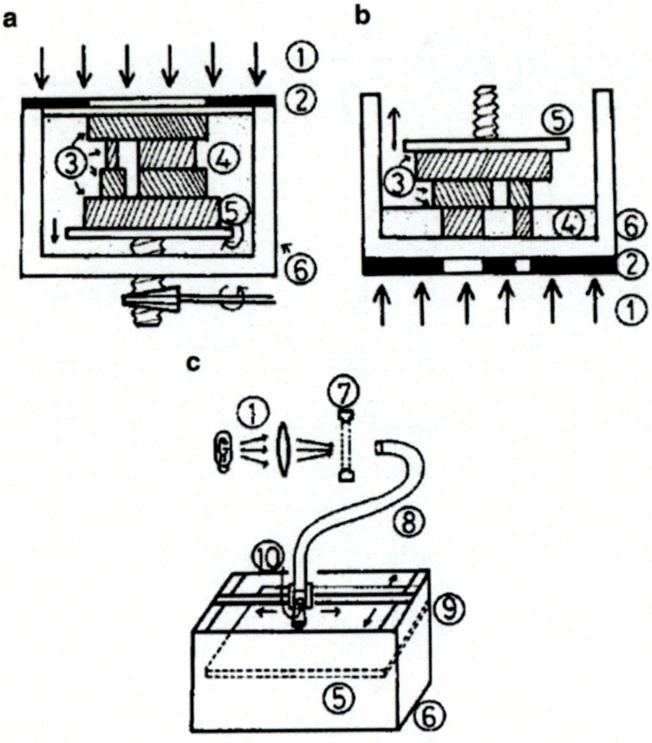

Fig. 2.4 Schematics of the three systems studied by Kodama [45]

Fig. 2.5 Herbert's apparatus for construction of solids of revolution [46]

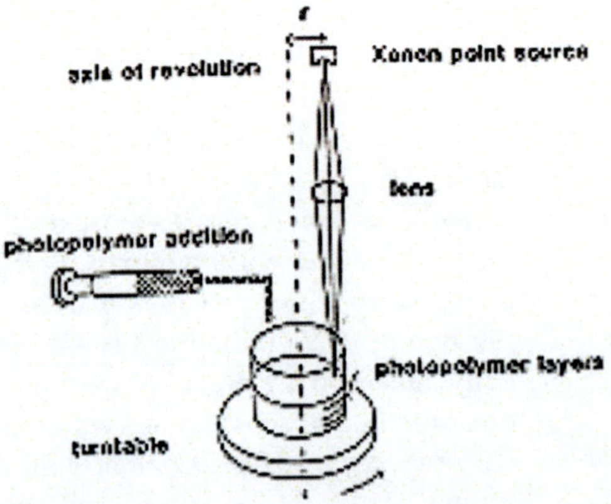

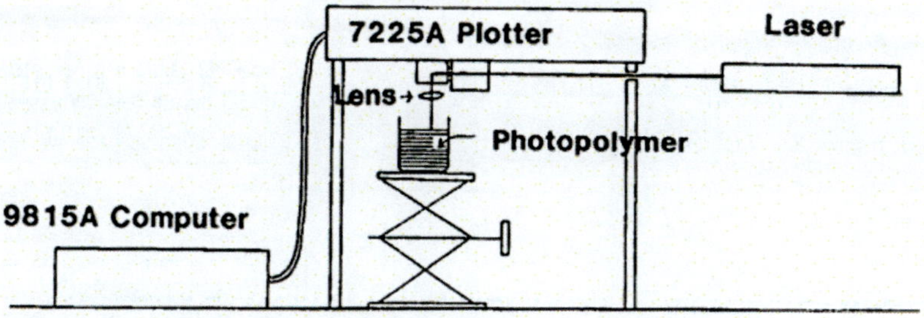

Fig. 2.6 Herbert's apparatus for generating models by polymerisation [46]

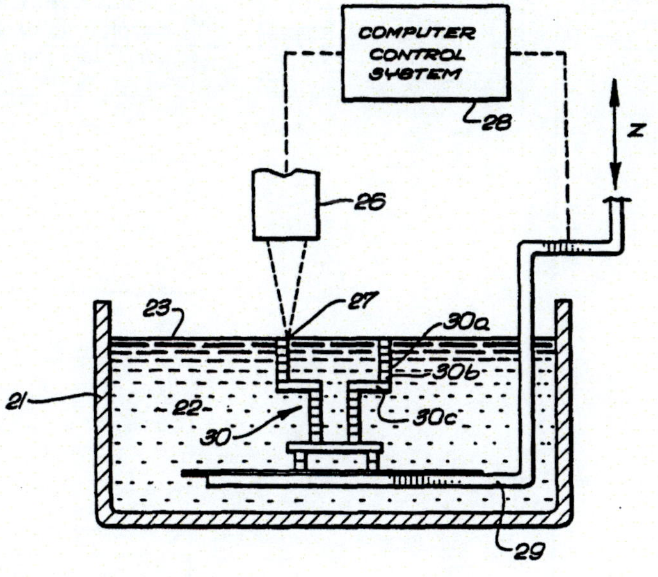

Fig. 2.7 Hull's stereolithography process [47]

The entire process of conceiving a model using stereolithography (the different phases of the building process are shown in Fig. 2.8) comprises the following steps [2, 3, 16, 47–49]:

1. Create a solid or surface model on a CAD system
2. Export the CAD model
3. Add support structures
4. Specify the build style variables and parameters necessary for slicing
5. Slice the computer model to generate the information that controls the SL apparatus

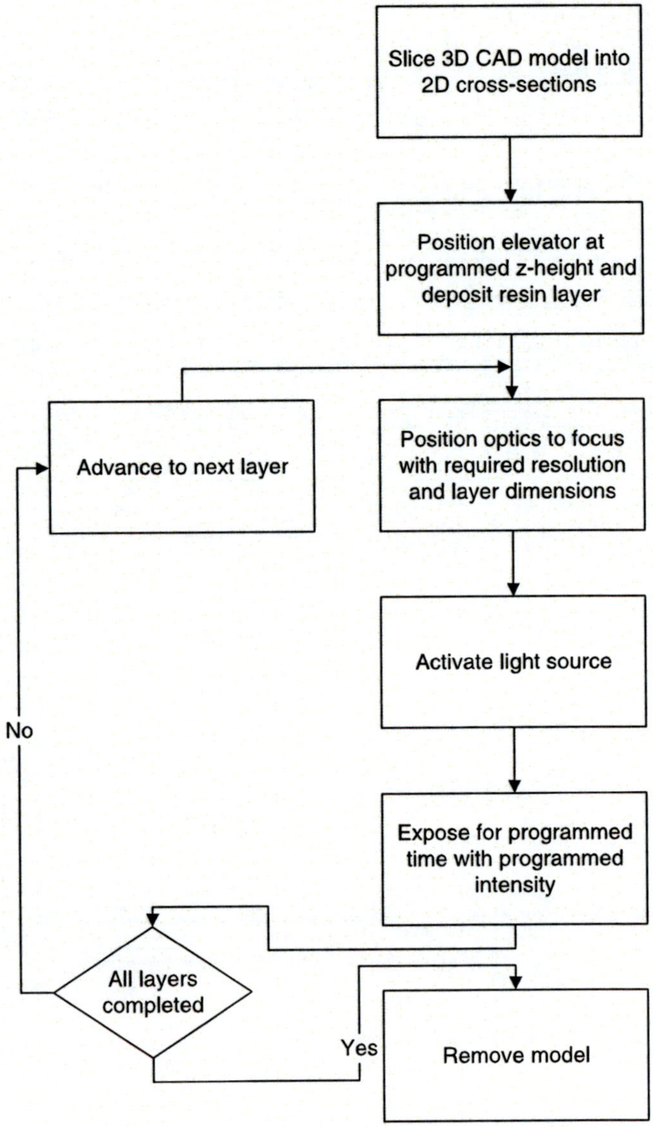

Fig. 2.8 Flowchart of the stereolithography building process [47]

6. Build the model using the slice file
7. Post-process and clean the part
8. Post-curing to complete the cure process.

The block diagram of the stereolithography system as proposed by Hull [47, 48] is shown in Fig. 2.9.

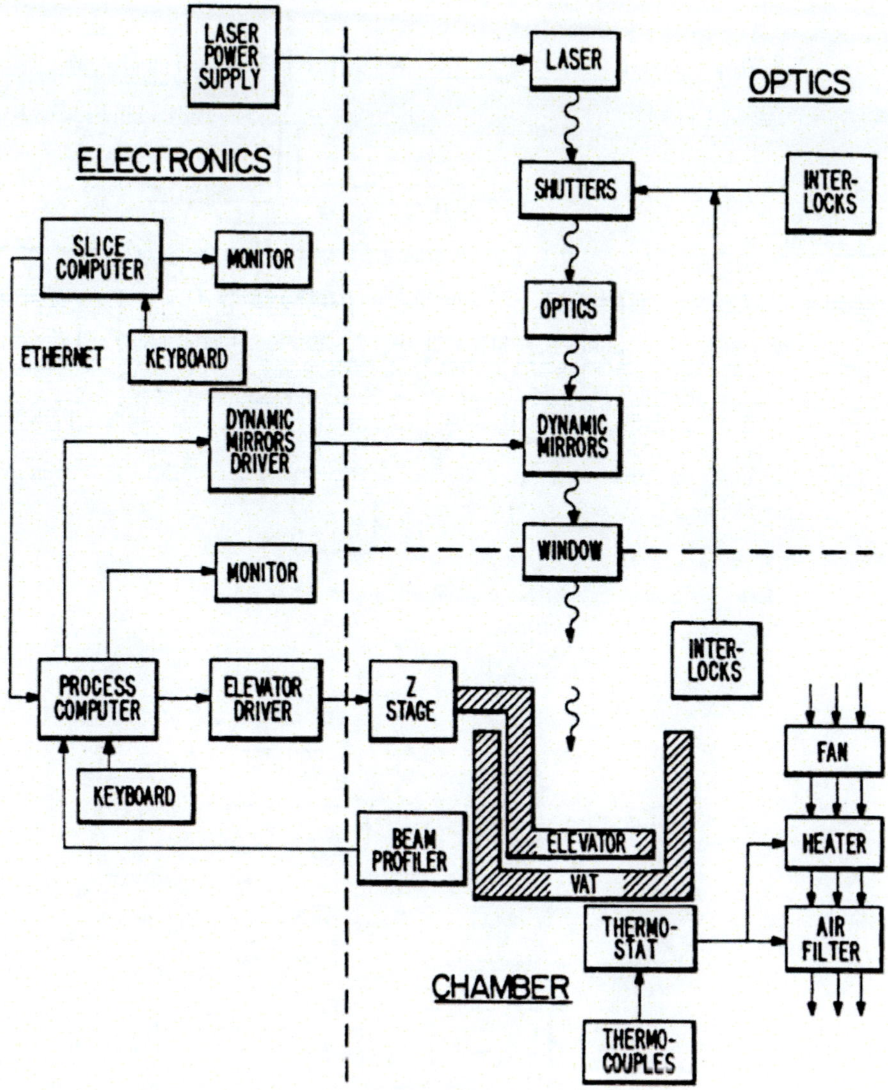

Fig. 2.9 The block diagram of the stereolithography system [48]

Hull also proposed other stereolithographic strategies as shown in Fig. 2.10. In this system the physical object is pulled up from the liquid resin, rather than down and further into the liquid photopolymeric system [47]. The radiation passes through a UV transparent window.

In order to minimize the amount photopolymerisable material required for the fabrication process, Murphy et al. [52] proposed a stereolithographic method and apparatus in which a membrane separates two liquid phases. The system (Fig. 2.11)

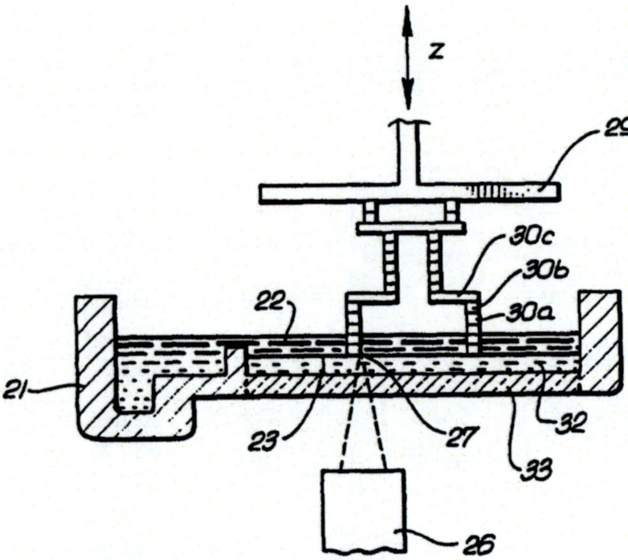

Fig. 2.10 The ascending fabrication platform proposed by [47]

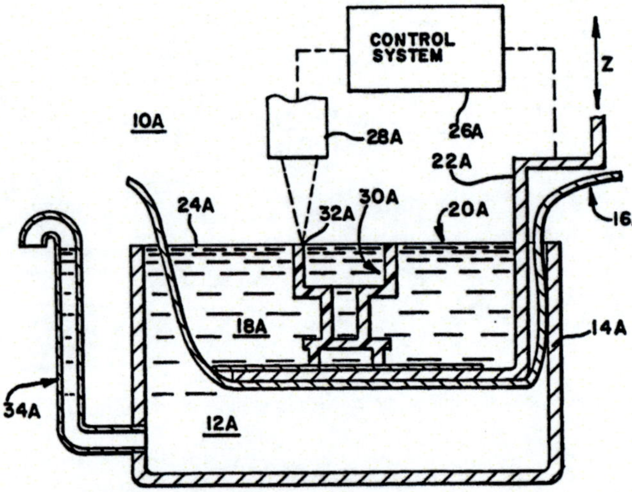

Fig. 2.11 Stereolithographic apparatus with membrane [52]

comprises a nonpolymerisable fluid phase, an impermeable movable membrane positioned on top of the fluid phase, a photopolymerisable liquid resin positioned on top of the membrane and a radiation source positioned above the polymerisable material [52].

Almquist and Smalley [53] proposed the concept of thermal stereolithography that uses a solid material, instead of a liquid one, which is flowable when subjected to light.

Marutani [54, 55] has proposed a new stereolithographic system that polymerises a liquid resin inside the vat rather than at the surface. In this system a UV laser beam penetrates through a pipe into the vat containing the liquid resin and solidifies it, thereby eliminating the need for successive layer deposition as in conventional stereolithography.

An important evolution step in the stereolithography domain is the so-called color stereolithography [56]. This process uses a clear liquid resin containing additives that color upon exposure to high doses of UV radiation. Through this process, each layer is cured in the usual way, using a dose of UV radiation sufficient for curing but not for coloring. When the "writing" process of each layer is completed, the laser rescans the area required to be colored at a lower speed, delivering in this way a much higher dose of UV radiation. This provides a means to highlight features in a model and, since the uncoloured stereolithography resins are transparent, to show features that may be embedded inside the encompassing solid object. A more laborious coloring strategy was proposed by Im et al. [57] (Fig. 2.12).

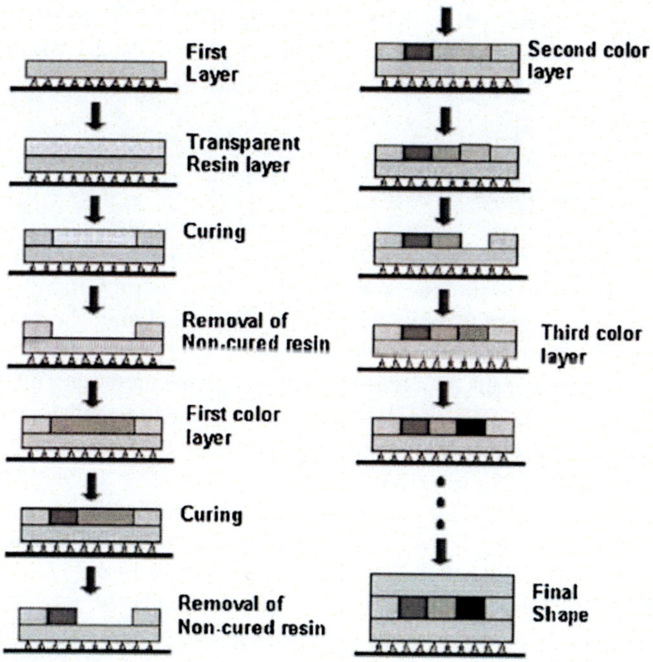

Fig. 2.12 Colouring stereolithography [57]

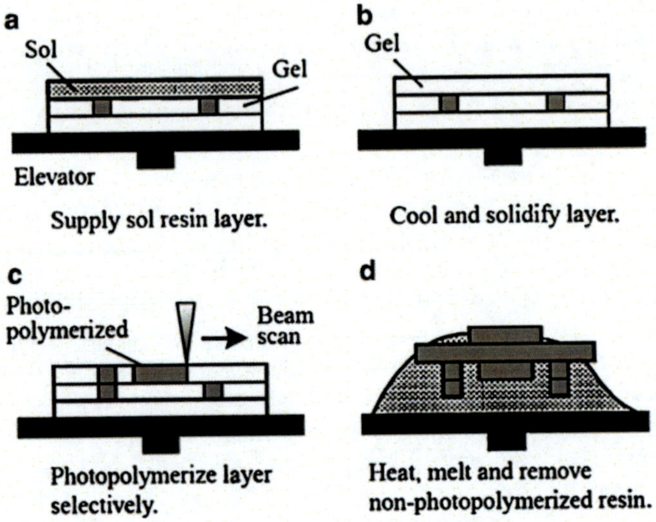

Fig. 2.13 Stereolithography using a solgel transformable photopolymer [60]

In order to eliminate the need of support structures, Murakami [58, 59] proposed a stereolithographic approach for fabricating solid (cured) objects from solid (non-cured) photopolymeric resin (Fig. 2.13). First, a liquid resin is supplied to form a new layer, cooled to the gel state (between −50 and −10°C), and then selectively photopolymerised. The final object is revealed by heating the gel resin block [58–60].

Other important inventions are listed on Table 2.1.

2.4.1.2 Photomask Systems

Pomerantz [61, 62] proposed a photomask system to produce 3D models (Fig. 2.14). The steps of his technique, currently known as Solid Ground Curing (SGC), are: deposit a thin layer of polymer; illuminate the polymer through a xerographically produced mask having geometry of a single cross section; removal by suction of the uncured material; fill the areas vacated by the uncured material with water or wax; cure or freeze the rest of the layer; grind the surface to establish a uniform layer; repeat the earlier steps until the model is complete.

Fudim [63, 64] developed a technique similar to the Pomerantz method. His technique [63, 64] involves the illumination of a photosensitive polymer, with UV radiation through masks and a piece of flat material transparent to the radiation that remains in contact with the liquid layer being formed. The method is simpler than the SGC technique, but requires an operator to create and manually position each mask.

Photomasking systems generally require the generation of many masks, and precise mask alignments. One solution to this problem is to use a liquid crystal

Table 2.1 Other relevant laser lithography patents

Inventors	Topic	Patent
Hull et al.	Discloses various removable support structures for stereolithography	US Patent 4999143
Modrek et al.	Presents techniques for post processing objects produced by stereolithography	US Patent 5076974
Spence et al.	Proposes the use of multiple wavelengths in the exposure of a stereolithographic medium	US Patent 5182056
Hull et al.	Discloses a program called Slice and various techniques for converting 3D object data into data descriptive of cross-sections	US Patent 5184307
Allison et al.	Proposes various build/exposure styles and various techniques for reducing object distortion	US Patent 5256340
Almquist et al.	Proposes various recoating techniques for stereolithography. Presents techniques such as (1) an inkjet dispensing device, (2) a fling recoater, (3) a vacuum applicator, (4) a stream recoater, (5) a counter-rotating roller recoater, and (6) a technique for deriving sweep extents	US Patent 08/790005
Partanen et al.	Proposes the application of solid-state lasers to stereolithography	US Patent 08/792347
Partanen et al.	Discloses the use of a pulsed radiation source in stereolithography	US Patent 08/847855
Bloomstein et al.	Presents a stereolithographic patterning system with variable size exposure areas	US Patent 633234

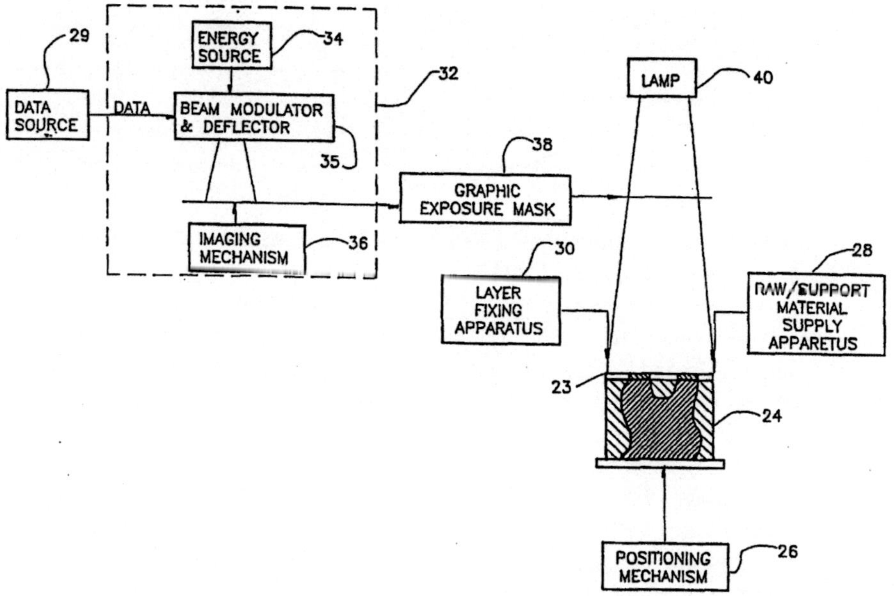

Fig. 2.14 The photo-fabrication system proposed by Pomerantz [61]

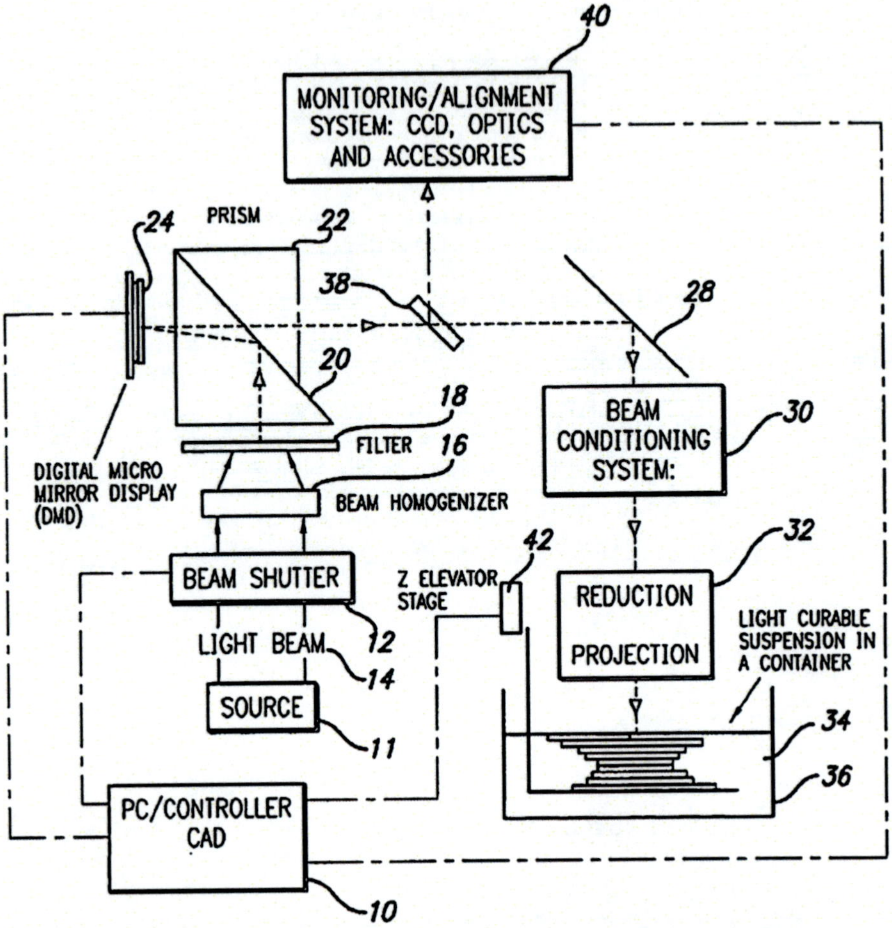

Fig. 2.15 Dynamic mask projection stereo micro lithography proposed by Zhang [65]

display (LCD) or a digital processing projection system as a reconfigurable mask [16]. Through this process, the CAD model is converted to a grayscale contour map, and the LCD mask modulates the light intensity distribution according to the gray-scale contour map of the model. However, due to the large pixel size and very low transmission in UV, the device's resolution is limited and contrast is poor. Therefore, several stereolithographic systems using a Digital Micromirror Device as a dynamic mask have been proposed (Fig. 2.15) [65, 66].

Recently Murakami [60] from the University of Tokyo, Japan, proposed a new stereolithographic system involving the separate use of a liquid photo-initiator and a photopolymer without photoinitiator. In this process (Fig. 2.16), the resin without photoinitiator is supplied as a layer, and then a mask pattern is drawn onto the surface with photoinitiator by inkjet printing. When the surface is exposed to UV light, only the pattern drawn with the photoinitiator, which acts as a positive mask, is cured.

Fig. 2.16 Stereolithography process using positive direct-mask exposure [60]

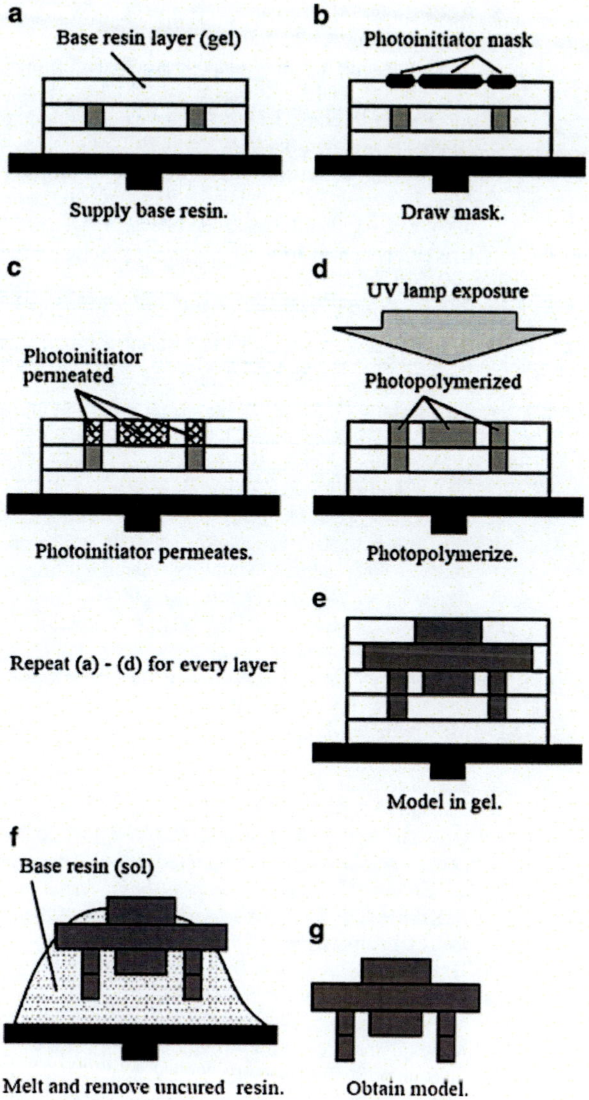

References

1. J. Wilson, J.F.B. Hawkes. *Lasers, principles and applications*. Prentice Hall, New York, 1987
2. I. Gibson, D.W. Rosen, B. Stucker. *Additive manufacturing technologies: rapid prototyping to direct digital manufacturing*. Springer, New York, 2009
3. C.K. Chua, K.F. Leong, C.S. Lim. *Rapid prototyping – principles and applications*. World Scientific Publishing, Singapore, 2003
4. I. Gibson. Rapid prototyping from product development to medicine, *Virtual and Physical Prototyping*. 1, 31–42, 2006

5. B. Bidanda, P.J. Bartolo. *Virtual prototyping & bio-manufacturing in medical applications*. Springer, New York, 2008
6. S. Kumas, J.P. Kruth. Composites by rapid prototyping technology. *Materials and Design*, 31, 850–856, 2010
7. J.P. Kruth, G. Levy, T.H.C. Childs. Consolidation phenomena in laser and powder-bed based layered manufacturing. *CIRP Annals – Manufacturing Technology*, 35, 730–759, 2007
8. P.J. Bartolo, J. Gaspar. Metal filled resin for stereolithography metal part. *CIRP Annals – Manufacturing Technology*, 57, 235–238, 2008
9. J. Cryzewski, P. Burzynski, K. Gawel, J. Meisner. Rapid prototyping of electrically conductive components using 3D printing technology. *Journal of Materials Processing Technology*, 209, 5281–5285, 2009
10. T. Grimm. *User's guide to rapid prototyping*. Society of Manufacturing Engineers, Dearborn, 2004
11. N. Tolochko, S. Mozzharov, T. Laoui, L. Froyen. Selective laser sintering of single- and two-component metal powders. *Rapid Prototyping Journal*, 9, 68–78, 2003
12. P.J. Bartolo, G. Mitchell. Stereo-thermal-lithography: a new principle for rapid prototyping. *Rapid Prototyping Journal*, 9, 150–156, 2003
13. G.N. Levy, R. Schindel, J.-P. Kruth. Rapid manufacturing and rapid tooling with layer manufacturing (LM) technologies, state of the art and future perspectives. *CIRP Annals – Manufacturing Technology*, 52(2), 589–609, 2003
14. D.L. Bourell, J.B. Beaman, M.C. Leu, D.W. Rosen. A brief history of additive manufacturing and the 2009 roadmap for additive manufacturing: looking back and looking ahead. *US-Turkey Workshop on Rapid Technologies*, 2009
15. J. Zhao, R. Xia, W. Liu, H. Wang. A computing method for accurate slice contours based on na STL model. *Virtual and Physical Prototyping*, 4, 29–37, 2009
16. P.J. Bartolo. *Optical approaches to macroscopic and microscopic engineering*, PhD Thesis, University of Reading, UK, 2001
17. M. Greulich, M. Greul, T. Pintat. Fast functional prototypes via multiphase jet solidification. *Rapid Prototyping Journal*, 1, 20–25, 1995
18. N.P. Karapatis, J.P.S. Van Griethuysen, R. Glardon. Direct rapid tooling. *Rapid Prototyping Journal*, 4, 77–78, 1998
19. G.A. Hindson, A.K. Kochhar, P. Cook. Procedures for effective implementation of simultaneous engineering in small to medium enterprises. *Proceedings of the Institution of Mechanical Engineers, Part B: Journal of Engineering Manufacture*, 212, 251–258, 1998
20. P. Krouwel, in *Research in design thinking*, Edited by N. Cross, C. Dorst and N. Rootenbury, Delft University Press, Delft, 1992
21. D. Kochan, C.K. Chua. State-of-the-art and future trends in advanced rapid prototyping and manufacturing. *International Journal of Information Technology*, 1, 173–184, 1995
22. S. Azernikov, A. Fischer. Emerging non-contact 3D measurement technologies for shape retrieval and processing. *Virtual and Physical Prototyping*, 3, 85–91, 2008
23. S. Rianmora, P. Koomsap, D.P.V. Hai. Selective data acquisition for direct integration of reverse engineering and rapid prototyping. *Virtual and Physical Prototyping*, 4, 227–239, 2009
24. F. Laroche, A. Bernard, M. Cotte. Advanced industrial archaeology: a new reverse-engineering process for contextualising and digitising ancient technical objects. *Virtual and Physical Prototyping*, 3, 105–122, 2008
25. N.M.F. Alves, P.J.S. Bartolo. Automatic 3D shape recovery for rapid prototyping. *Virtual and Physical Prototyping*, 3, 123–137, 2008
26. W. De Vries. *Analysis of material removal processes*. Springer-Verlag, New York, 1992
27. S. Kalpakjian, S.R. Schmid. *Manufacturing engineering and tooling*. Prentice Hall, New Jersey, 2000
28. E.M. Malstrom. *Manufacturing cost estimating*. Marcel Dekker, New York, 1981

29. I. Madrazo, C. Zamorano, E. Magallón, T. Valenzuela, A. Ibarra, H. Salgado-Ceballos, I. Grijalva, R.E. Franco-Bourland, G. Guízar-Sahagún. Stereolithography in spine pathology: a 2-case report. *Surgical Neurology*, 72, 272–275, 2009
30. V. Dedoussis, V. Canellidis, K. Mathioudakis. Aerodynamic experimental investigation using stereolithography fabricated test models: the case of a linear compressor blading cascade. *Virtual and Physical Prototyping*, 3, 151–157, 2008
31. V. Dedoussis, J. Giannatsis. Stereolithography assisted redesign and optimisation of a dishwasher spraying arm. *Rapid Prototyping Journal*, 10, 255–260, 2004
32. J. Giannatsis, V. Dedoussis, D. Karalekas. Architectural scale modelling using stereolithography. *Rapid Prototyping Journal*, 8, 200–207, 2002
33. P.J. Bártolo, C.K. Chua, H.A. Almeida, S.M. Chou, A.S.C. Lim. Biomanufacturing for tissue engineering: present and future trends. *Virtual and Physical Prototyping*, 4, 203–216, 2009
34. M. Burns. *Automated fabrication – improving productivity in manufacturing*. Prentice Hall, New Jersey, 1993
35. Y. Morita, T. Noikura, R. Petzold, M. Blank, W. Kalender, S. Hiura, A. Okubo, K. Sugihara, T. Kamiinaba, Y. Izumi. Rapid prototyping for dentistry in Japan. *Proceedings of the eighth international conference on rapid prototyping*, Edited by T. Nakagawa, Y. Marutani, M. Imamura, M. Agarwala, A. Lightman, D. Klosterman, and R.P. Chartoff, University of Dayton, 2000
36. C.C. An, R.H. Chen. The experimental study on the defects occurrence of SL mold in injection molding. *Journal of Materials Processing Technology*, 201, 706–709, 2008
37. S. Rahmati, P. Dickens. Rapid tooling analysis of stereolithography injection mould tooling. *International Journal of Machine Tools & Manufacture*, 47, 740–747, 2007
38. G. Kakarala, A.D. Toms, J.H. Kuiper. Stereolithography models for biomechanical testing. *The Knee*, 13, 451–454, 2006
39. H. Gernsheim, A. Gernsheim. *The history of photography: from the camera obscura to the beginning of the modern era*. McGraw Hill, New York, 1969
40. W.K. Swainson. *Method, medium and apparatus for producing three-dimensional figure product*, US Patent 4041476, 1977
41. R.E. Schwerzel, V.E. Wood, V.D. McGinniss, C.M. Verber. 3D photochemical machining with lasers, Applications of lasers to industrial Chemistry. *SPIE*, 458, 90–97, 1984
42. W.K. Swainson, S.D. Kramer. *Three-dimensional pattern making methods*, US Patent 4333165, 1982
43. W.K. Swainson, S.D. Kramer. *Three-dimensional patterned media*, US Patent 4466080, 1984
44. W.K. Swainson, S.D. Kramer. *Method and media for accessing data in three dimensions*, US Patent 4471470, 1984
45. H. Kodama. Automatic method for fabricating a three-dimensional plastic model with photo-hardening polymer. *Review of Scientific Instruments*, 52, 1770–1773, 1981
46. A.J. Herbert. Solid object generation. *Journal of Applied Photographic Engineering*, 8, 185–188, 1982
47. C.W. Hull. *Apparatus for production of three-dimensional objects by stereolithography*, US Patent 4575330, 1986
48. C.W. Hull, S.T. Spence, D.J. Albert, D.R. Smalley, R.A. Harlow, P. Steinbaugh, H.L. Tarnoff, H.D. Nguyen, C.W. Lewis, T.J. Vorgitch, D.Z. Remba. *Method and apparatus for production of three-dimensional objects by stereolithography*, US Patent 5059359, 1991
49. C.W. Hull, S.T. Spence, D.J. Albert, D.R. Smalley, R.A. Harlow, P. Stinebaugh, H.L. Tarnoff, H.D. Nguyen, C.W. Lewis, T.J. Vorgitch, D.Z. Remba. *Method and apparatus for production of high resolution three-dimensional objects by stereolithography*, US Patent 5184307, 1993
50. J.C. André, M. Cabrera, J.Y. Jezequel, A. Méhauté. French Pat. 2583333, 1985
51. J.C. André, A. Méhauté, O. Witthe. *Dispositif pour realiser un module de piece industrielle*, French Pat. 8411241, 1984
52. E.J. Murphy, J.J. Krajewski, R.E. Ansel. *Stereolithographic method and apparatus in which a membrane separates phases*, US Patent 5011635, 1991

53. T.A. Almquist, D.S. Smalley. *Thermal stereolithography*, US Patent 5672312, 1997
54. Y. Marutani, T. Kamitani. 3-Dimensional exposure using an air bubble in the resin. *Proceedings of the seventh international conference on rapid prototyping*, Edited by A.J. Lightman and R.P. Chartoff, University of Dayton, 1997, 213
55. J.-P. Kruth, M.C. Leu, T. Nakagawa. Progress in additive manufacturing and rapid prototyping. *CIRP Annals – Manufacturing Technology*, 47, 525–549, 1998
56. B. Swaelens, W. Vancraen. Laser photopolymerisation models based on medical imaging: a development improving the accuracy of surgery. *Proceedings of the seventh international conference on rapid prototyping*, Edited by A.J. Lightman and R.P. Chartoff, University of Dayton, 1997, 250
57. Y.G. Im, S.I. Chung, J.H. Son, Y.D. Jung, J.G. Jo, H.D. Jeong. Functional prototype development: inner visible multi-color prototype fabrication process using stereo lithography. *Journal of Materials Processing Technology*, 130–131, 372–377, 2002
58. T. Murakami, A. Kamimura, N. Nakajima. Refrigerative stereolithography using sol-gel transformable photopolymer resin and direct masking. *Solid Freeform and Additive Manufacturing – 2000*, Edited by S.C. Danforth, D. Dimos, and F. Pritz, Materials Research Society, Warrendale, 2000
59. T. Murakami, A. Kamimura, N. Nakajima. Refrigerative stereolithography using direct masking. *Proceedings of the eigth International Conference on Rapid Prototyping*, Japan Society of Die and Molds Technology, Tokyo, Japan, 2000
60. T. Murakami, T. Yada, G. Kobayashi. Positive direct-mask stereolithography with multiple-layer exposure: layered fabrication with stair step reduction. *Virtual and Physical Prototyping*, 1, 73–81, 2006
61. I. Pomerantz, J. Cohen-Sabban, A. Bieber, J. Kamir, M. Katz, M. Nagler. *Three dimensional modelling apparatus*, US Patent 4961154, 1990
62. I. Pomerantz, S. Gilad, Y. Dollberg, B. Ben-Ezra, Y. Sheinman, G. Barequet, M. Katz. *Three dimensional modelling apparatus*, US Patent 5519816, 1996
63. E.V. Fudim. *Method and apparatus for production of three-dimensional objects by photo-solidification*, US Patent 4801477, 1989
64. E.V. Fudim. *Method and apparatus for production of three-dimensional objects by photo-solidification*, US Patent 4752498, 1988
65. X. Zhang. *Dynamic mask projection stereo micro lithography*, US Patent 2005/0259785 A1, 2005
66. C. Sun, N. Fang, D.M. Wu, X. Zhang. Projection micro-stereolithography using digital micro-mirror dynamic mask. *Sensors and Actuators A*, 121, 113–120, 2005

Chapter 3
Infrared Laser Stereolithography

André Luiz Jardini Munhoz and Rubens Maciel Filho

3.1 Introduction

Infrared laser stereolithography is a new process of stereolithography, substantially different from those already existents; that uses an infrared laser (carbon dioxide laser) to localize the curing reaction of a thermo-sensitive materials building the desired three-dimensional (3D) shape. Being developed by Scarparo and collaborators at the University of Campinas (UNICAMP) in Brazil, since 1992 [1], and patented in 2002, this technique is to have commercial value, with thermo-sensitive material that must have useful properties for applications in rapid manufacturing and tooling, medicine, or other important fields. Certainly, further advances and process optimization are a part of the ongoing activities that are in consonance with a large number of programs around the world of very interesting work towards improvement of stereolithography process.

Conventional stereolithography uses ultraviolet (UV) lasers, typically operated at a wavelength of 352 nm, to cure photosensitive polymers, by computer control of the direction and power of the laser beam to fabricate complex structures [2, 3]. Using computer control of the optical and mechanical elements as well as the thermal and chemical conditions, it has proved to be possible to fabricate complex structures with high spatial resolution.

It is known that the study and perfecting of new industrial parts are now made easier and faster by using the stereolithography. It is believed that the application of IR laser radiation to thermo-sensitive materials, such as epoxy resins, may provide a novel, reliable, and cost-effective means of manufacturing industrial prototypes [4]. It has been reported in the literature, for example, that the automobile industry plans to incorporate new light weight materials into automobiles. Plastic models built with stereolithography resin present very good surface finish and strength and may also be translucent.

A.L. Jardini Munhoz (✉)
School of Chemical Engineering, State University of Campinas, Campinas, São Paulo, Brazil
e-mail: jardini@feq.unicamp.br

P.J. Bártolo (ed.), *Stereolithography: Materials, Processes and Applications*,
DOI 10.1007/978-0-387-92904-0_3, © Springer Science+Business Media, LLC 2011

3.2 Potential Applications of Lasers Sources in Stereolithography

The internationally recognized term LASER is an acronym standing for *Light Amplification by Stimulated Emission Radiation.* The noun laser represents an item of equipment utilizing the principle above to produce a light beam having remarkable properties [5].

The laser, as an identifiable item of equipment, made its first appearance with the work of Maiman (1960). However, a number of key advances during preceding century contributed to its development. In the mid-nineteenth century the behavior of light was quite successfully explained by the laws of Ampere, Gauss and Lenz, which are better known today in combined form as Maxwell's Equations. The interactions on an atomic scale between theory and experiment were eventually explained by the German physicist Plank, who postulated a particulate or quantum nature of light itself. In 1913, the Danish physicist Bohr extended Plank's ideas to atomic energy structure and was able to explain the absorption and emission spectra of atoms in terms of transactions between atomic energy levels.

This was followed by Einstein's (1917) pivotal paper on the interaction between atomic energy levels and light. This effect, now called Stimulated Emission, is central to the operation of all laser devices. The first device utilizing stimulated emission was reported by Gordon, Zeiger, and Townes (1955). This device achieve *Microwave Amplification by Stimulated Emission Radiation* (hence the acronym MASER). Following several years of intense debate concerning the possibility of this microwave effect occurring at optical or visible wavelengths, the matter was finally put to rest in 1960 with the invention of the ruby laser by Maiman of the Hughes Research Laboratory. Once the principle had been established, that the optical maser and laser was in fact a possible device, there followed a decade of intense activity during which most of the laser types were established. In particular, and in chronological order, the Helium–Neon laser was investigated and developed by Javan et al. (1961), Neodymium: YAG by Geusic et al. (1964), and Carbon Dioxide by Patel (1964).

The five major characteristics of the laser beam in determining the type of laser to use for a particular application are the beam's power, wavelength, temporal and spatial mode, and focal spot size.

- Laser Power
  Output power is the most basic characteristic of a laser. The amount of laser power required is determined by examining the optical and thermal properties of the material to be irradiated. The thermal properties determine the amount of energy required to melt or vaporize the material that include *heat capacity*, *latent heat*, and *thermal diffusivity* that determine the energy transmitted to the surrounding material during processing. The optical properties affect the surface of the material irradiated by laser beam. Of the optical material properties, the *absorption coefficient* determines the fraction of the energy that is absorbed by the material in function of wavelength of laser beam, surface roughness, temperature, and phase of the material. The remainder of the beam energy is reflected back into the

environment. Estimation of power requirements can be performed by using numerical models of the laser process to relate the material properties, operating parameters such as laser power, focal spot size, and depth of focus.

- Wavelength
The wavelength is the characteristic spatial length associated with one cycle of vibration for a photon in the laser beam. The wavelength of a laser is determined by the properties of its active medium (gas, liquid, solid, or semiconductor) and, to a certain extent, by how that medium is excited. The range of lasing wavelength is bracketed between the UV and infrared region of the electromagnetic spectrum. The absorption of materials can be highly dependent on the wavelength of incident light, and thus certain lasers will be more suitable for the processing of polymeric materials used in stereolithography process.
- Temporal Mode
Lasers can operate in either a continuous wave (CW) mode or a pulsed beam mode. In CW mode, the laser beam is emitted without interruption. In the pulsed mode, the laser is emitted periodically. The selection of a laser and an operating mode will depend strongly on the desired machining operation. Pulsed operation is usually best for deep penetration processes. The concentration of energy in each pulse leads to a small percentage of energy lost through conduction into the material or dissipation to the environment. Pulsed operation is used to minimize the heat affected zone in materials that are sensitive to elevated temperature, such as polymer used in stereolithography.
- Spatial Mode
The beam profile can be characterized by its *Transverse Electromagnetic Mode* (TEM). In a TEM the electric and magnetic fields are each perpendicular to the direction of travel of the wave. TEM modes are normally written in form of TEM_{nm}. Physically, the subscripts n and m denote the number of times the electric (or magnetic) field crosses the x, y, and z axes, respectively. That is, the subscript indicates the number of nodes in directions orthogonal to the beam propagation. TEM_{00} has a Gaussian spatial distribution and is usually considered the best mode for laser machining because the phase front is uniform and there is a smooth drop off of irradiance from the beam center. This minimizes diffraction effects during focusing and allows the generation of small spot sizes.
- Focal Spot Size
In materials processing, power density (power per unit area) of the laser beam at the material surface is of prime importance [6]. Power density great enough to solidify or melt any material such as sintering or curing process can be generated by focusing a laser beam. Some lasers are capable of being focused to produce what is called a diffraction-limited beam spot diameter.

The maximum power density is obtained at the focal point of a lens, where the beam is at its smallest diameter (Fig. 3.1); the location of this minimum diameter is called the *focal spot size*. Several factors influence focal spot size. First, focal spot size is directly related to the quality of the incoming beam, which can be quantified by the divergence can be focused to a smaller spot than a beam with high divergence. Second, focal spot size is influenced by diffraction. When focusing a diffraction-limited laser beam with a lens, a longer focal length or high f-number corresponds

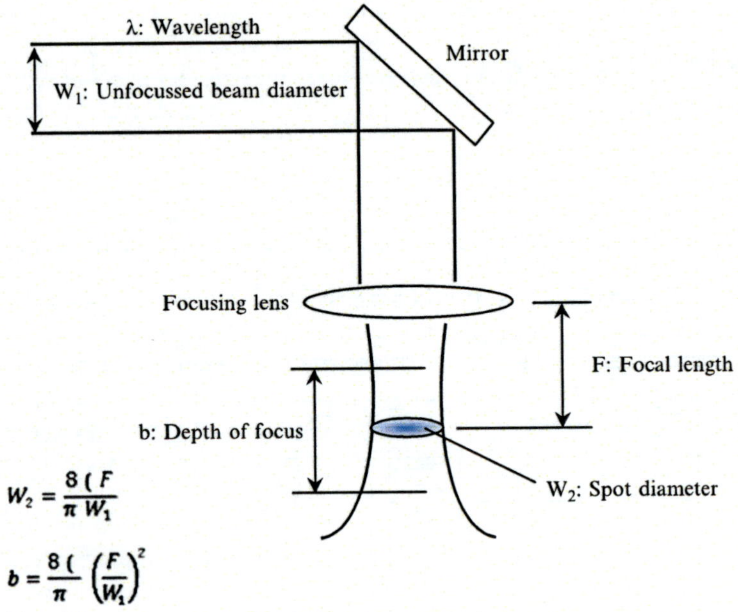

Fig. 3.1 Calculation of focal spot diameter and depth of focus

to a larger focused spot diameter. Finally, the diameter of the incoming laser beam affects the focal spot size.

3.3 Basic Chemistry of Laser Curing

The main advantage of using radiation of laser to cure is that it takes only a fraction of a second to achieve an extensive through-cure of the resin to a highly localized region. In consideration of the intrinsic characteristics of the laser emission, these powerful light sources present many advantages which make them very attractive for curing applications. First, the laser emission provides a great directivity to the laser beam that can be focused down to a tiny spot of extremely high intensity. With such a sharp "light pencil," it becomes then possible to directly write complex relief patterns, at micrometric resolution, by scanning the thermo-sensitive plate at very large speeds. Another consequence of this high directivity is that the light intensity remains essentially constant along the laser beam; by contrast to conventional light sources, the light intensity is thus not decreasing as the source-object distance increases, so that a uniform illumination of large 3D samples can be carried out by an even scanning with the laser beam. Owing to the temporal coherence of the laser emission which occurs at a well-defined wavelength, the cure penetration can be better controlled than with polychromatic light sources. Besides, the narrow bandwidth of the laser emission line will reduce the extent of the undesirable secondary chemical reactions, while at the same time eliminating the energy wastage in the nonabsorbing parts of the spectrum and the related heating of the sample that are common to all the industrial laser sources.

In conventional stereolithography system the UV laser emission is used to produce electronically molecules photosensitive materials leading to a highly cross-linked and insoluble material in the three dimensions.

In Infrared laser stereolithography, the reaction is initiated thermally. Infrared photons (light particles) do not contain enough energy to directly initiate the chemical curing reactions. Instead, energy from the infrared laser is absorbed in the resin as heat, and it is the heat energy that induces polymerization. The principle of the initiation step is as follows: an absorbing reactant locally transfers the energy of the light to the medium.

$$\text{Absorbing reactant} \xrightarrow{\text{IR energy}} \text{localized heating}$$

Via infrared laser curing, the viability of a thermal localized curing in the thermo-sensitive material by precisely confining the laser energy to the volume irradiated is reported. Thermal energy of IR lasers can be effectively used for activating a cross-linking polymer. The interaction of the infrared laser with thermo-sensitive materials lead to a thermal process in which controls of the output power, dwell time, scan speed, spot size, and other thermal and optical properties of materials are the principal features to be considered in study of this nature.

Ultraviolet lasers also generate heat in the resin, but the infrared experiments used much higher powered lasers than are usually used in UV laser, and so generated much higher levels of heat. The anticipated advantages of thermal curing over direct photo-curing are:

(a) Improved depth penetration when the resin contains filled particles that block the passage of UV light,
(b) The lower cost of infrared lasers.

3.4 Thermo-Sensitive Materials

A set of materials may be used, but the thermo-sensitive materials considered in this study for the Infrared laser stereolithography is composed of epoxy resin, in this case, diglycidyl ether of bisphenol A (DGEBA), a curing agent, to know, diethylenetriamine (DETA) and silica powder.

Curing reaction occurs by means of cationic polymerization between epoxy resin and DETA reactants. Thermo-sensitive resins are normally made from relatively low molecular weight, semiliquid substances. The localized curing becomes critical when the mixture of the sample components is not appropriate. In inappropriate proportions, the curing between reagents will be incomplete, hindering the

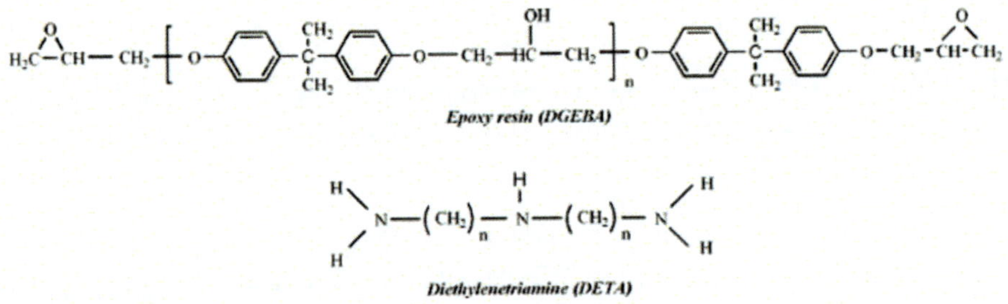

Fig. 3.2 Chemical structures of the thermo-sensitive materials

formation of the structure of high molecular weight and high density of cross-linking [7]. The chemical structures of the materials are given in Fig. 3.2.

The interest for epoxy resins is due to the variety of chemical reactions in which they can take part and to the diversity of materials that can be used as agents of cure. The kinetic characterization of the epoxy resin as well as the determination of the optimal proportion resin/agent of cures is important, not only for a better understanding of the relation structure-properties, but it is also fundamental in the optimization of the process conditions with impact in the product end quality. The curing reaction of DGEBA and DETA system is studied by Dynamic Scattering Calorimetry (DSC). The method for determination of the optimum ratio is based on the search for the maximum enthalpy change.

Silica plays an important key in controlling the localized curing. The amount of silica has a decisive influence in the process of localized curing. If there is no silica participation in the process or the amount is very small, the curing is not localized though it may occur, since the heat diffusion between the reagents involved in the process is fast. If the amount is large, the contact between the reagents is blocked up, which makes the connection between them difficult. Using the right amount of silica, the diffusion of heat through adjoining areas and the contact blockade between reagents does not happen, hence curing occurs locally. The type and the amount of silica were found to be critical to confine the curing process to a localized volume.

3.5 Theoretical Model

A physical theoretical model has been worked out aiming at the exact characterization of every physical phenomenon that occurred in the process of local curing. The model describes the energy flow deposited by the laser in terms of the control of the operational parameters and the behavior of the resin, aiming local curing.

The determination of the released energy distribution is essential for obtaining the local curing. The model takes into account the thermal and the optical characteristics of the resin as well as the absorption depth, thermal conductivity, and diffusivity. Initially, in order to obtain the power associated to the laser beam, the "dwell time" of the beam in the sample was determined. The local curing was achieved by CW CO_2 laser repeatedly over a circular trajectory on the sample's surface with a scan speed υ. By dividing the beam diameter 2ω by the scan speed, one obtains the "dwell time":

$$\tau_d = \frac{2\omega}{\upsilon} \tag{3.1}$$

concerning the time of interaction laser/resin at a surface point.

As the resin is highly absorptive at the CO_2 laser wavelength (10.6 μm), it is assumed that, during the "dwell time," nearly all the beam energy has to go into the inner part of the sample at a distance from the surface equivalent to the absorption depth δ. The absorption depth can be determined by measuring the transmittance of a non-cured sample, at the band of the CO_2 (10.6 μm) laser emission spectrum. It is assumed that energy E has been absorbed in the small cylindrical volume V during the "dwell time." This volume is defined as:

$$V = \pi \omega^2 \delta \tag{3.2}$$

The energy released in V is the product of the laser power by "dwell time":

$$E = P \tau_d \tag{3.3}$$

The approach of (3.2) is reasonable, inclusive with experimental evidence, because the sample absorbs at very small depths. Energy absorption in materials is quite critical concerning the depths they reach. In materials which do not absorb the energy strongly, the absorption depth may exceed the focus depth of the beam. As a consequence the confinement of energy at the surface of the model is not maintained. By the mean energy E it is possible to determine the variation of temperature, which is proportional to the deposited energy concerning the specific heat C_p and mass m of the material contained in volume V, according to the following equation:

$$E_p - C_p m \Delta T \tag{3.4}$$

The mass of the heated volume may be calculated by using the mass density of the sample.

A numerical solution is applied based on the method of finite elements, developed to the general equation for conduction of heat [8]. If it is assumed that nearly all the flow of energy deposited by the laser beam is absorbed every moment the laser passes at a point on the surface of the sample, it follows that the irradiated volume will undergo a temperature increase which is determined by the expression:

$$\Delta T = \frac{E_p}{m C_p} = \frac{P \tau_d}{\pi \omega^2 \delta \rho C_p} \tag{3.5}$$

As the laser induces a linear heating rate, the general equation for conduction of heat dependent on time is applied as follows:

$$\nabla^2 T = \frac{1}{D}\frac{\partial T}{\partial t} - \frac{G}{K} \tag{3.6}$$

where, D stands for the thermal diffusivity of the sample, K is the thermal conductivity, and G describes the rate of energy generated by the laser source. In Gaussian profile [9] of intensity of the laser beam, the term of source G generated by the CO_2 laser may be expressed by:

$$G = \frac{P}{\pi\,\omega^2\,\delta}\exp^{\left(-r^2/\omega^2\right)}\exp^{\left(\frac{-z}{\delta}\right)}\,S(\tau_d) \tag{3.7}$$

For $0 \leqslant t \leqslant \tau_d, S(\tau_d) = 1; t > \tau_d, S(\tau_d) = 0$ where $S(\tau_d)$ is the function which considers the term of source in the general equation, r is the distance from the beam center, z is depth of sample surface, δ is the absorption depth and P is the output power of the laser.

3.5.1 Optical Characterization

The optical characterization of the properties of the thermo-sensitive resin will depend on the sample composition, which in proper conditions allows to obtain information about the absorption depth δ. Therefore, these analyses have proved to be of great use for study of the samples behavior in terms of the silica absorption of energy, since each sample was submitted to a variation in the amount of silica in the composition. The absorption depth, as defined in the theoretical physical model, is essential for the definition of the volume of the cured sample. If the absorption depth determines the infra red radiation depth in the sample, it follows that the amount of silica in the composition may determine the dimension of the cured volume, seeing that as the amount of silica changes, there occurs a variation in the absorption depth. In order to determine the absorption depth in each sample, the semiliquid samples are inserted between two KBr crystals (transparent to infrared light).

The equipment used for the transmittance measurements is the Infrared Spectrum Analyzer, Model IR-700. Through an analytical solution, by applying Lambert–Beer Law ($I_T = I_0 \exp^{-x/\delta}$), a satisfactory estimate is established. Considering that the intensity is partly reflected and attenuated by the crystals, through the analytical solution, it is possible to determine the depth of energy absorption by the sample. The determination of δ is given by:

$$\delta_s = -\frac{x_s}{\alpha_c(x_{c1} + x_{c2}) + \ln\left[T\,\frac{1}{(1+R)^2}\right]} \tag{3.8}$$

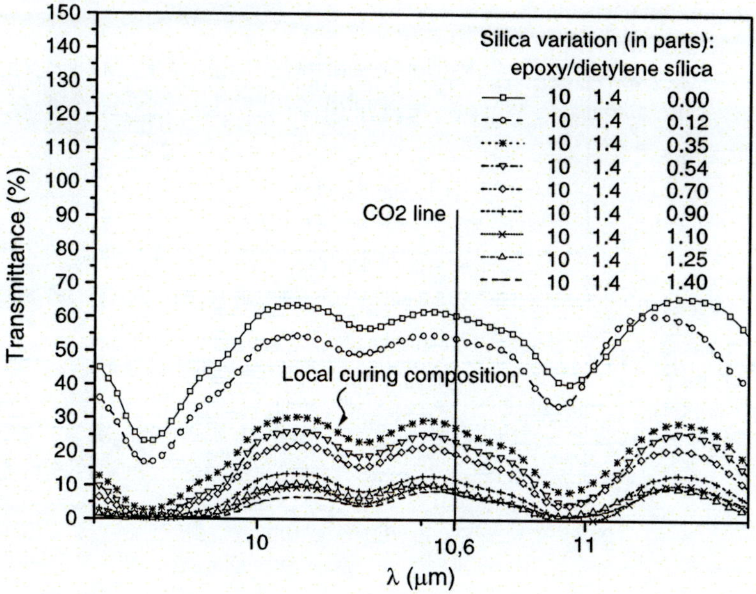

Fig. 3.3 Spectrums of CO_2 laser infrared radiation obtained from semiliquid samples in terms of silica variation in the composition

where, $I_T/I_0 = T$ is the sample transmittance, R is the crystal reflectance, x_c is the KBr crystal thickness, α_c is the optical absorption coefficient of the KBr crystal, and δ_s is the sample absorption depth.

Figure 3.3 shows different transmittance behavior as silica is added to the sample, along the CO_2 line (at 10.6 μm). From this experiment, it is possible to establish the relationship between the amount of silica variation and the depth of the optical absorption (Fig. 3.4).

Considering the importance of silica in the formation of the appropriate composition of the thermo-sensitive resin, it is necessary to analyze the effect since it is considered to be one of the main parameters in the physical interpretation and development of the process to obtain local curing. A further concern is to create a theoretical model with the laser operational parameters which could determine the boundary conditions of the laser application on the sample for the obtainment of the localized curing.

3.5.2 *Thermal Characterization*

The usual way of studying the cure kinetics of thermoset resins using isothermal and/or dynamic experiments with DSC, consists in obtaining the experimental reaction rate cure. The dynamic study carried out show that activation energy varies

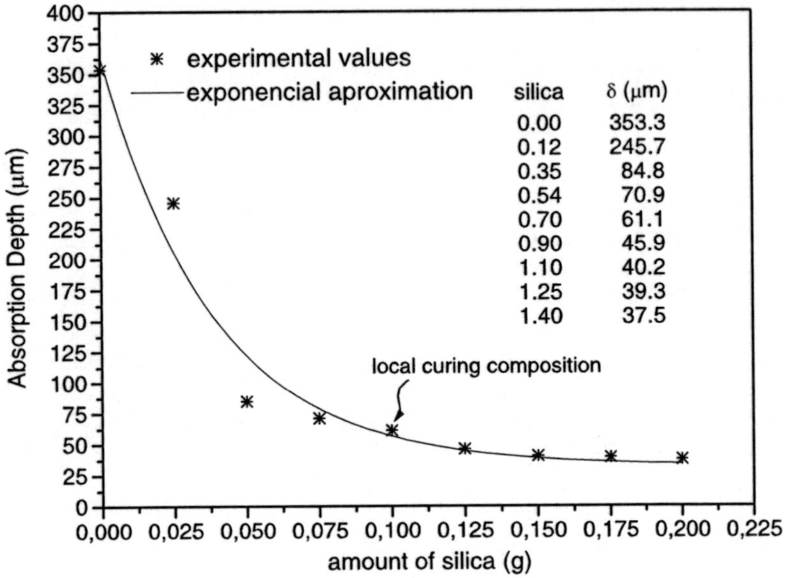

Fig. 3.4 Depth of optical absorption in terms of silica variation

according to the degree of conversion during curing and it raises the question concerned with the usual isothermal method that characterize the complete curing processes with a single activation energy. Both methods are used to obtain kinetic parameters of the curing system DGEBA and DETA. In order to identify the empirical kinetic method that produces the best results, an extensive kinetic study should be made performing different kinetics analysis using diverse empirical method.

3.5.2.1 Stoichiometric Epoxy/Curing Agent Ratio

The determination of this parameter is of vital importance when using a commercial system of two components (resin and hardener). This parameter is determined by dynamic experiments to a heating rate of 10°C/min using different concentrations (phr of amine) of DETA with the aim of produce changes in the heat of reaction. Phr is defined as parts by weight per 100 parts resin.

To obtain optimum properties with epoxy resin and reactive agents, it is desirable that the epoxy resin and curing agent react at approximately stoichiometric quantities. Stoichiometric ratio can be calculated as follows:

- Calculate the amine hydrogen equivalent weight of the curing agent using the following equation:

$$\text{AHEW} = \frac{\text{Molecular Weight of Amine}}{\text{No of Active Hydrogens}} \tag{3.9}$$

where, the molecular weight of the amine of DETA is 103 and the number of active hydrogens are 5, it is that directly bonded to the nitrogen of the amine molecule. Therefore:

$$\text{Amine AHEW DETA} = \frac{103}{5} = 20,6 \tag{3.10}$$

- Calculate the phr of the amines (parts by weight per 100 parts resin) has been made using the following equation, where the epoxy equivalent weight of diglycidyl ether of bisphenol A resin is 179.

$$\text{phr of Amine} = \frac{\text{Amine AHEW} \times 100}{\text{Epoxy Equivalent Weight}} \tag{3.11}$$

$$\text{phr DETA} = \frac{20,6 \times 100}{179} = 11,5$$

3.5.2.2 Differential Scanning Calorimetry

The calorimetric measurements are made with a Mettler Toledo DCS823e calorimeter. Before the measurements on the sample, indium standard were used to calibrate the temperature and energy of the DSC apparatus. The isothermal curing is performed at a temperatures range of 80–130°C in nitrogen atmosphere. The weight of the samples 5–10 mg of binary mixture DGEBA/DETA prepared before performing each experiment and hermetically sealed in aluminum DSC pans. After the isothermal curing, a dynamic scan of −50 to 250°C is made at a heating rate of 10 K/min to determine the residual heat. Immediately after inserting, the sample was sent to the calorimeter temperature to −50°C. Nitrogen with a flow rate of 45 ml/min was used as a purge gas.

3.5.2.3 Isothermal Kinetic Analysis

For isothermal cure reaction, the kinetic study can be described by an equation that relates the degree of conversion with the consumption of reagents, expressed as:

$$\frac{d\alpha}{dt} = k(T)f(\alpha) \tag{3.12}$$

In the thermosetting cure resin, although several simultaneous reactions occur during the curing process, some simple models have been developed assuming that only a unique reaction may represent the whole curing process. This model is expressed by a rate equation of nth order:

$$\frac{d\alpha}{dt} = k(1-\alpha)^n \tag{3.13}$$

where, n represents the order of the reaction, t the time, α the conversion, and k is the rate constant obeying dependence Arrhenius temperature, as follows:

$$k = A\,e^{-\frac{E_a}{RT}} \tag{3.14}$$

where, T is the temperature, A is a preexponential factor, R is the universal gas constant, and E_a represents the activation energy for the reaction.

For an isothermal reaction, Eq. 3.13 predicts the maximum of reaction rate at time $t = 0$, which is not the case for autocatalytic cure processes of system epoxy-amino where hydroxyl groups generated by the reaction serves as an additional catalyst in the reaction. Moreover, different events (phase separation, gelation, vitrification, etc.) occur in the course of the reaction which makes the cure kinetics more difficult.

A typical isothermal plot is shown in Fig. 3.5. In this Figure the initial increase in power flow represent the energy required to heat the sample to the desired temperature. Integrating the difference of the nonreactive sample curve and the reaction sample curve from $t = 0$ to t s gives the amount of energy released by the reaction prior to t. As $t \to \infty$ the energy released in the isothermal scan approach ΔH.

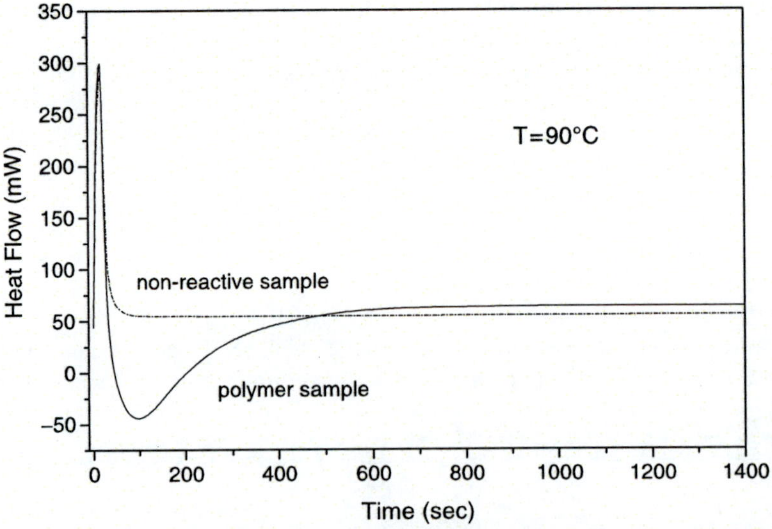

Fig. 3.5 DSC isothermal scan of the curing process at 90°C

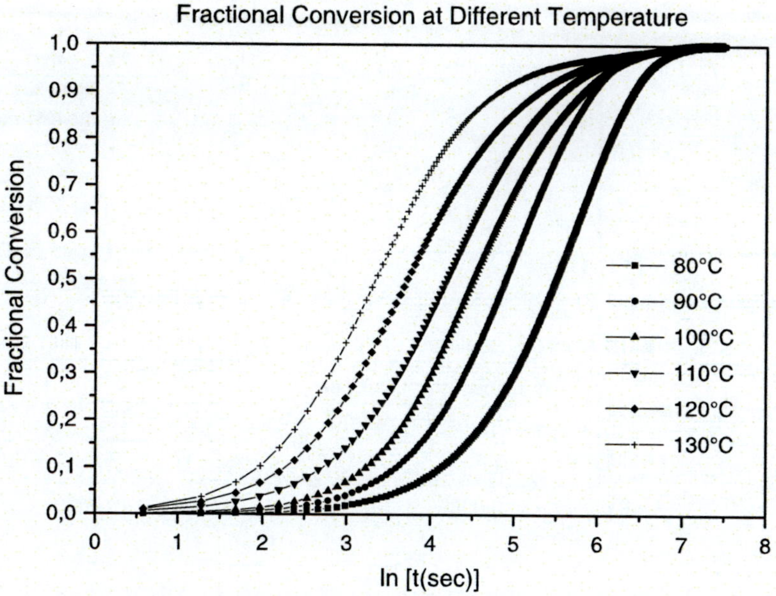

Fig. 3.6 Plot of fractional conversion of the reaction for various temperatures

A standard procedure [10] to obtain E_A (activation energy) using (3.13) is given by:

$$\ln(t_{1\beta}) - \ln(t_{2\beta}) = \frac{E_A}{R}\left[\frac{1}{T_1} - \frac{1}{T_2}\right] \tag{3.15}$$

where, $t_{1\beta}$ and $t_{2\beta}$ are the times required to reach a given fractional conversion at two different temperatures T_1 and T_2, and R is the molar gas constant. Several isothermal scans of the thermo-sensitive material are performed and the semilog plots of β versus $\ln(t)$ are shown together in Fig. 3.6.

The standard Arrhenius plot corresponding to the Fig. 3.6 is shown in Fig. 3.7, where each of the lines represents one specific fractional conversion β. The results show that the activation energy for the curing process is $E_A = 50.1$ KJ/mol.

3.5.2.4 Optimum Epoxy/Curing Agent Ratio

The properties of the epoxy systems can be controlled by changes both in stoichiometric formulation and cure conditions. Therefore, the best properties are achieved when using the appropriate proportions of the reactants. In this condition, the system gets the network structure fullest because it reaches the maximum conversion. To evaluate the optimum proportion of resin/agent cure

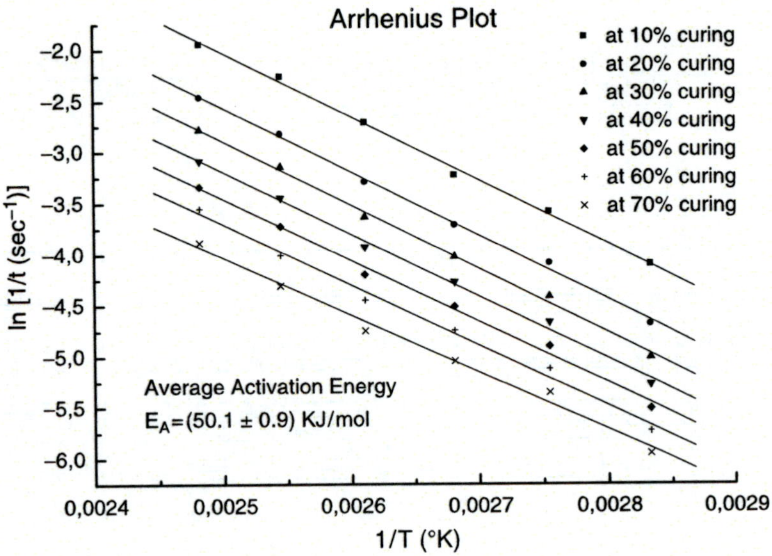

Fig. 3.7 The Arrhenius plots of the curing process taken from the results of Fig. 3.6

necessary for the subsequent kinetic study, a series of dynamic experiments is performed varying the amount of hardener. The methodology consists in determining the maximum enthalpy (ΔH) as a result of the formation of different network structures with different proportions of reagents in the mixture, as shown in Fig. 3.8. In Table 3.1, the corresponding values of enthalpy used for different proportions are shown.

Based on the experiences made with different dynamic proportions, the greater enthalpy corresponded to the proportion 100 g DGEBA and 14 g of DETA with a value of 521.54 J/g. Despite the stoichiometric calculations resulted in a ratio of 11.5 parts of agent cure for 100 parts resin epoxy, according to the previous result, better results are obtained with the proportion 100:14 of DGEBA/DETA. This is probably, because major amounts than cure agent must be added to all the epoxy rings are consumed in the reaction.

Figures 3.9 and 3.10 show the thermograms corresponding to the isothermal cures and dynamic post-cure. Figure 3.9 depicts the kinetic effect of the temperature. Thus, at higher temperatures, the curing times are lower and the reaction rates higher. The existence of a residual heat after the isothermal cures indicates that at the experimental conditions cures are not completed, as show Fig. 3.10. This may be because the material has vitrified and the movement of the existing reactive species in the reaction medium is restricted. The curing temperature influences last cure reached before vitrification, since at a higher curing temperature the heat detected in a dynamics post-cure is lower. Figures 3.9 and 3.10 have been used to calculate the signal by integrating the calorimetric heats, to know the isothermal, residual, and dynamic heats of reaction expressed in Table 3.2.

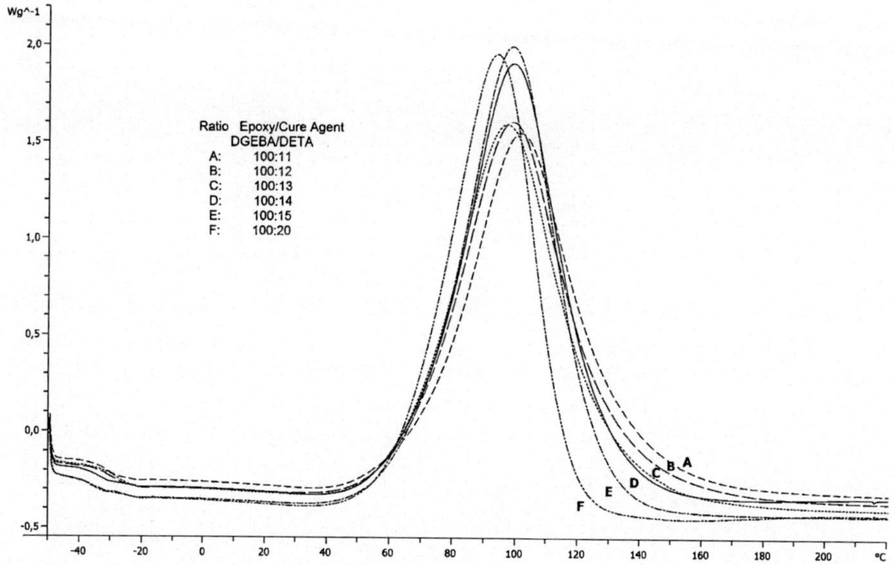

Fig. 3.8 DSC thermograms for different amounts of hardener

Table 3.1 Enthalpy reaction to the different ratios of epoxy resin and curing agent used at heating rate of 10°C/min

Ratio DGEBA/DETA	ΔH (J/g)
100:11	486,38
100:12	499,37
100:13	505,42
100:14	521,54
100:15	515,92
100:20	480,46

As the curing temperature increases, ΔH_{iso} increases and ΔH_{res} decreases. It is also interesting to see that the heat obtained by dynamic procedure is always higher than ΔH_{iso}+ΔH_{res}. Therefore the dynamic reaction heat obtained at 10°C/min of 521.54 J/g was taken as the total reaction heat.

3.5.2.5 Dynamic Kinetic Analysis

This dynamic kinetic analysis estimates the activation energy, E_{α}, as a function of the conversion fraction, α, regardless the kinetic equation based on data from the DSC by conducting dynamic experiments to deferent heating rates. Samples with 100 parts DGEBA epoxy resin and 14 parts hardener DETA, is cured in dynamic conditions for different heating rates: 2.5, 5, 10, 15, and 20°C/min. The heat flux

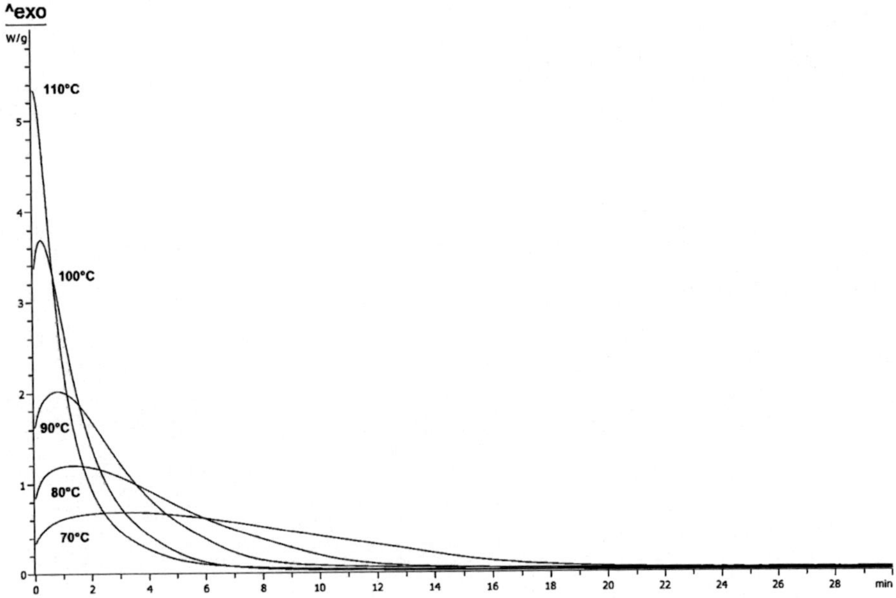

Fig. 3.9 Isothermal DSC thermograms for different curing temperatures

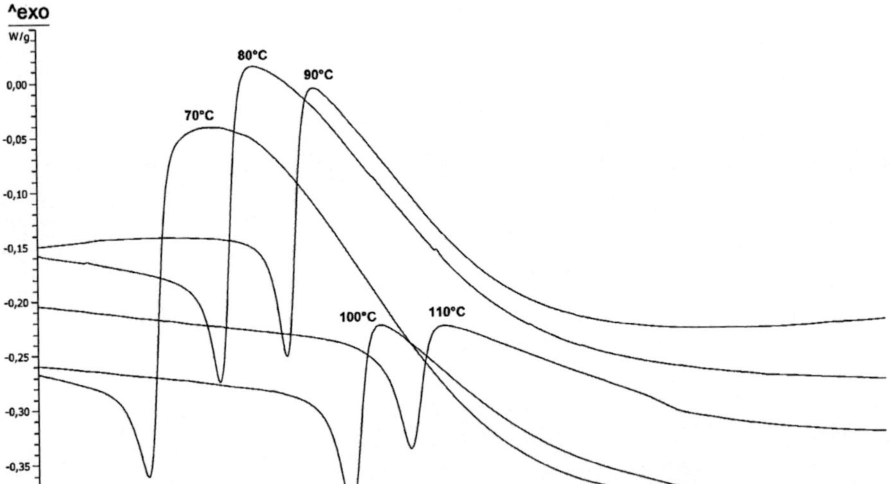

Fig. 3.10 Dynamic thermograms performed to 10°C/min after the isothermal scanning for different temperatures

Table 3.2 Experimental reaction heat obtained from isothermal, dynamic residual, and dynamic curves

Isothermal experiments				Dynamic experiment
T (°C)	ΔH_{iso}	ΔH_{res}	$\Delta H_{iso} + \Delta H_{res}$	ΔH_{din}
70	438,17	69,45	507,62	521,54
80	451,31	52,84	504,15	
90	466,24	31,70	497,94	
100	473,66	24,27	497,93	
110	481,49	19,42	500,91	

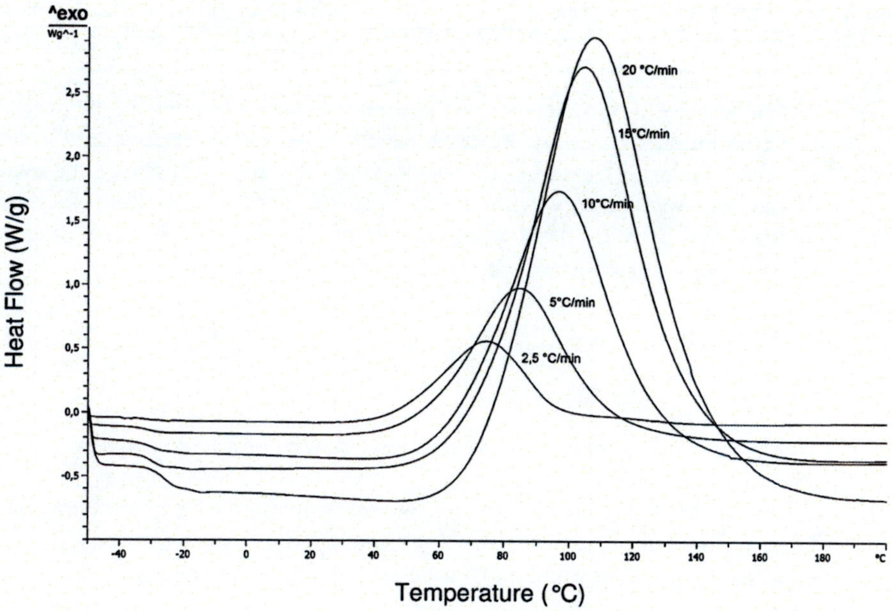

Fig. 3.11 Dynamic DSC thermograms obtained from different heating rates (K/min) for the system DGEBA/DETA

standard with the mass of the samples (W/g) against the temperature for different heating rates is show in Fig. 3.11. The great advantage of the dynamic kinetic analysis compared with the isothermal one is that it is possible to determine the activation energy for a given degree of conversion without needing to know $f(\alpha)$, since it has only been assumed that, for a given degree of conversion, $f(\alpha)$ takes the same form regardless of the curing temperature. Different thermal analysis experiments have shown different activation energies during the course of the cure reaction for similar epoxy systems. Furthermore, the dynamic kinetic analysis can be applied to different degree of conversion to see how the reactive process evolves and whether single activation energy can describe the whole curing process, as it is assumed in the isothermal model.

The kinetics of cure thermal processing, for non isothermal conditions when the temperature varies with time, with a constant heating rate, $\beta = \mathrm{d}T/\mathrm{d}t$, Eq. 3.12 is represented as follows:

$$\frac{\mathrm{d}\alpha}{\mathrm{d}T} = \left(\frac{A}{\beta}\right) \mathrm{e}^{-\frac{E_a}{RT}} f(\alpha) \tag{3.16}$$

where, A is the preexponential factor, E_a is the activation energy, and R is the gas constant. The rate (3.17) in integral form can be expressed as:

$$g(\alpha) = \int_0^{\alpha} \frac{\mathrm{d}\alpha}{f(\alpha)} = \left(\frac{A}{\beta}\right) \int_0^{T} \mathrm{e}^{-\frac{E_a}{RT}} \mathrm{d}T \tag{3.17}$$

Vyazovkin (1996) reported that the isoconversional methods offer greater accuracy in determining the curing kinetics, where it is necessary to make a series of measures under different heating rates. Thus, this approach allows the evaluation of the activation energy without the need to know the reaction kinetic model. Writing Eq. 3.16 in the logarithmic form for different heating rates β_i and considering it at a constant conversion, it possible to obtain:

$$\ln\left[\beta_i\left(\frac{\mathrm{d}\alpha}{\mathrm{d}T}\right)_{\alpha,i}\right] = \ln[A_\alpha f(\alpha)] - \frac{E_{a_\alpha}}{RT_{\alpha,i}} \tag{3.18}$$

where, subscript α refers to the value related to a considered conversion and i to a given heating rate.

Assuming A, E_a, and the kinetic model in relation to a given conversion, the kinetic parameters are the same for the temperature changes; it can be equated expressions (Eqs. 3.19 and 3.20) in relation to a given conversion:

$$g(\alpha) = t_a A_a \mathrm{e}^{-\frac{E_a}{RT_0}} \tag{3.19}$$

$$g(\alpha) = \left(\frac{A_\alpha}{\beta}\right) \int_0^{T_\alpha} \mathrm{e}^{-\frac{E_a}{RT}} \mathrm{d}T \tag{3.20}$$

where, hold under isothermal (T_0 = constant) and non-isothermal (β = constant) conditions. Simultaneous solution of Eqs. 3.19 and 3.20 to t_α yields:

$$t_\alpha = \left[\beta\, \mathrm{e}^{-\frac{E_{a_\alpha}}{RT_0}}\right]^{-1} \int_0^{T_\alpha} \mathrm{e}^{-\frac{E_{a_\alpha}}{RT}} \mathrm{d}T \tag{3.21}$$

where, T_α is an experimental value of the temperature corresponding to a given conversion at the heating rate β. Equation 3.21 enables the estimation of the time at which a given conversion will be reached at an arbitrary temperature T_0.

Similarly, Eq. 3.20 corresponding to different heating rates can be equated at the same conversion, leading to:

$$\left(\frac{1}{\beta_1}\right)\int_0^{T_\alpha} e^{-\frac{E_{a_\alpha}}{RT}} dT - \left(\frac{1}{\beta_0}\right)\int_0^{T_{\alpha,0}} e^{-\frac{E_{a_\alpha}}{RT}} dT = 0 \tag{3.22}$$

where, β and T_α are experimental values. $T_{\alpha,0}$ found as a solution in Eq. 3.22 is the temperature at which a given conversion will be reached at an arbitrary heating rate, β_0.

Solving Eqs. 3.21 and 3.22 for different conversions, it is possible to predict a dependence of α on t at an arbitrary temperature and/or a dependence of α on T at an arbitrary heating rate.

Equation 3.17 can be used to calculate the activation energy associated, through data obtained from the dynamic curves at different heating rates. Figure 3.12, represents the calculation of the activation energy.

Figure 3.12 shows that once the reaction has started the activation energy decreases with the increase in the degree conversion, then decreases slightly, and for the final stages of conversion (80–90%) begins to increase. This behavior has been observed in previous works. The decrease in activation energy is due to the autocatalytic effect which increases the reaction rate, generating an increase in the mobility of reactive species. The increase in the activation energy observed for high conversions (90–100%) is due to the increase in the density of cross-linking

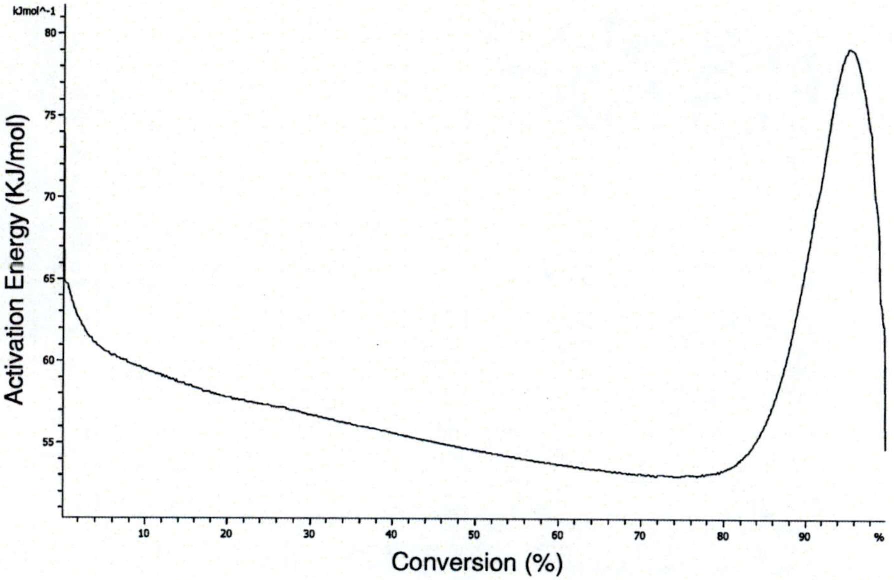

Fig. 3.12 Activation energy obtained by dynamics (kinetic) analysis

in the chemical structure, leading to a restriction of mobility chains (diffusion phenomenon), in addition, the shortage of reactive species.

3.5.2.6 Infrared Laser Stereolithography Process

Infrared laser stereolithography system involves the building of physical parts through a point by point cure process (Fig. 3.13). With this system, a laser beam is focusing on sample surface, curing a small volume. The basic system is made up of a computer, serving as the equipment control unit, a movable platform containing thermo-sensitive resin, an infrared laser (CO_2 laser) and an optical system (lens and mirror) to direct and focus the laser beam, as shown in Fig. 3.7. The strategy in building models is according to the specifications in the sliced model. The laser irradiates the resin surface curing a layer, which correspond to a given transversal section of the object. After finishing that section, the elevator platform drops a distance equal to the thickness of the next layer and the whole process is repeated. In this way, the system needs a leveler that reduces model surface stress and stabilizes the resin surface, minimizing the processing time of each layer. The diagnostic tests of the laser scanning parameters to obtain the localized cure in thermo-sensitive process are performed using the apparatus showed in Fig. 3.7. The laser is applied direct to the sample over a convenient substrate. Prototype constructions are obtained by scanning laser over several

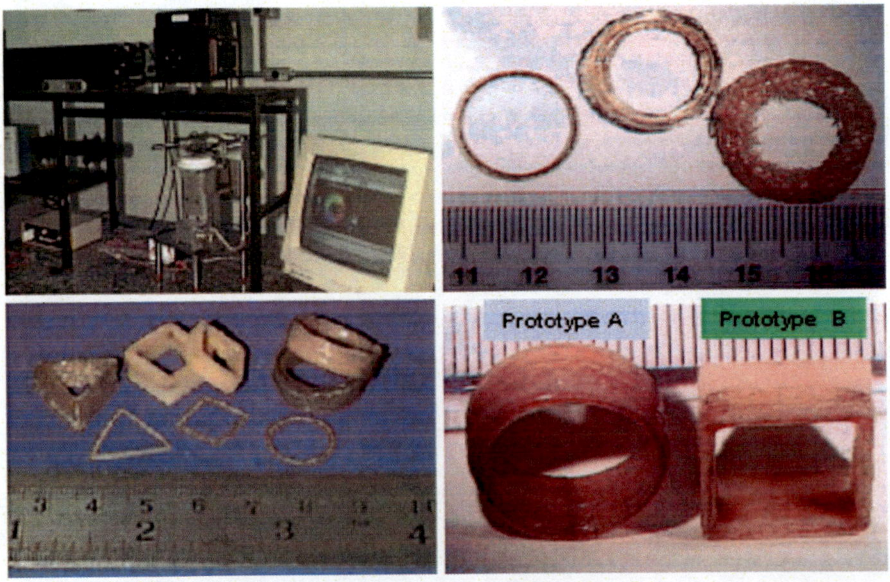

Fig. 3.13 Infrared laser stereolithography system and basic products

Table 3.3 Diagnostic test: laser parameter and composition of sample

	Values	
Parameters	Prototype A	Prototype B
Energy [mJ]	9.5	8.7
Power [W]	20	20
Laser beam diameter [mm]	0.8	0.8
Scan speed [mm/s]	1,595.4	1,755.0
Dwell time [μs]	476	433
DGEBA [parts]	10	10
DETA [parts]	1.4	1.4
Silica [parts]	0.7	0.54
Lateral layer width [mm]	0.9	1.2
Sample width [mm]	0.2	0.2
Sample height [mm]	5.5	5.8
Sample length [mm]	16.0	13.5

different paths on the sample surface, varying the laser speed and the laser revolution time. Several stoichiometric compositions of the sample were taken into account, as well as the laser parameters, involved in the curing process (Table 3.3). In the top left side are shown the experimental apparatus already described; a prototype monolayer obtained from a correct stoichiometric sample composition (including silica) and correct laser parameters is compared with two other monolayers constructed using incorrect sample composition and incorrect laser parameters. Prototypes of different geometries and sizes (layer by layer constructed) are shown below. In the right side, the final products are shown with different spatial resolution.

3.6 Conclusion

Stereolithography is perhaps the oldest rapid prototyping technology known. 3D Systems has been manufacturing stereolithography equipment and they have a large number of rapid prototyping machines in the work force today. Parts built with stereolithography resins possess excellent surface finish and strength, and are also translucent. The QuickCast technology allows them to be used in casting process. This advanced technology has been applied using best efforts with the help of technologies and funding from the US government to develop a tripled-efficient "clean car" within a decade, and a year later they reported an encouraging progress. The key to a super-efficient car is to cut all the losses by making the car very light and aerodynamically slippery and then recovering most of its braking energy. Unfortunately, this cost is high.

The application of IR laser radiation to the thermo-sensitive materials, such on sample, may provide a novel, reliable, and cost-effective means to manufacturing industrial prototypes. The result can be a hard, fairly, tough, and well formed plastic. The great advantage that has been observed between initial products (IR) and those using UV lasers is that the final product shows no marked contractions after curing and requires no post-cure treatment. The CO_2 laser is considerably cheaper than the UV lasers at the necessary power levels.

In general, thermosets possess good dimensional thermal stability and can be used in appliance housings, automotive parts, clothing textile treating, electrical wiring boards, and medical and dental instruments and fillings. This investigation has presented the results of the spatial selective solidification of semiliquid sample composed by epoxy, DETA and silica powder, using CO_2 laser as a localized heat source. The localized cure occurs only for optimized processing using a scanning system along with a careful control of the laser parameters. The final product obtained with those special controls is hard table, showing no significant shrinkage and requiring no post-cure treatment. Basic comparative results among the resins used in stereolithography is shown, pointing out the hit points of both photosensitive and thermo-sensitive materials, along with the advantage of each method.

Spatial resolution is determined by the thickness of the uncured semi-liquid resin and laser parameters. In order to adapt the infrared laser curing process to different geometry, one must account for the effect of several experimental parameters, including laser power, pulse repetition rate, scanning frequency and scan speed, size of laser beam, thermal conductivity of the sample, and the mixture of reactants.

Three-dimensional parts have been constructed using those mentioned parameters setting with smaller CO_2 laser beam size. The technique is able to produce structures with lateral spatial revolution of tens of microns. Localizing the curing process to a presumably small region in all the three dimensions requires that the laser energy be deposited in a well-defined volume so that heat will not be conducted away to regions where we do not wish to cure the sample. The thermo-sensitive materials supported by an optimum mix of the components and with special role of the silica power, and relatively low thermal conductivity, strongly absorb the CO_2 laser irradiation such that the curing process is restricted to a desired region. Improvement of those mentioned parameters may reduce the curing laterally, within roughly the diameter of the laser beam and the absorption depth.

As a result, the infrared laser stereolithography leads to reduced manufacturing cost, shorter lead time, and improve control and prediction of the properties of the finished product.

References

1. Barros, M.L., Scarparo, M.A.F., Gerck, E., Kiel, A., Hurtack, J.J., (1994) Stereolithography with thermosensitive resins using a CO_2 laser, *Journal of Applied Polymer Science*, vol 54, pp. 1575–1578.
2. Belforte, D.A., (1990) Argon ion laser speed prototype modeling, *Machine Design*, June 21.
3. Kaplan, R., (1990) Stereolitography: a marriage of technologies, *Photonics Spectra*, June.

4. Jardini A.L., Maciel Filho R., Scarparo M.A.F, Andrade S.R., Moura L.F.M., (2004) Advances in stereolithography: new experimental technique in production 3D plastic model using infrared laser radiation, *Journal of Applied Polymer Science*, Vol. 92, pp. 2387–2394.
5. Crafer, R.C., Oakley, P.J., (1993) *Laser processing in manufacturing*, Chapman & Hall, London.
6. Ready, J., (1979) *Lasers: Their Unusual Properties and Their Influence on Applications*, Lasers in Modern Industry, Society of Manufacturing Engineers Marketing Services Dept., Dearborn, MI, pp. 17–38.
7. May, C. A., (1988) *Epoxy resins – chemistry and technology*, Second Edition.
8. Carslaw, H.S. and Jaege, J.C., (1959) *Conduction of heat in solids*, 2nd ed. 10 (*Oxford University*), New York.
9. Sanders, D.J., (1984) Temperature Distributions Produced by Scanning Gaussian Laser Beams, *Applied Optics*, v. 23, no.1.
10. Wisanrakkit G. and Gilham J.K., (1990) *Polymer Characterazation*, Advances in Chemistry Series, v. 227.

Chapter 4
Microstereolithography

Arnaud Bertsch and Philippe Renaud

4.1 Introduction

Microstereolithography is a technology that is based on the same manufacturing principle as stereolithography. Three-dimensional (3D) objects are built by the superimposition of many layers, each being produced by a light-induced space-resolved photopolymerization of a liquid resin. As the resolution of microstereolithography is far better than other rapid prototyping techniques, this technique creates interest in both the rapid prototyping domain, where it can be used to produce high-resolution prototypes, but also in the microengineering field, as it is clearly the microfabrication process that can produce small objects with the most complicated shapes and intricate details.

Many different designations (IH process, spatial forming, 3D optical modeling, micro-photoforming, microstereophotolithography, optical forming, etc.) have been used by the research teams who published the first scientific reports on this technique, to account for the variations in the design of the apparatuses built. The name "Microstereolithography" is now commonly accepted by almost all users and developers of this technology, as it clearly states the connection of the process both to the stereolithography and the microfabrication fields.

The first developments of the microstereolithography technique started in 1993 and different research teams have devised strategies for improving both the vertical and the lateral resolution of the stereolithography process, which resulted in the fabrication of a variety of machines that can be classified into three categories:

- Scanning microstereolithography machines are based on a vector by vector tracing of every layer of the object with a light beam.
- Integral microstereolithography processes are based on the projection of the image of the layer to be built on the surface of the resin with a high resolution.

A. Bertsch (✉)
Ecole Polytechnique Fédérale de Lausanne, Lausanne, Switzerland
e-mail: arnaud.bertsch@epfl.ch

P.J. Bártolo (ed.), *Stereolithography: Materials, Processes and Applications*,
DOI 10.1007/978-0-387-92904-0_4,

- Sub-micron microstereolithography processes allow the polymerization of the layers composing the object directly inside the reactive medium and no longer on its surface.

All microstereolithography machines are based on the space-resolved photopolymerization of a liquid resin that should have very specific characteristics, as it is directly related to the resolution of the components that will be produced.

Microstereolithography is starting to be a commercially available manufacturing process. As the market for miniaturized products expands, there is an increasing need for high-resolution small size prototype parts. If the production of small mechanical components is the first commercial application of microstereolithography, this technology can also produce useful components for the microrobotic, microfluidic, microsystems, and biomedical fields. Very often the use of microstereolithography in the microsystems field requires the usage of a variety of materials, either by the combination of components made by microstereolithography and by other methods that provide a special function to the assembly. Components in composites materials have also been produced by adapted microstereolithography machines, which allows to fabricate 3D microcomponents in metal or ceramics after a sintering step.

4.2 From Stereolithography to Microstereolithography

Microstereolithography has evolved from stereolithography, a rapid prototyping technique that has a typical resolution in the range of 150 μm in the three directions of space [1, 2]. This evolution has significantly improved both the lateral and the vertical resolution of stereolithography. As the fabrication method of this technique relies on the space-resolved and light induced polymerization of a resin, the improvement of its resolution is mostly related to the reduction of the interaction volume between light and matter.

By the nature of the stereolithography process, based on the stacking of partially cured layers, resolution, precision, and surface roughness are inherently anisotropic. Consequently, it is useful to distinguish between vertical (along the build axis) and lateral (in-plane) resolution.

4.2.1 Thickness of the Polymerized Layer

Photopolymerization is a chemical reaction that starts when a given quantity of photons of adequate energy is absorbed per volume unit of photosensitive medium. This creates reactive species (radicals or cations, in most cases), which induce the polymerization of the liquid monomers and oligomers composing the resin into a solid polymer, by a chain reaction. In general, the resulting polymer is cross-linked, and does not dissolve in the monomer bath.

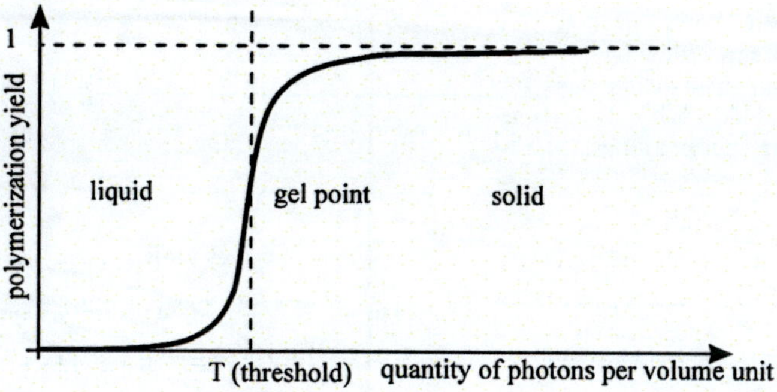

Fig. 4.1 Polymerization as a threshold system

The liquid/solid phototransformation can be approximated by a threshold system (Fig. 4.1). The threshold T is reached when a certain quantity of photons has been absorbed per volume unit of reactive medium:

$$T = I_a t_0 \tag{4.1}$$

where, T is the threshold value (photons m^{-3}), I_a the light absorption velocity (photons m^{-3} s^{-1}), and t_0 is the irradiation time at the threshold (s).

Time-resolved IR spectroscopy can be used to evaluate experimentally the polymerization kinetics of a photosensitive resin, and the results are in good agreement with the threshold system approximation, as proven by Zissi et al. (Fig. 4.2) for the photopolymerization of a resin containing diacryl 101 (2,2′-bis [4-(methacryloxy-ethoxy)phenyl] propane) as monomer and DMPA (α,α' dimethoxy-2-phenyl acetophenone) as photoinitiator [3].

In an isotropic and homogeneous medium containing a substance absorbing at a given wavelength λ, the transmitted photon flux F_T is given by Beer–Lambert law:

$$F_T = F_0 \exp(-\alpha\, c\, l) \quad \text{with} \quad \alpha = 2.3\, \varepsilon \tag{4.2}$$

where F_0 is the incident flux (photons m^{-3} s^{-1}), c the concentration (mol l^{-1}), l the depth (cm), ε the coefficient of molar extinction (l mol^{-1} cm^{-1}), α the Napierian coefficient of molar extinction (l mol^{-1} cm^{-1}).

The light absorption velocity along the z axis is given by the following relation:

$$I_a(z) = F_0\, \alpha\, c \exp(-\alpha\, c\, l) \tag{4.3}$$

Combining (4.1) and (4.3) gives the temporal evolution of the polymerized thickness:

$$e = \frac{1}{\alpha c} Ln\left(\frac{t}{t_0}\right) \quad \text{with} \quad t_0 = \frac{T}{\alpha\, c\, F_0} \tag{4.4}$$

where, t is the irradiation time (s).

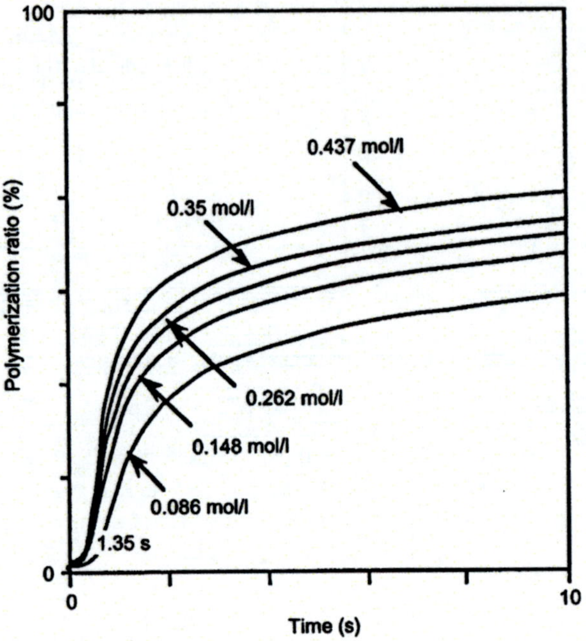

Fig. 4.2 Polymerization kinetics measured by time-resolved IR spectroscopy (monomer: diacryl 101, photo-initiator: DMPA) from [3]

This relation shows that the evolution of the polymerized thickness "*e*" with the irradiation time is logarithmic. This simple model is in good agreement with experimental results in the case of resins that do not undergo changes in their absorption coefficient during the polymerization process. When there is a bleaching phenomenon associated with the polymerization process, the evolution of the polymerized depth with time is no longer logarithmic, but it can be obtained by numerical simulation [4].

4.2.2 Reducing the Polymerized Thickness

The vertical resolution of the stereolithography process depends on the polymerized thickness of the layers created, and not on the thickness of the layer of fresh liquid resin spread on the surface of the object being built. Making thinner layers by depositing less resin on the surface of the component being manufactured is not sufficient to control the vertical resolution. If the light is not confined to the surface, it will penetrate deep in the chemical medium and result in parasitic polymerizations in the already polymerized parts of the object. The control of the light penetration in the resin and the control of the polymerization depth are consequently the parameters of importance for the improvement of the vertical resolution of the stereolithography process [3, 5–8].

By understanding the contribution of the different terms of (4.4) to the polymerized thickness, various ways of improving the resolution can be deduced as shown:

(a) Irradiating the resin close to the polymerization threshold.
Thin polymerized layers can be obtained by reducing the irradiation time of the chemical medium, such that the energy received by the resin is close to the critical energy *Ec* required to start the polymerization process. In that case, (4.4) can be simplified to:

$$e \cong \frac{1}{\alpha c}\frac{t - t_0}{t_0} \tag{4.5}$$

With this method, the layer thickness can be controlled by the choice of the irradiation time, but as the incident energy is close to the critical energy required to start the polymerization process, the resulting polymer network has in general poor mechanical properties. Additionally, small variations in the irradiation energy induce large changes in the polymerized thickness, which is a major problem for vector-by-vector fabrication processes. In particular, when two polymerized vectors are secant, the intersection point is irradiated with twice the energy of other points, which results locally in very large changes of the polymerized thickness and makes it impossible to control the polymerization process accurately (Fig. 4.3).

Many research teams have tried to use directly stereolithography resins in microstereolithography machines, and reduced the irradiation time for the production of thinner layers. By doing so, they use conditions close to the threshold and have to face the drawbacks of these operating conditions, which include poor mechanical properties of the built objects, overcure errors, and a difficult control of the layer thickness.

(b) Using highly absorbing reactive media
For large values of the irradiation energy compared to the critical energy, (4.4) can be written as:

$$e \cong \frac{R_1}{\alpha c} \quad \text{where} \quad R_1 = Ln\left(\frac{\alpha c F_0 \tau}{T}\right) \tag{4.6}$$

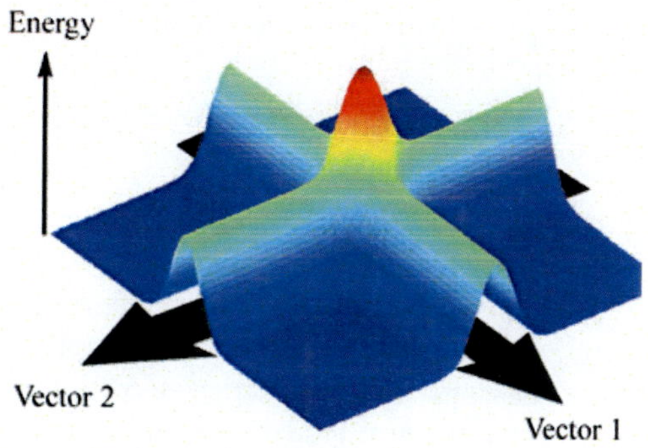

Fig. 4.3 In vector-by-vector processes, when two vectors are secant, the intersection point receives a double energetic dose. This can lead to local changes in the polymerized thickness, in particular when the irradiation is performed in conditions close to the polymerization threshold

There are only very small variations of R_1 with small variations of the irradiation time τ, when $\tau >> t_0$. In this case, the layer thickness is strongly related to the absorption of the reactive medium, which limits the penetration of the light in the resin. The polymerized thickness can be reduced by using reactive resins with low value of the optical thickness $\mu = 1/\alpha c$. Such resins show a strong absorption at the irradiation wavelength and often a strong reactivity. They generally contain high concentrations of photoinitiators or photosensitizers having high extinction coefficients.

The use of such resins in conditions far from the irradiation threshold is suitable for vector-by-vector fabrication as small variations of the irradiation time only induce a limited increase of the polymerized thickness. In the case of two secant vectors, the intersection point still receives a double energetic dose, but the resulting change of thickness is limited, as the light is confined at the surface by the absorption of the resin.

(c) Using neutral absorbers.
This solution is in fact a combination of both previous cases: It consists in adding a nonreactive chemical that will strongly absorb the light at the irradiation wavelength and dissipate the captured energy by ways that do not interfere with the photopolymerization process. When using a neutral absorber, (4.4) becomes:

$$e = \frac{1}{\alpha c + \alpha_N c_N} Ln\left(\frac{t}{t'_0}\right) \quad \text{with} \quad t'_0 = \frac{(\alpha c + \alpha_N c_N)T}{\alpha^2 c^2 F_0} \tag{4.7}$$

where, c_N is the concentration and α_N is the napierian coefficient of molar extinction of the neutral absorber.

The use of a neutral absorber consequently reduces the optical thickness of the resin, but at the same time less energy is available for starting the polymerization reaction, which also reduces the polymerized thickness. The drawback of neutral absorbers is that their use reduces the reactivity of the resin. Zissi et al. used Tinuvin P as neutral absorber in microstereolithography resins, as this molecule has a high absorption coefficient at 364 nm. The addition of this neutral absorber at relatively low concentrations reduces strongly the polymerized thickness of the stereolithography resin as presented on Fig. 4.4.

4.2.3 The z-Overcure Error

The z-overcure error is a phenomenon that induces the solidification of a certain amount of additional polymer present below the layer being built, due to the penetration of the light across already polymerized layers (this phenomenon is known as "print-through" in stereolithography). This z-overcure error is a known source of inaccuracy in the vertical direction in stereolithography and slice algorithms have been produced to compensate for this effect.

Fig. 4.4 Effect of an increasing concentration of Tinuvin P used as neutral absorber on the polymerized thickness of a microstereolithography resin (diacryl 101 with DMPA as photoinitiator) from [3]

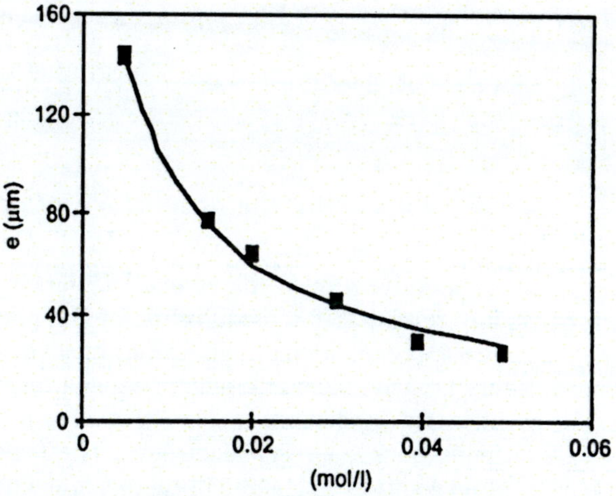

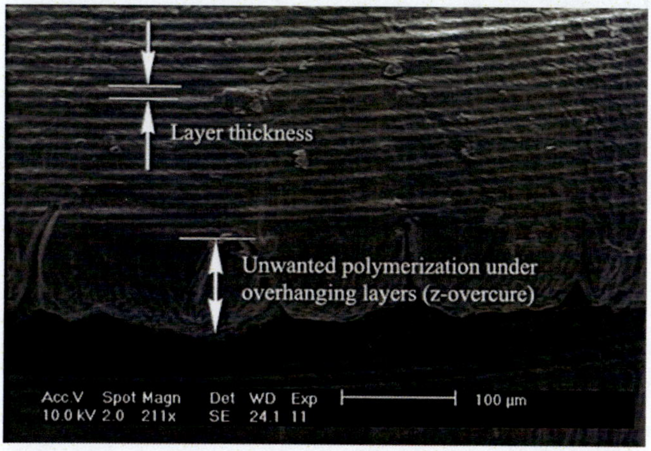

Fig. 4.5 Example of z-overcure error. An unwanted polymerization has occurred under overhanging layers in an object made by microstereolithography

The z-overcure error can have dramatic effects on the degradation of the vertical resolution in microstereolithography as presented on Fig. 4.5, and result in unwanted polymerizations that can deteriorate features in the already polymerized layers of the object being built. In fact this often happens when irradiating the resin close to the polymerization threshold, or when resins with low reactivity and low absorption are used. In these cases, when the first layer of an overhanging structure is built, the resin underneath is sensitized but does not polymerize as the threshold energy needed to start the polymerization process has not been reached locally. As photoinitiation is a cumulative process, more energy will be added in this already sensitized zone if more layers are added on top of the overhanging area. When the energy in the sensitized zone reaches the threshold, the polymerization process starts and creates unwanted polymer structures.

4.2.4 *Improving the Lateral Resolution of Stereolithography*

If the improvement of the vertical resolution of the stereolithography process is mainly related to the modification of the photosensitive resins used in order to reduce the penetration of light in the chemical media, such modifications only have a very limited influence on the horizontal resolution. In fact when the light beams used in the irradiation process have a nonuniform energetic repartition (as for example when Gaussian beams are used), the reduction of the polymerized thickness related to chemical modifications of the photopolymerizable resin also induce a small reduction of the dimensions in the lateral direction.

However, major improvements in the horizontal resolution of the stereolithography process are not related to the chemistry of the process, but to the resolution of the scanning system used to irradiate the surface of the resin to create the layers, and to the quality and stability of the light beam.

The laser diameter of a SLA-250, the most widely used stereolithography machine commercialized by 3D systems, is 0.25 mm. The laser beam diameter sets the limit of the smallest feature that can be manufactured, and consequently the resolution of the system. Given a stable laser beam profile, accurate correction factors for the width of the scanning beam and resin shrinkage, a horizontal precision in the order of the half of the laser beam width, as a rule of thumb, can be routinely achieved.

(a) Small-spot stereolithography
The horizontal resolution of the stereolithography process can be significantly improved by using a single mode instead of the common multimode HeCd laser on a conventional SLA-250 machine. The spot size at the focal point can be reduced from 0.25 mm to about 0.08–0.1 mm, which pushes down the lower limit of feature feasibility [9, 10]. This technique has found applications in the area of watchmaking, electronics, [11, 12] and in the electro- and medico-technical industry (hearing aids), and is commercially available, as a few service bureaus worldwide produce high-resolution prototypes made with the small-spot technology.

(b) Microstereolithography
To improve the horizontal resolution of the stereolithography process the light beams used to irradiate the surface of the reactive medium simply need to have a smaller diameter in order to reduce the surface of interaction between light and matter.

In the stereolithography technique, a laser beam is focused and scanned on the surface of the resin. A lateral improvement of resolution can be obtained by focusing more precisely, the beam on the surface, such that the spot size is reduced to only a few micrometers in diameter. This requires to adapt the optics of the system, but also to accurately and continuously measure the position of the surface of the liquid photopolymer on which the beam is scanned and to dynamically focus it with a sufficient precision. A first kind of microstereolithography processes based on this principle has been developed, for which every layer of the object being built is obtained by a vector-by-vector scanning of the resin surface as it is the case in the stereolithography process.

To avoid the problems related to the focusing of the light beam on the resin surface, another family of microstereolithography processes has been defined. These processes, named "integral microstereolithography" or "projection microstereolithography" take advantage of the relatively large depth of focus that can be achieved while projecting a high-resolution image on a surface, and are consequently relatively simple machines.

Finally, the improvement of the lateral resolution can be obtained by a radical change of the stereolithography concept. Building the layers that compose the object inside the reactive medium rather than on its surface allows very small increments in both the vertical and lateral directions, and result in processes that can have a submicron resolution. In these processes, the energetic density of the light beam has to be maximal under the surface, which is not something easy to achieve.

4.3 Scanning Microstereolithography Machines

4.3.1 Constrained Surface Techniques

The first scientific papers published on microstereolithography were issued in 1993, one by Takagi et al. at the MHS conference [13] and the second by Ikuta et al. at the MEMS conference [14]. The machines presented in these papers, which are more extensively presented in later publications are relatively similar [15–17]. They use a vector-by-vector tracing of each layer of the object with an UV light beam that is focused on the resin surface through a transparent window. The focus point of the apparatus remains fixed during the fabrication steps, while X-Y translation stages are used to move either all the optical system focusing the light beam on the resin surface or the photoreactor in which the object is made. The light beam is occulted by a shutter during translations without polymerization or when a new layer is made (Table 4.1). Figure 4.6 shows a schematic diagram of such a microstereolithography machine.

Using a fixed focused beam together with an X-Y translation stage to solidify the layers rather than galvanometric mirrors (as it is the case in the stereolithography process), solves the problems of defocusing of the beam occurring when scanning a light beam with mirrors and allows to achieve a small and well-controlled focal point. However, this solution is slow and does not allow for building complex objects quickly.

Table 4.1 Characteristics of the microstereolithography machines described by Takagi and Ikuta in 1993

	Takagi et al. (1993)	Ikuta et al. (1993)
Name of process	Photo forming	IH process
Light source	He–Cd laser, UV (325 nm)	Xenon lamp, UV
Constrained surface with	Quartz window+PFA tape	Transparent window
Irradiation	From bottom	From top
Maximum size of structure	20 × 20 × 20 mm	10 × 10 × 10 mm
Announced resolution	5 × 5 × 3 μm (x,y,z)	60 μm, up to 8 μm
Resin type	Acrylic	Not specified

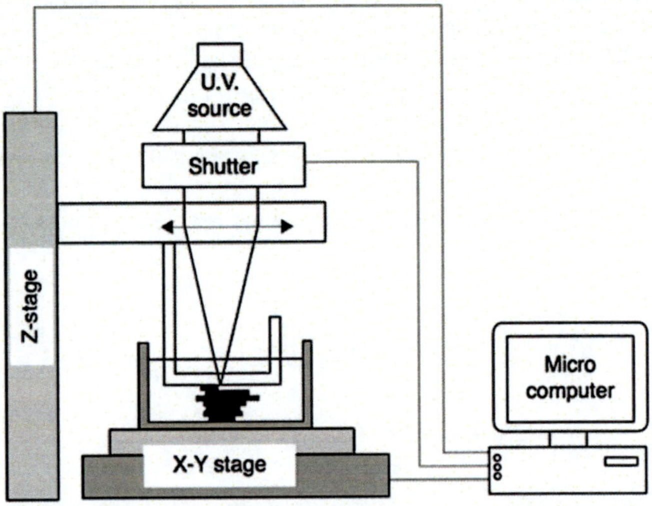

Fig. 4.6 Schematic diagram of a scanning microstereolithography machine using the constrained surface configuration

The use of a glass window to push on the liquid avoids the problems related to spreading the fresh resin on the already polymerized part of the object and allows to build layers of well-controlled thickness. However, there is a major disadvantage in polymerizing through a transparent window, as the formed polymer sticks to it, which often induces partial or total destruction of the component manufactured during its fabrication.

4.3.2 Free Surface Technique

As the adhesion of the polymer to the glass window is a major limitation of the constrained surface microstereolithography processes [18], various research

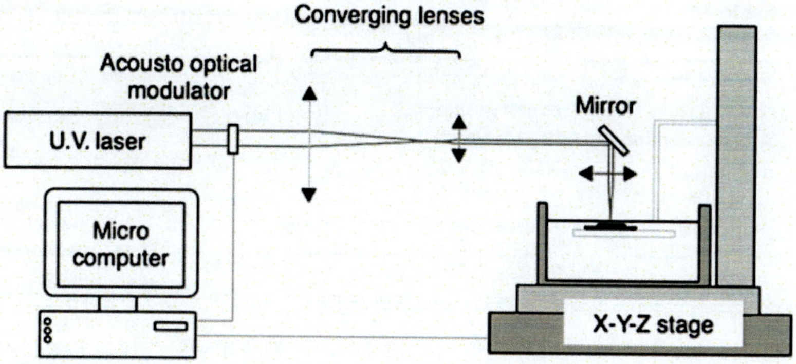

Fig. 4.7 Schematic diagram of a scanning microstereolithography apparatus using a free-surface configuration

teams developed scanning processes with a free surface technique. Again, the vector-by-vector fabrication of each layer is obtained by moving the photoreactor with an X-Y-Z motorized translation stage rather than by moving the light beam with galvanometric mirrors (Fig. 4.7).

Although some problems are avoided by the free-surface configuration, others need to be addressed. The control of the thickness of the deposited layer of liquid is difficult, as only gravity forces intervene for leveling the surface after fresh liquid is spread on top of the object being manufactured. The time required to obtain the desired thickness of fresh resin depends on the rheological properties of the resin as well as on the geometry of the last layer built. Low viscosity photopolymerizable resins have to be used in such apparatuses.

The first paper describing a scanning microstereolithography process with a free surface technique was published by Zissi et al. in 1994 [19]. They used an acousto-optical modulator as shutter, and an IR laser diode to monitor the position of the surface of the resin during the vector-by-vector scanning of each layer, which allows for a precise focusing of the light beam on the surface. Acrylate-based phototpolymerizable resins were developed specifically for this project, using neutral absorbers to reduce the polymerized thickness and polymerization inhibitors to control the reactivity of the resins.

In 1998, Zhang et al. also presented a free-surface scanning microstereolithography machine they designed [20]. They used a laser rather than a lamp to obtain a smaller spot size and more stable beam intensity. The UV curable resins developed for this apparatus contained HDDA (1,6 hexanediol diacrylate) as a low viscosity monomer, and a conventional UV initiator (Table 4.2).

Instead of waiting for gravity to level the surface, different authors have investigated improved scrapping devices to spread the resin on the object being built. Kobayashi et al. obtained a 10 μm resolution in the deposited material by adapting the angle of the scraping device and by coating one of its surfaces with fluorocarbon [18, 21].

Table 4.2 Characteristics of the microstereolithography machines described by Zissi and Zhang

	Zissi et al. (1994)	Zhang et al. (1998)
Name of process	Microstereophotolithography	Microstereolithography
Light source	Argon ion laser	Argon ion laser
Surface monitoring	IR laser diode	CCD camera
Announced resolution	30 × 30 × 20 μm (*x,y,z*)	Spot has 1–2 μm
Resin type	Acrylate based resin containing unreactive absorbers and polymerization inhibitors	HDDA monomer containing 4 wt % of benzoin ethly ether as photoinitiator

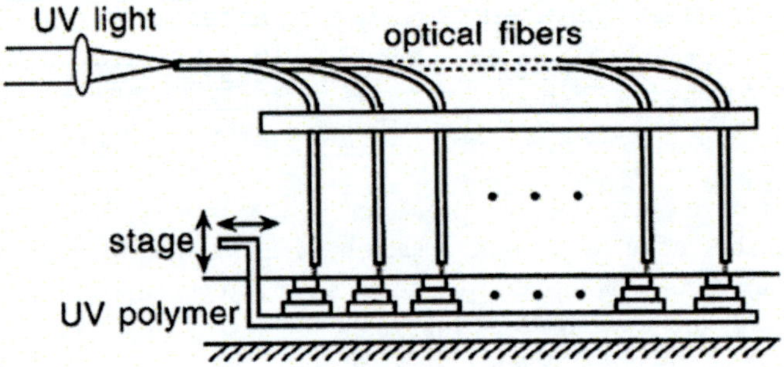

Fig. 4.8 Schematic diagram of collective manufacturing by scanning microstereolithography using an array of optical fibers. From [22] © 1996, IEEE. Reprinted by permission

4.3.3 *Collective Manufacturing with Optical Fibers*

Most conventional microfabrication techniques are wafer-based and allow a collective fabrication of a large number of identical components on the same substrate. This is not the case of most microstereolithography machines developed up to now, for which objects are generally built one at a time. Ikuta et al. tried to overcome this major limitation in 1996, by developing a collective microstereolithography process using an array of five single-mode optical fibers to focus the UV light of a xenon lamp in five different locations on the free surface of a photopolymerizable resin (Fig. 4.8). As for most scanning microstereolithography processes, the use of an X-Y-Z translation stage moving the photoreactor allows the solidification of the vectors composing each layer [22].

4.4 Integral Microstereolithography Processes

Integral microstereolithography processes (also named projection microstereolithography) are based on the projection of an image created by a dynamic mask on the surface of the photosensitive resin, and consequently, the object to be built has to be

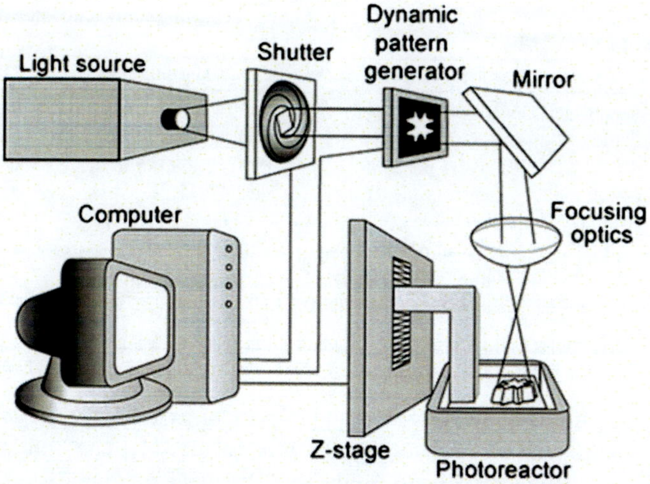

Fig. 4.9 Schematic diagram of an integral microstereolithography apparatus

described by a series of black and white bitmap files rather than by vectors as it is the case in scanning methods. These bitmap files are then used to shape the light beam using a dynamic mask, such that the image of each layer can be projected on the surface of the liquid resin after being reduced and focused by an appropriate optical system. The projected image induces a solidification of the irradiated areas, and a shutter controls the duration of the irradiation step. The superimposition of many layers of different shapes allows the fabrication of a complex object in a similar way as for conventional rapid prototyping techniques (Fig. 4.9).

Integral microstereolithography processes are inherently faster than scanning techniques, as the irradiation of a complete layer is done in one step, regardless to its complexity. Additionally, the light flux density on the surface of the photopolymerizable resin is much smaller when projecting an image than when focusing a light beam in one point, which allows to avoid problems related to unwanted thermal polymerization.

4.4.1 Liquid Crystal Displays as Dynamic Mask

The first integral microstereolithography machine, developed by Bertsch et al. in 1995 used a liquid crystal display (LCD) developed for rear-projection applications as dynamic mask [23]. Every pixel of a LCD is a small cell that can be set either to its transparent or opaque state by changing the orientation of the molecules it contains. This allows to modulate the light beam in the microstereolithography machine.

At the time the first integral microstereolithography machines were built, LCDs were not compatible with UV light, which required the development of specific photosensitive resins reacting to irradiation in the visible spectrum. As visible

Table 4.3 Characteristics of the integral microstereolithography machines described by Bertsch, Chatwin and Loubère

	Bertsch et al. (1995)	Chatwin et al. (1998)	Loubère et al. (1998)
Light source	Ar^+ laser (515 nm)	Ar^+ laser (351.1 nm)	Halogenure lamp
LCD display	260 × 260 pixels	600 × 800 pixels	640 × 480 pixels
Speed	90 layers per hour	<60 layers per hour	60 layers per hour
Announced resolution	5 × 5 × 5 μm (*x*,*y*,*z*)	50 μm (z)	5 × 5 × 10 μm (*x*,*y*,*z*)
Resin	PETIA with EosinY as photosensitiser and amine as co-initiator	Cibatool 5180 (low viscosity epoxy-based resin)	PETIA with EosinY as photosensitiser and amine as co-initiator

wavelengths are less energetic than UV radiations, there are very few photoinitiators available to start the polymerization process directly, and multi-components polymerizable resins have been used. The chemical medium developed for this first integral microstereolithography apparatus contained Eosin Y as a photosensitising dye, N-methyl diethanolamine as coinitiator and pentaerythritol triacrylate as monomer. This photopolymerizable resin fulfils the requirements to be used in the designed machine but has a few drawbacks: It has a high shrinkage and a photobleaching of the dye is observed during the photopolymerization process, which makes it difficult to control accurately the polymerized thickness. The maximum of reactivity of this resin corresponds to the maximum of absorption of the dye; consequently, an argon-ion laser emitting at 515 nm was used as light source together with a complex optical system aimed at destroying the spatial coherence (which eliminates the speckle effect) and redistributing the irradiance of the laser from a Gaussian to a flat-top profile [4, 24–26]. Other authors replicated this first integral microstereolithography machine, and replaced the light source by a broadband halogen lamp [27, 28].

In 1998, Chatwin et al. presented an integral microstereolithography machine operating in the UV. They used a special LCD component (a polysilicon thin film twisted nematic LCD manufactured by CRL Smectic Technologies Ltd.), transparent to wavelengths above 350 nm. This spatial light modulator was used in conjunction with the 351.1 nm line of an argon-ion laser and an optical system using diffractive elements to obtain a flat-top profile in the light beam [29–31]. Epoxy-based stereolithography resins were used with this machine [32–34] (Table 4.3).

4.4.2 Digital Micromirror Device (DMD™) as Dynamic Mask

The Digital Micromirror Device made by Texas Instruments, is widely used in video projection applications and has been used as dynamic mask in integral microstereolithography machines [35, 36]. This component is in fact an array of micromirrors actuated by electrostatic forces, and is used as a light switch in the microstereolithography apparatus: Each 16 μm square mirror can be independently actuated either to reflect the incident light beam either into or out of the pupil of the optical system, such that the corresponding pixel of the projected image appears bright or dark. For video projection applications, gray levels and colored images can be produced by combining the rapid on/off movement of the mirrors with a rotating color wheel, but these features are not used in the microstereolithography apparatus.

The first microstereolithography machine using the DMD™ component as dynamic mask was developed by Bertsch et al. and used a metal halide lamp combined with optical filters to select a band of visible wavelength for the irradiation of the resin [37, 38]. Different photopolymerizable resins were developed for this apparatus. In a first step, a high resolution resin reacting at 530 nm was used. It allowed the manufacture many components, but showed important curl deformations. In a second step, resins reacting at 410 nm were developed. Their composition is a lot closer to conventional acrylate-based stereolithography resins, even if they do not react with UV light, but in the blue part of the visible spectrum.

A second machine was later built by the same research team with an improved resolution (XGA: 1,024 $\times$ 768 pixels), and an irradiation in the UV [39]. An acrylate-based resin was formulated specifically for this machine, which was used in an industrial context to produce high-resolution prototypes.

More recently, other authors have also investigated such type of integral microstereolithography processes [40, 41].

4.5 Sub-Micron Processes

Since the first research started on the stereolithography process, the teams investigating this technique have dreamt of manufacturing the object directly inside the reactive medium rather than layer by layer on the surface. There are of course many advantages at creating the object directly in the resin, without superimposing the layers; no support parts are needed. No time is spent spreading the liquid on the part being manufactured, which potentially can speed-up the process significantly. Freely movable structures can be fabricated without the need of sacrificial layers. Many attempts have been done to develop such kind of processes, but they have never been successful, except for the fabrication of very small components by microstereolithography because nonlinear phenomena can be exploited when working at the micro-scale [42].

4.5.1 *Two-Photon Microstereolithography*

Two-photon absorption is a nonlinear phenomenon that occurs in all materials when the irradiance is sufficiently high and when the combined energy of two photons matches the transition energy between the ground state and an excited state. The rate

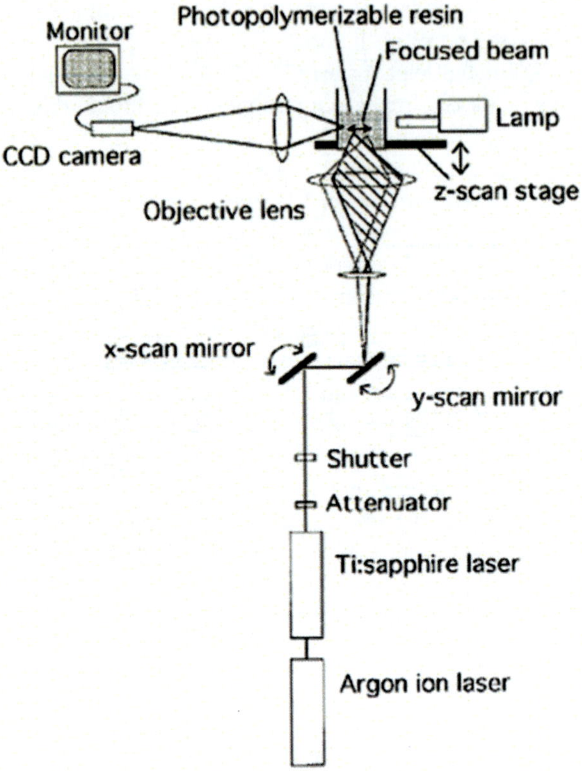

Fig. 4.10 Schematic diagram of a microstereolithography machine based on the two-photon absorption. From [43] © 1998, IEEE. Reprinted by permission

of two-photon absorption is proportional to the square of the incident light intensity. The quadratic dependence of the two-photon absorption rate on the light intensity confines this phenomenon to the area at the focal point, which allows to confine the photopolymerization phenomenon in submicron volumes.

Maruo et al. developed a two-photon microstereolithography process in 1996, and used a mode locked Ti:Sapphire laser emitting at 770 nm [43–46]. The beam power at peak in the resin was about 3 kW with a repetition rate of 76 MHz and a pulse width of 130 fs. The resin was a mixture of urethane acrylate monomers/oligomers and UV photoinitiators. It was transparent to IR light at 770 nm and did not attenuate the light beam, which could be focused inside the resin without polymerization at the surface by one-photon absorption. The lateral deflection of the light beam was obtained with scanning mirrors, whereas the object was manufactured in the vertical direction by moving the photoreactor (Fig. 4.10). Very small objects have been built with this process with a submicron resolution, like a microtube having an outside diameter of less than 3 μm and an inside diameter of 1.8 μm.

Kawata et al. developed a microstereolithography apparatus based on the same principle with a 120 nm resolution, which is smaller than the diffraction limit (the nonlinear effects allow to exceed it). To demonstrate the very high resolution

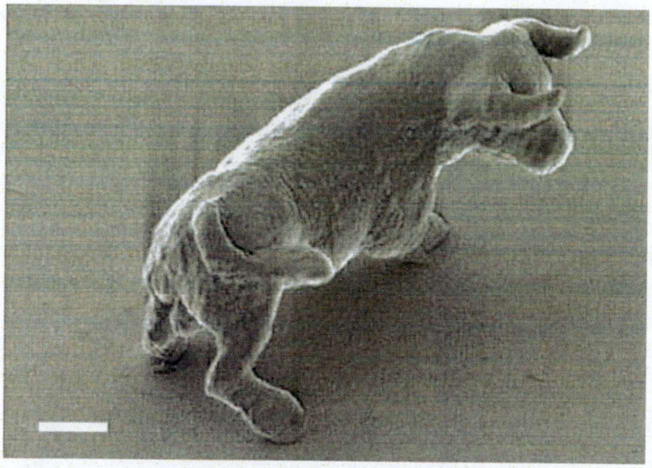

Fig. 4.11 Microbull. The scale bar corresponds to 2 μm [48]. Reprinted by permission from Macmillan Publishers Ltd: Nature, © 2001

obtained with this process, Kawata et al. built a micro-bull (Fig. 4.11), which is 10 μm long and 7 μm high and about the size of a red blood cell [47–50].

Other research groups have also investigated the same domain [51–54], and new photoinitiators having an enhanced two-photon sensitivity have been synthesized [55, 56].

4.5.2 *Single Photon Process Exploiting Nonlinearities of the Photopolymerization Reaction*

A radiation that is weakly absorbed by the photosensitive resin can be tightly focused inside the reactive medium such that the light intensity is sufficient to solidify the resin near the focal point, but not in the out-of-focus region. Appropriate conditions for exploiting this phenomenon can be found when using resins for which the dissolved oxygen molecules inhibit the polymerization reaction.

Maruo et al. presented a microstereolithography process based on the single photon under surface polymerization at the MEMS conference in 1998 [57, 58]. The single photon photopolymerization is stimulated not by a pulsed laser but by a continuous wave blue He–Cd laser, as the blue wavelength at 441.6 nm is weakly absorbed by conventional UV initiated resins. The resolution of this process has been evaluated by manufacturing polymerized threads at a constant laser scanning speed while varying the laser power [58, 59]. Measuring their width and thickness with a scanning electron microscope allowed the measurement of the process resolution, which is close to the one of the two-photon technique; the best lateral and depth resolution attained were 0.43 and 1.4 μm respectively. With this system, the lateral resolution is nearly equal to the diffraction limit of light.

4.6 Other Processes Related to Microstereolithography

The development of microstereolithography machines has contributed to the development of other techniques also based on the multilayer microfabrication approach, which allow the manufacturing of small objects having complex shapes, and are representative of possible ways to push further the microstereolithography concept. Research in this field aims mainly at widening the choice of materials used in the microstereolithography process to composites, ceramics, and metals and to further increase the resolution of the process to push this technologies to the nano-dimensions.

4.6.1 Microstereolithography with Materials Other than Polymers

Ceramic particles embedded in a photopolymerizable polymer matrix have already been used in the stereolithography field, to create big-size prototype parts in composite materials and to obtain ceramic parts to be used either as functional component or as mold for investment casting. According to the close connections between rapid prototyping technologies and microstereolithography, the results obtained in formulating resins highly loaded in powder ceramic material was the starting point in the study of microstereolithography of ceramics.

The insertion of ceramic powders in photopolymerizable media for microstereolithography was studied by a few research teams. Zhang et al. used a scanning microstereolithography machine to study the manufacturing of components from aqueous and nonaqueous alumina suspensions [20, 60, 61]. Moneret et al. used an integral microstereolithography machine to manufacture composite components from a dispersion of alumina particles in HDDA (a low viscosity acrylate monomer) [62–64].

Bertsch et al. adapted the integral microstereolithography machine they developed to use photopolymerizable media containing very high loads of alumina nanoparticles (up to 80 wt%) [65, 66]. The photosensitive media are no longer liquids, but pastes containing four components: alumina nanoparticles, a low viscosity monomer (Polyethylene glycol 400 diacrylate), a UV photoinitiator (DMPA), and a silane (3-glycidoxypropyltrimethoxysilane). The latest component is used to stabilize the formulation and prevent alumina particles to agglomerate. The obtained photosensitive pastes showed a high reactivity, similar to the conventional stereolithography resins, and were used to manufacture objects made of a large number of layers of 10 μm in thickness without major loss of resolution related to the light scattering of the reactive medium. Spreading the photosensitive paste on top of the component being built required the use of an automated scraping device made of a blade, synchronized with the other operations of the microstereolithography machine. Objects in composite materials are obtained when using photosensitive pastes and an adapted microstereolithography machine, and pure ceramic components can be manufactured from them, placing them in a furnace and applying an

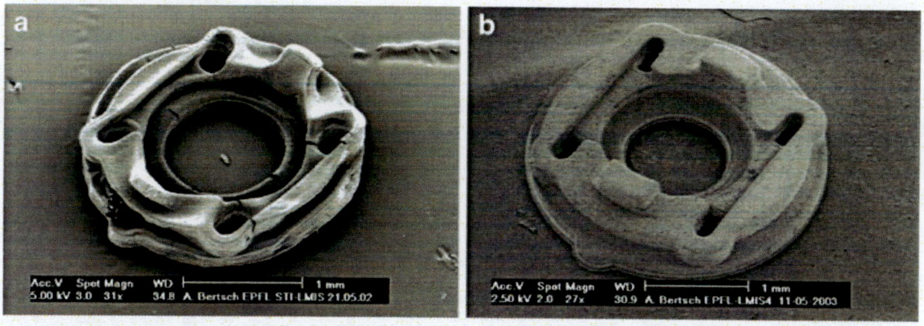

Fig. 4.12 Two identical components sintered from composites of different ceramic loadings. (**a**) From 50 wt% composite (27% shrinkage, 30% porosity), shows deformation and cracks. (**b**) From 75 wt% composite (22% shrinkage, 8% porosity) has kept its original shape

appropriate temperature ramp for debinding and sintering the alumina particles. Once sintered the objects show some shrinkage and a residual porosity. When using a high content in ceramic particles, the components keep their original shape and show no deformations or cracks (Fig. 4.12).

4.6.2 The EFAB Process

The EFAB process is a layer-by-layer manufacturing technique based on an in-situ patterning by electrodeposition that uses prefabricated, unbounded microfabricated masks [67, 68]. EFAB is in fact an acronym for Electrochemical FABrication. First, a series of masks in an insulator material are produced by photolithography on a flat copper substrate that will be used as anode. These masks correspond to the different layers of the object to be built. By pressing the insulator mask against the substrate, copper can be deposited by electroplating through the openings in the insulator mask. Once the current layer has been electroplated, the mask is removed. Nickel is then blanket-deposited onto the openings left in the copper, by electrodeposition, but without a mask. After both materials are deposited, the surface of the layer is planarized by lapping the surplus material to achieve the chosen layer thickness and obtain a flat surface for starting the next layer fabrication. As both the support and the building material are electrically conductive, new layers can be electrodeposited on top of the previously created layers without restrictions. By repeating the masking, electrodeposition, and planarization steps, complex 3D metallic structures will be formed. The selective dissolution of copper by etching with acid allows to release the final components.

Components made by the EFAB process can be obtained from MicroFabrica Inc. (www.microfabrica.com). Structures can be produced with a 2 μm accuracy and repeatability, with 5 μm features. As it is the case in stereolithography, complex mechanisms can be made without assembly.

4.6.3 Spatial Forming

Spatial forming is a more interesting experiment than a real technology development as this technique has been studied in 1991, but did not induce further investigations. This technique uses a modified offset printing press and two customized inks containing, respectively, high loads of ceramic or metal powders [69]. Half a micrometer of ceramic-based ink is deposited by offset printing on a ceramic substrate and cured by UV light. This process is repeated 20–30 times, before the non-image voids are filled with ink containing metal powder and cured with UV radiations. The ink-deposition process is then repeated until all layers have been deposited. The components are introduced in controlled atmosphere furnaces where they are debinded and sintered in conditions such that the ceramic material crumbles away and the finished metallic part separates from the substrate. In principle, spatial forming can be used to fabricate small 3D metallic components, and mass production is possible.

4.6.4 Applications

The first domain of application of the microstereolithography process is to produce small size prototype components when other rapid prototyping techniques reach their limits. Microstereolithography can be successfully used when small prototypes have to be built with dimensions of only a few millimeters or less, with small openings or intricate details, and avoid to manually polish the surface of the manufactured components, which can be very tedious, if not impossible, when dealing with small parts [70].

In many cases, the small spot stereolithography technique, which is available from a few service bureaus worldwide, is sufficient for producing components with a resolution higher than conventional rapid prototyping machines. However, some applications demand a substantially higher resolution and precision and can benefit from the potential of microstereolithography.

The first company that started to sell prototype parts made by microstereolithography is microTEC, in Duisburg, Germany (www.microtec-d.com) [71]. If the production of small mechanical components is the only current commercial application of microstereolithography, all universities and research institutes who investigated this technology manufactured interesting objects and try to find potential applications for this process.

4.6.5 Objects with Complex Geometry

All research teams who investigated microstereolithography validated the concept they developed by manufacturing components demonstrating the freedom in the

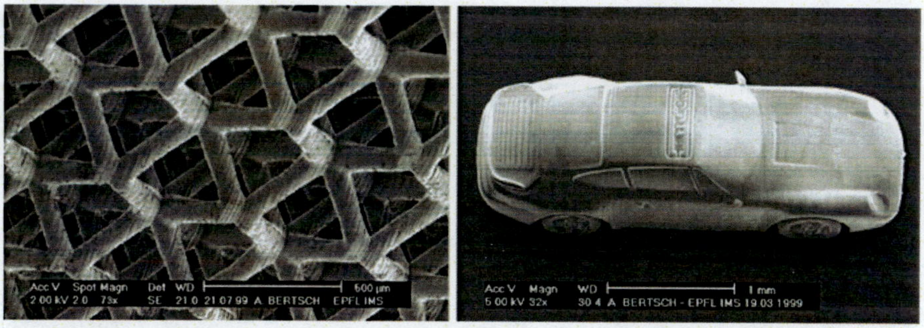

Fig. 4.13 3D network (*left*) and scale model of a car (*right*) made by integral microstereolithography

shapes this technology can produce. Most of these components have no real use, but their fabrication allows to identify and solve the problems encountered in the process development phase, and to refine the process and the resins. Typical components are: bending pipes [14], microtubes [43], micro coil springs [14, 30, 72, 73], spiral structures [45], micro cones [20] and pyramids [40], microgears [20, 25, 26, 28, 30, 31, 40, 58, 74], parts with overhanging regions [28], freely movable microstructures [46, 57, 58, 75], micromechanisms [51, 75–81], 3D networks [82], and scale models of larger components (small car, statue of liberty, micro turbines, Christmas tree, etc.) [9, 38]. Figure 4.13 shows examples of such components, which are extremely difficult or impossible to be built with conventional microfabrication techniques.

4.6.6 Small Mechanical Components

Based on the principle of layer-by-layer manufacturing, microstereolithography belongs to the rapid-prototyping family. Consequently, a first domain of application of this technique is the manufacturing of small high-resolution components that other rapid prototyping techniques fail to produce with adequate detail. There is a growing demand of such components, in particular, in domains connected to the medical field. Medical probe heads in which optical and chemical sensors could be embedded will be less invasive if they could be created with a smaller size [9]. Hearing-aid manufacturers try to design lightweight products small enough not to be detected, comfortable, with rounded shapes to be close from the natural geometry of the ear canal. This requires to prototype small mechanical components with intricate details (Fig. 4.14). In the medical domain, 3D models made by microstereolithography can also be used to train the surgeons before a difficult operation. Microstereolithography can also be used to produce small connectors and many other prototype parts.

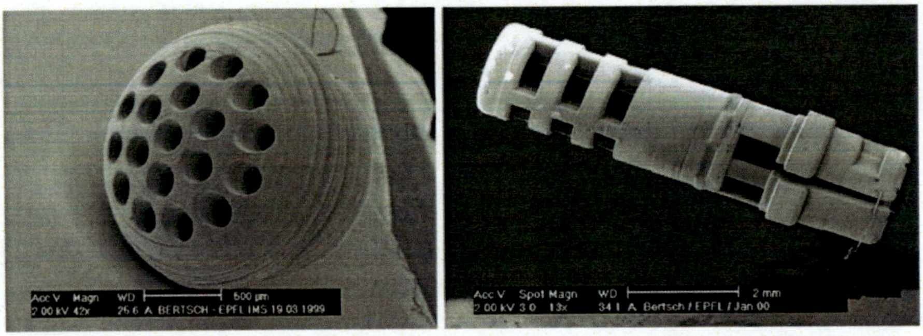

Fig. 4.14 Mushroom-shaped cap of a hearing aid (*left*). Component of a medical probe (*right*)

4.6.7 Microsystems

Many research teams have investigated the use of microstereolithography components in the Microsystems field, and in particular the combination of microstereolithography with structures and devices made by conventional silicon-based micromachining technologies. Microstereolithography does not allow wafer-level manufacturing, but its 3D capability is, however, an attractive feature for the microtechnology field. To combine it with other microfabrication technologies, microstereolithography can be applied to grow the 3D component directly on the part previously patterned by conventional microfabrication technologies, by inserting it in the photoreactor and aligning it with the light beam of the microstereolithography machine [37, 83].

- Microstereolithography has been combined with UV-LIGA (thick resist UV lithography), in order to add both the advantages of UV-LIGA (wafer-level processing, micrometer accuracy, smooth and vertical walls) to those of microstereolithography (complex geometries, curved and conical structures). To demonstrate this combination (Fig. 4.15), a conical axle made in a post-processing step by microstereolithography was added on a piece of gearing made in SU-8 (a negative epoxy resist).
- By building a polymer structure on top of a piezoelectric actuator, Takagi et al. realized a micro clamping tool [15].
- Microactuators having multiple degrees of freedom have been fabricated by inserting shape memory alloy (SMA) wires in a mechanical structure made by microstereolithography, in which clamping areas are provided (Fig. 4.15). Some SMA wires are submitted to an initial stress to store mechanical energy during the assembly step, whereas other passive wires have just a stiffening action. All wires are glued to the structure. When heating, the wires are initially submitted to mechanical stress by Joule effect and the whole structure deforms. It returns to its original position when the heating is stopped. Microactuators having a distributed elasticity or elastic hinges have been designed and tested [84–86].

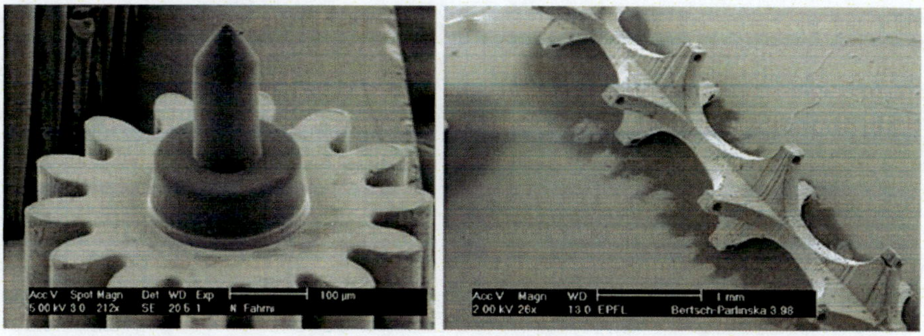

Fig. 4.15 Conical axle added by microstereolithography on a SU-8 micro gear (*left*). Structure of microactuator before assembly with SMA wires (*right*)

- A few research teams tried to immerse silicon wafers on which sensors and other components have been patterned by conventional microfabrication processing, in the resin bath of a stereolithography machine, for in situ packaging of these sensors by microstereolithography [87, 88]. Results were however rather limited, as the sensing components, once packaged, are almost impossible to clean from the photosensitive resin used in the stereolithography step. Moreover, dicing a 3D structure bound to a silicon wafer is certainly not straightforward.

Microstereolithography has also been used in the field of Microsystems as an inexpensive prototyping technique in the case of the fabrication of distributed MEMS phase shifters for microwave and millimeter wave applications. In that case, common silicon surface micromachining is used for all processing steps except the last one, defining the MEMS bridge, which is added on top of the structure by microstereolithography of a SU-8 layer. This technique allows to test MEMS bridges of various dimensions without the need of producing a large number of photomasks [89].

Hybrid polymer structures made by microstereolithography can also lead to the integration of various functions in microsystems. For example, the combination of two polymers with different refractive indices can be used for the manufacturing of optical wave-guides. This has been investigated by Maruo et al., who made such multi-polymer structures by adapting their microstereolithography machine such that the photosensitive resin could be changed for each layer of the manufactured object [90]. The integration of optical wave-guides in biochemical reactors made with the micro-stereolithography technology could lead to the creation of optical sensors embedded in chemical and biochemical microsystems.

4.6.8 *Passive Microfluidic Components*

The 3D shapes that can be generated by microstereolithography make it possible to create microfluidic components much more similar to the ones of the macro-world.

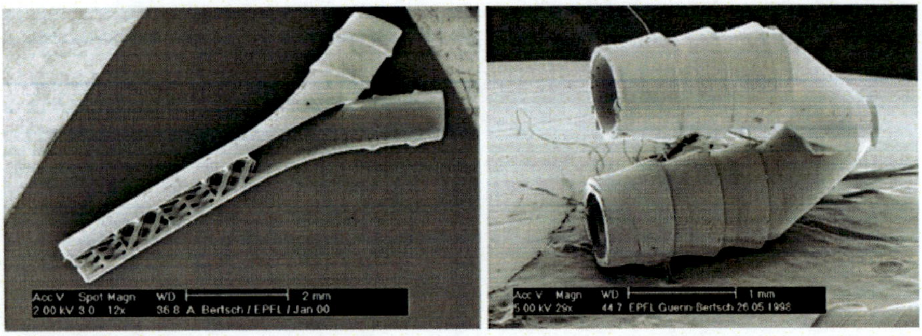

Fig. 4.16 Cut-out view of a micromixer based on the geometry of large scale industrial static mixers (*left*). Fluidic connector (*right*)

It is relatively easy to create networks of pipes having shapes that are impossible to obtain by conventional microfabrication techniques. Various research teams used microstereolithography to define a large selection of passive microfluidic components: bended pipes, a micro-venous valve [14], microfluidic channels [18, 91, 92], micro-connectors (Fig. 4.16) [38], and 3D micro mixers having shapes similar to large scale industrial static mixers (Fig. 4.16) [93–95], etc.

Passive microfluidic components can also be obtained by inserting membranes or other components during the microstereolithography fabrication step. A large variety of such passive microfluidic systems have been demonstrated and reported in the literature:

- Ikuta et al. built a microreactor around a photosensor, and demonstrated cell-free Luciferase (the photo-protein of firefly) synthesis in this fully integrated biochemical chip [96]. They later synthesized this substance from DNA in another microreactor obtained by assembling together different modules [97, 98], each of them producing a simple operation (valve, reactor, reservoir, sensor, etc.) [99].
- A prototype of micro-osmotic valve for insulin injection was made by building a 3D structure around a cellulose film used as semi-permeable membrane and a rubber component used as a diaphragm. In this device, when water is transferred through the semipermeable membrane by osmotic forces, a deformation of the diaphragm opens the valve, allowing the insulin injection [100].
- A structure made by microstereolithography containing a microreactor chamber and fluidic channels was used to perform the separation of Luciferase from other reagents. An ultrafiltration membrane and a photodiode, both inserted in the device during its fabrication, provide the function of the microreactor and allow to monitor the evolution of the concentration of the separated components by photoluminescence [101].
- A 10-way switching valve made for multidirectional flow control was obtained by combining different components made by microstereolithography with a steel ball used to reduce friction, a solenoid used for switching, and rubber valves built into the polymer housing. This device was used to switch fluid flows between different outlets, the inlet being located at the center of the device [102].

- Polymer tubing and a reaction chamber were grown on top of chemical gas sensors made with conventional silicon micromachining. If the 3D components were successfully produced, the cleaning step of the sensors embedded in the device could not be performed in a satisfying way [88].

4.6.9 Active Microfluidic Components

Three-dimensional components made by microstereolithography can be combined with piezoelectric actuators or shape memory alloy components to produce active microfluidic elements, such as pumps, valves, and active micromixers. Many research papers report investigations on this subject:

- A micropump was manufactured by Dario et al. using conventional stereolithography to form the pump body with internal channels and a valve housing. It is actuated by a piezoelectric ceramic disk, glued to the polymer components [103, 104]. The pump has been tested with water and the flow rate can be chosen between 1 and 50 μl/min by changing the frequency of the sinusoidal voltage applied to the actuator.
- A micropump based on the principle of a deformable chamber has been built by Ikuta et al., using microstereolithography to build the pump body and fluidic channels, and inserting rubber valves during the microstereolithography step [105, 106]. The upper part of the pump chamber is closed by a rubber sheet, which can be deformed by a shape memory alloy actuator (Fig. 4.17). Flow rates of up to 10 μl/min were achieved when pumping water with this device.
- Still using shape memory alloy actuators, Ikuta et al. produced a switching valve that can be used to direct a flow of liquid to either of the two outlets. Switching times in the range of 0.5 s have been measured [107, 108].

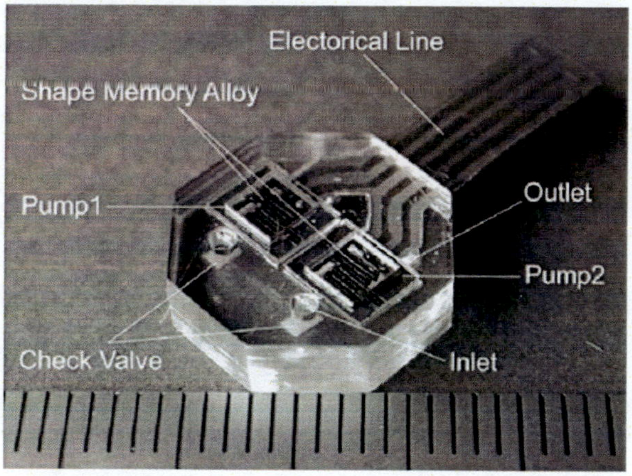

Fig. 4.17 Micropump chip made by Ikuta et al. from [106] © 2001, IEEE. Reprinted by permission

- Using the two-photon microstereolithography process, Maruo et al. produced an optically-driven micropump in which two rotors are moved in opposite direction by a laser tweezer [75]. Fluid is trapped between the two rotors and is carried from inlet to outlet by their movement, providing the pumping mechanism. Flow rates in the range of less than 1 pl/min were achieved.
- An ultrasonic homogenizer was made by Ikuta et al. by inserting a piezoelectric element in a part having a horn shape, which allowed to disrupt the cells. The resulting solution was used for separating the proteins by gel electrophoresis [108].

Complex fluidic systems can be obtained by assembling active and passive microfluidic components. Ikuta et al. developed a simple method for holding and connecting fluidic components to obtain chemical and biochemical chips [97, 98, 109, 110]. All individual components have an octagonal shape and their function can be assessed separately. Their interconnection and coupling without leakage is obtained by the insertion of a molded thin silicone rubber film between the individual components. The holder unit allows to stack the different components, which are held together by a spring and separated by a silicon rubber film that prevents fluid leakage. This versatile fluidic device has been named "Biochemical IC" [96, 99, 110–112].

4.7 Conclusion

Microstereolithography is a technology that is not yet commercially available, although a few service bureaus are using it for manufacturing small objects that are impossible to be produced by conventional rapid prototyping technologies. Microstereolithography has been mostly studied by various research teams in an academic context, each team developing its own apparatus, implementing specific features, and using specially developed resins. This technique has been mostly used as complement for the existing techniques in the Microsystems field, allowing to produce components with complicated geometries for microrobotics, microfluidics, and micromechanics applications. Its compatibility with conventional microfabrication technologies has been evaluated in specific cases, but it is not yet a wafer-level collective fabrication technique; this is a limitation in the microsystems field.

Objects made by microstereolithography have a 3D geometry. Most of them cannot be molded easily, which implies that in some applications, microstereolithography is not a prototyping, but a manufacturing technology. In this case, the materials in which the objects are produced need to have adequate mechanical, physical or chemical characteristics. Current developments of the microstereolithography process are related to the extension of the materials that can be used in this process to a variety of polymers, biocompatible or biodegradable resins [113, 114] as well as ceramic, metal, and composite materials, contributing to the further general evolution of micro and nanosystems towards the biomedical domain.

Acknowledgements The authors would like to thank all the research teams who have been involved in the microstereolithography field since this subject has started to be studied, in particular Prof. Koji Ikuta and Prof. Jean-Claude André who had a pioneer vision of this scientific domain.

References

1. Hull, C.W., *Apparatus for production of three dimensional objects by stereolithography*. 1986: US Patent.
2. André, J.C., A. Le Méhauté, and O. De Witte, *Dispositif pour réaliser un modèle de pièce industrielle*. 1984: French Patent.
3. Zissi, S., et al., *Stereolithography and Microtechniques*. Microsystem Technologies, 1996. **2**(2): p. 97–102.
4. Bertsch, A., J.Y. Jézéquel, and J.C. André, *Study of the spatial resolution of a new 3D microfabrication process: the microstereophotolithography using a dynamic mask-generator technique*. Journal of Photochemistry and Photobiology. A, Chemistry, 1997. **107**: p. 275–281.
5. Yamaguchi, K. and T. Nakamoto. *Consideration on the accuracy of fabricating microstructures using UV laser induced polymerization*. In *the 6th International Symposium on Micro Machine and Human Science (MHS'94)*. 1994.
6. Yamaguchi, K. and T. Nakamoto. *Consideration on the optimum conditions to produce micromechanical parts by photo polymerization using direct focused beam writing*. In *6th International Symposium on Micro Machine and Human Science (MHS'95)*. 1995. Piscataway, NJ, USA.
7. Nakamoto, T. and K. Yamaguchi. *Consideration on the producing of high aspect ratio micro parts using UV sensitive photopolymer*. In *the 7th International Symposium on Micro Machine and Human Science (MHS'96)*. 1996.
8. Nakamoto, T., et al., *Manufacturing of three-dimensional micro-parts by UV laser induced polymerization*. Journal of Micromechanics and Microengineering, 1996. **6**(2): p. 240–253.
9. Bertsch, A., et al., *Rapid prototyping of small size objects*. Rapid Prototyping Journal, 2000. **6**(4): p. 259–266.
10. Tönshoff, H.K., et al. *Qualification of different standard photo resins and new laser sources for micro-stereolithography*. In *3rd micro materials conference MICROMAT2000*. 2000. Berlin, Germany.
11. Bernhard, P., *Proform devises method for adding a second small spot laser to its SLA 250/40*. Rapid prototyping report, 1997. Dec. 1997.
12. Inverson, N.J., *Why not use stereolithography as a manufacturing process?* Rapid prototyping report, 1998. Mars 1998.
13. Takagi, T. and N. Nakajima. *Photoforming applied to fine machining*. In *4th International Symposium on Micro Machine and Human Science (MHS'93)*. 1993.
14. Ikuta, K. and K. Hirowatari. *Real three dimensional micro fabrication using stereo lithography and metal molding*. In *6th IEEE Workshop on Micro Electro Mechanical Systems (MEMS'93)*. 1993.
15. Takagi, T. and N. Nakajima. *Architecture combination by micro photoforming process*. In *7th IEEE Workshop on Micro Electro Mechanical Systems (MEMS'94)*. 1994. Oiso, Japan.
16. Takagi, T. and N. Nakajima, *Photoforming applied to fine forming*. JSME International Journal – Series C, 1995. **38**(4): p. 811–817.
17. Ikuta, K., K. Hirowatari, and T. Ogota. *Three dimensional micro integrated fluid systems (MIFS) fabricated by stereo lithography*. In *the IEEE Micro Electro Mechanical Systems Workshop (MEMS'94)*. 1994. Oiso, Japan.

18. Kobayashi, K. and K. Ikuta. *Development of free surface microstereolithography with ultra high resolution to fabricate hybrid 3-D microdevices*. In *IEEE International symposium on Micro-Nano Mechatronics and Human Schience*. 2005.
19. Zissi, S., et al. *Limites de la stéréolithographie pour des applications microtechniques*. In *3e Assises Européennes du Prototypage Rapide*. 1994. Paris, France.
20. Zhang, X., X.N. Jiang, and C. Sun, *Micro-stereolithography for MEMS*. Micro Electro Mechanical Systems (MEMS) ASME, 1998. **66**: p. 3–9.
21. Kobayashi, K. and K. Ikuta. *Advanced free-surface microstereolithography with 10-micrometer resolution for hybrid microstructures*. In *IEEE/ASME International Conference on Advanced Intelligent Mechantronics*. 2007.
22. Ikuta, K., et al. *Development of mass productive micro stereo lithography*. In *8th IEEE Workshop on Micro Electro Mechanical Systems (MEMS'96)*. 1996. Piscataway, NJ, USA.
23. Bertsch, A., et al. *Nouveau procédé de microstéréolithographie utilisant des filtrages dynamiques*. In *4èmes assises européennes du prototypage rapide*. 1995. Paris, France: Remark S. A. Ed., Paris.
24. Bertsch, A., *Microstéréophotolithographie par masquage dynamique*. 1996, Ph. D. Thesis in Process Engineering, INPL-ENSIC. p. 168.
25. Bertsch, A., et al. *Manufacture of multilayers microparts by microstereophotolithography*. In *3rd France-Japan Congress and 1st Europe-Asia Congress on Mechatronics*. 1996. Besançon, France.
26. Bertsch, A., et al., *Microstereophotolithography using a liquid crystal display as dynamic mask-generator*. Microsystem Technologies, 1997. **3**(2): p. 42–47.
27. Loubère, V., S. Monneret, and S. Corbel. *Microstereolithography using a mask-generator display*. In *the 4th Japan-France Congress and 2nd Asia-Europe Congress on Mechatronics*. 1998. Kitakyushu, Japan.
28. Monneret, S., V. Loubère, and S. Corbel. *Microstereolithography using a dynamic mask generator and a non-coherent visible light source*. In *SPIE Symposium on Design, Test and microfabrication of MEMs/MOEMs*. 1999. Paris, France.
29. Chatwin, C.R., et al., *UV microstereolithography system that uses spatial light modulator technology*. Applied Optics, 1998. **37**(32): p. 7514–7522.
30. Farsari, M., et al., *A novel high-accuracy microstereolithography method employing an adaptive electro-optic mask*. Journal of Materials Processing Technology, 2000. **107**: p. 167–172.
31. Farsari, M., et al., *Microfabrication by use of a spatial light modulator in the ultraviolet: experimental results*. Optics Letters, 1999. **24**(8): p. 549–550.
32. Farsari, M., et al. *Holographic Measurements of Photopolymers for Microstereolithography Application*. In *Lasers and Electro-Optics Europe (1998 CLEO/Europe)*. 1998.
33. Farsari, M., et al., *Four wave mixing studies of UV curable resins for microstereolithography*. Journal of Photochemistry and Photobiology. A, Chemistry, 1998. **115**: p. 81–87.
34. Chatwin, C.R., et al., *Characterization of epoxy resins for microstereolithographic rapid prototyping*. The International Journal, Advanced Manufacturing Technology, 1999. **15**: p. 281–286.
35. Hornbeck, L.J. *Digital Light Processing (TM) for high-brightness high-resolution applications*. In *Electronic Imaging (EI'97) – Projection displays III*. 1997. San Jose, California, USA.
36. Nelson, W.E. and R.L. Bhuva. *Digital micromirror device imaging bar for hard copy*. In *the SPIE color hard copy and graphics arts IV*. 1997. San Jose, California, USA.
37. Bertsch, A., H. Lorenz, and P. Renaud. *Combining microstereolithography and thick resist UV lithography for 3D microfabrication*. In *11th IEEE Workshop on Micro Electro Mechanical Systems (MEMS'98)*. 1998. Heidelberg, Germany.
38. Beluze, L., A. Bertsch, and P. Renaud. *Microstereolithography: a new process to build complex 3D objects*. In *SPIE Symposium on Design, Test and microfabrication of MEMs/MOEMs*. 1999. Paris, France.
39. Bertsch, A., et al., *Microstereolithography: a review*. Materials Research Society Symposia Proceedings, 2003. **758**: p. LL1.1.1–LL1.1.12.

40. Choi, J.W., et al., *Fabrication of 3-dimensional microstructures using dynamic image projection*. Key Engineering Materials, 2007. **339**: p. 473–478.
41. Anonymous, *Projection microstereolithography creates 3-D microstructures*. Laser Focus World, 2000. **36**(12): p. 11.
42. Sun, H.-B. and S. Kawata, *Two-photon photopolymerization and 3D lithographic microfabrication*. 2004: Springer-Verlag. 268.
43. Maruo, S. and S. Kawata, *Two-photon-absorbed near-infrared photopolymerization for three-dimensional microfabrication*. Journal of Microelectromechanical Systems, 1998. **7**(4): p. 411–415.
44. Maruo, S., O. Nakamura, and S. Kawata. *Three dimensional microfabrication with two-photon absorbed photopolymerization*. In *SPIE – Optics for Science and New Technology*. 1996.
45. Maruo, S., O. Nakamura, and S. Kawata, *Three-dimensional microfabrication with two-photon-absorbed photopolymerization*. Optics Letters, 1997. **22**(2): p. 132–134.
46. Maruo, S. and K. Ikuta. *Movable microstructures made by two -photon three dimensional microfabrication*. In *International symposium on micromechatronics and human science (MHS'99)*. 1999.
47. Kawata, S. *Three-dimensional micro-fabrication with two-photon and single-photon polymerization*. In *4th Pacific Rim Conference on Lasers and Electro-Optics, 2001 (CLEO/Pacific Rim 2001)*. 2001.
48. Kawata, S., et al., *Finer features for functional microdevices*. Nature, 2001. **412**: p. 697–698.
49. Tanaka, T. and S. Kawata. *Three-dimensional fabrication and observation of micro-structures using two-photon absorption and fluorescence*. In *Micro and Nano-photonic materials and devices conference, SPIE*. 2000.
50. Tanaka, T. and S. Kawata. *Three-dimensional microfabrication by two-photon initiated photopolymerization*. In *Conference on Lasers and Electro-Optics (CLEO '01)*. 2001.
51. Galajda, P. and O. Pál, *Complex micromachines produced and driven by light*. Applied Physics Letters, 2001. **78**(2): p. 249–251.
52. Miwa, M., et al., *Femtosecond two-photon stereo-lithography*. Applied Physics A, 2001. **73**: p. 561–566.
53. Teh, W.H., et al., *Effect of low numerical aperture femtosecond two-photon absorption on (SU-8) resist for ultrahigh aspect ratio microstereolithography*. Journal of Applied Physics, 2005. **97**: p. 054907-1–054907-11.
54. Sun, H.-B., M. Maeda, and K. Takada, *Experimental investigation of single voxels for laser nanofabrication via two-photon photopolymerization*. Applied Physics Letters, 2003. **83**(5): p. 819–821.
55. Cumpston, B.H., et al., *Two-photon polymerization initiators for three-dimensional optical data storage and microfabrication*. Nature, 1999. **398**: p. 51–54.
56. Yu, T., et al., *Chemically amplified positive resists for two photon three-dimensional microfabrication*. Advanced Materials, 2003. **15**(6): p. 517–521.
57. Ikuta, K., S. Maruo, and S. Kojima. *New micro stereo lithography for freely movable 3D micro structure – Super IH Process with submicron resolution*. In *11th IEEE Workshop on Micro Electro Mechanical Systems (MEMS'98)*. 1998. Heidelberg, Germany.
58. Maruo, S. and K. Ikuta, *Three dimensional microfabrication by use of single photon absorbed polymerization*. Applied Physics Letters, 2000. **76**(19): p. 2656–2658.
59. Maruo, S. and K. Ikuta, *Submicron stereolithography for the production of freely movable mechanisms by using single-photon polymerization*. Sensors and Actuators: A, 2002. **100**: p. 70–76.
60. Varadan, V.K., X. Jiang, and V.V. Varadan, *Microstereolithography and other fabrication techniques for 3D MEMS*. 2001: Wiley. 260.
61. Zhang, X., X. Jiang, and C. Sun, *Micro-stereolithography of polymeric and ceramic microstructures*. Sensors and Actuators A, 1999. **77**: p. 149–156.

62. Provin, C. and S. Monneret, *Complex ceramic-polymer composite microparts made by microstereolithography*. IEEE transactions on electronics packaging manufacturing, 2002. **25**(1): p. 59–63.
63. Provin, C., et al., *Three-dimensional ceramic microcomponents made using microstereolithography*. Advanced Materials, 2003. **15**(12): p. 994–997.
64. Monneret, S., et al., *Microfabrication of freedom and articulated alumina-based components*. Microsystem Technologies, 2002. **8**: p. 368–374.
65. Bertsch, A., S. Jiguet, and P. Renaud, *Microfabrication of ceramic components by microstereolithography*. Journal of Micromechanics and Microengineering, 2004. **14**(2): p. 197–203.
66. Jiguet, S., et al. *Microstereolithography and ceramic composite three-dimensional parts*. In *Shaping II*. 2002. Gent, Belgium.
67. Cohen, A., et al. *EFAB: Rapid, low-cost desktop micromachining of high aspect ratio true 3-D MEMS*. In *12th IEEE International Conference on Micro Electro Mechanical Systems (MEMS'99)*. 1999. Orlando, Florida, USA.
68. Anonymous, *Microfabrication-rapid prototyping's killer application?* Rapid Prototyping Report, 1999. **9**(6): p. 1–5.
69. Taylor, C.S., et al. *'Spatial forming' a three dimensional printing process*. In *8th IEEE Workshop on Micro Electro Mechanical Systems (MEMS'95)*. 1995.
70. Bertsch, A., et al. *Fine features by integral microstereolithography*. In *4th International Workshop on High Aspect Ratio Micro Structure Technology (HARMST'01)*. 2001. Baden-Baden, Germany.
71. Bohlmann, H. and R. Götzen. *High aspect ratio components through RMPD*. In *4th International Workshop on High Aspect Ratio Micro Structure Technology (HARMST'01)*. 2001. Baden-Baden, Germany.
72. Varadan, V.K., V.V. Varadan, and S. Motojima. *Three dimensional polymeric and ceramic MEMS and their applications*. In *SPIE Conference on Smart Structures and Materials 1996 – Smart Electronics and MEMS*. 1996. San Diego, California, USA.
73. Yoshimoto, T., et al. *Micro stereo lithography system*. In *SPIE*. 2006.
74. Cabrera, M., et al., *Microfabrication of very small objects: Pushing the limits of stereophotolithography*. Molecular Crystals and Liquid Crystals, 1998. **315**: p. 223–234.
75. Maruo, S. and H. Inoue, *Optically Driven micropump produced by three-dimensional two-photon microfabrication*. Applied Physics Letters, 2006. **89**: p. 144101-1–144101-3.
76. Maruo, S., K. Ikuta, and K. Hayato. *Light-driven MEMS made by high-speed two-photon microstereolithography*. In *14th IEEE International Conference on Micro Electro Mechanical Systems (MEMS2001)*. 2001. Interlaken, Switzerland.
77. Maruo, S., K. Ikuta, and H. Korogi. *Optical drive of constrained micromechanisms produced by two-photon microstereolithography with 200 nm resolution*. In *4th International Workshop on High Aspect Ratio Micro Structure Technology (HARMST'01)*. 2001. Baden-Baden, Germany.
78. Maruo, S., K. Ikuta, and H. Korogi. *Remote Light-driven Micromachines Fabricated by 200nm Microstereolithography*. In *IEEE Nanoelectromechanical systems (NEMS) Conference (IEEE-NANO 2001)*. 2001.
79. Maruo, S., K. Ikuta, and H. Korogi, *Submicron manipulation tools driven by light in a liquid*. Applied Physics Letters, 2003. **82**(1): p. 133–135.
80. Maruo, S. and H. Korogi, *Focre-Controllable, optically driven micromachines fabricated by single-step two-photon microstereolithography*. Journal of Microelectromechanical Systems, 2003. **12**(5): p. 533–539.
81. Maruo, S., K. Ikuta, and H. Korogi. *Direct nanomanipulation tools for biological samples*. In *MicroTotal Analysis Systems (MicroTAS'02)*. 2002: Kluwer.
82. Lee, S.-J., et al., *Application of microstereolithography in the development of three-dimensional cartilage regeneration scaffolds*. Biomedical Microdevices, 2008. **10**: p. 233–241.

83. Bertsch, A., H. Lorenz, and P. Renaud, *3D microfabrication by combining microstereolithography and thick resist UV lithography*. Sensors and Actuators: A, 1999. **73**: p. 14–23.
84. Ballandras, S., et al., *Microstereophotolithography and shape memory alloy for the fabrication of miniaturized actuators*. Sensors and Actuators A, 1997. **62**: p. 741–747.
85. Bertsch, A., et al. *Conception and realization of miniaturized actuators fabricated by Microstereophotolithography and actuated by Shape Memory Alloys*. In *3rd France-Japan Congress and 1st Europe-Asia Congress on Mechatronics*. 1996.
86. Calin, M., et al. *Design and control of Compliant Microrobots*. In *5th IEEE Symposium on Emerging Technologies and Factory Automation ETFA*. 1996. Piscataway, NJ, USA.
87. Tse, A.L., P.J. Hesketh, and D.W. Rosen. *Stereolithography on silicon for microfluidics and microsensor packaging*. In *4th International Workshop on High Aspect Ratio Micro Structure Technology (HARMST'01)*. 2001. Baden-Baden, Germany.
88. Tse, A.L., et al., *Stereolithography on silicon for microfluidics and microsensor packaging*. Microsystem Technologies, 2003. **9**(5): p. 319–323.
89. Ji, T.S., K.J. Vinoy, and V.K. Varadan, *Distributed MEMS phase shifters by microstereolithography on silicon substrates for microwave and millimeter wave applications*. Smart Materials and Structures, 2001. **10**: p. 1224–1229.
90. Maruo, S., K. Ikuta, and T. Ninagawa. *Multi-polymer microstereolithography for hybrid opto-MEMS*. In *14th IEEE International Conference on Micro Electro Mechanical Systems (MEMS2001)*. 2001. Interlaken, Switzerland.
91. Ikuta, K. *Biomedical micro device fabricated by micro stereo lithography (IH process) "Metabiotic Device" – a synthetic approach for life science*. In *6th International Symposium on Micro Machine and Human Science (MHS'95)*. 1995.
92. Mizukami, Y., D. Rajniak, and M. Nishimura. *An integrated micro-electrophoretic chip fabricated using a new stereolithographic process*. In *13th IEEE International Conference on Micro Electro Mechanical Systems (MEMS 2000)*. 2000.
93. Bertsch, A., et al. *3D micromixers – Downscaling large-scale industrial static mixers*. In *14th IEEE International Conference on Micro Electro Mechanical Systems (MEMS2001)*. 2001. Interlaken, Switzerland.
94. Bertsch, A., et al., *Static micromixers based on large-scale industrial mixer geometry*. Lab on a Chip, 2001. **1**(1): p. 56–60.
95. Kim, D.S., et al., *A barrier embedded Kenics micromixer*. Journal of Micromechanics and Microengineering, 2004. **14**: p. 1294–1301.
96. Ikuta, K. *Biochemical IC using micro stereolithography-towards artificial cellular device*. In *International Microprocesses and Nanotechnology Conference*. 2000.
97. Hasegawa, T. and K. Ikuta. *Simple and effective method for hold and connet of chemical IC chip-set*. In *the Micro Total Analysis Systems 2001 Symposium (microTAS2001)*. 2001. Monterey, CA, USA.
98. Hasegawa, T. and K. Ikuta. *Silicone rubber coupling for microfluidic devices – Theoretical analysis of sealing properties for interconnection*. In *Micro Total Analysis Systems (MicroTAS'02)*. 2002. Kluwer Academic Publishers.
99. Ikuta, K., A. Takahashi, and S. Maruo. *In-chip cell-free protein synthesis from DNA by using biochemical IC chips*. In *14th IEEE International Conference on Micro Electro Mechanical Systems (MEMS2001)*. 2001. Interlaken, Switzerland.
100. Nagakura, T., et al. *Study of a micro-osmotic valve for insulin injection system in treatment of diabetes mellitus – Using the 3D optical modeling method-*. In *EMBEC'99*. 1999.
101. Ikuta, K., et al. *Micro concentrator with opto-sense micro reactor for biochemical IC chip family – 3D composite structure and experimental verification*. In *12th IEEE International Conference on Micro Electro Mechanical Systems (MEMS'99)*. 1999. Orlando, Florida, USA.
102. Hasegawa, T., K. Ikuta, and K. Nakashima. *10-way micro switching valve chip for multi-directional flow control*. In *7th International conference on miniaturized chemical and biochemical analysis systems (MicroTAS'03)*. 2003. Squaw Valley, California, USA.

103. Carrozza, M.C., et al., *A piezoelectric-driven stereolithography-fabricated micropump*. Journal of Micromechanics and Microengineering, 1995. **5**: p. 177–179.
104. Accoto, D., M.C. Carrozza, and P. Dario, *Modelling of micropumps using unimorph piezoelectric actuators and ball valves*. Journal of Micromechanics and Microengineering, 2000. **10**: p. 277–281.
105. Ikuta, K., et al. *Fluid drive chips containing multiple pumps and switching valves for biochemical IC family*. In *13th IEEE International Conference on Micro Electro Mechanical Systems (MEMS 2000)*. 2000.
106. Ikuta, K., T. Hasegawa, and T. Adachi. *The optimized SMA micro pump chip applicable to liquids and gases*. In *Thansducers'01 Eurosensors XV Workshop*. 2001. Munich, Germany.
107. Hasegawa, T., et al., *Multi-directional micro-switching valve chip with rotary mechanism*. Sensors and Actuators A, 2008. **143**: p. 390–398.
108. Ikuta, K., et al. *Micro ultrasonic homogenizer chip made by hybrid microstereolithography*. In *Micro Total Analysis Systems (MicroTAS'02)*. 2002: Kluwer.
109. Ikuta, K., et al. *Chemical IC chip family for on-chip cell-free protein synthesis from DNA*. In *Micro Total Analysis Systems 2001 Symposium (microTAS2001)*. 2001. Monterey, CA, USA: Kluwer academic publishers.
110. Ikuta, K., et al. *User-assembly, fully integrated micro chemical laboratory using biochemical IC chips for wearablw/implantable applications*. In *MicroTotal Analysis Systems (microTAS'02)*. 2002: Kluwer.
111. Ikuta, K., et al. *Biochemical IC chip toward cell free DNA protein synthesis*. In *11th IEEE Workshop on Micro Electro Mechanical Systems (MEMS'98)*. 1998. Heidelberg, Germany.
112. Ikuta, K., et al. *Biochemical IC chip for pretreatment in biochemical experiments*. In *MEMS 2003*. 2003.
113. Yamada, A., F. Niikura, and K. Ikuta, *A three dimensional microfabrication system for biodegradable polymers with high resolution and biocompatibility*. Journal of Micromechanics and Microengineering, 2008. **18**: p. 1–9.
114. Liska, R., et al., *Photopolymers for rapid prototyping*. Journal of Coating Technology and Research, 2007. **4**(4): p. 505–510.

Chapter 5
Polymeric Materials for Rapid Manufacturing

Fred J. Davis and Geoffrey R. Mitchell

5.1 Introduction

Rapid Manufacturing also known as solid free-form fabrication is a term describing a range of processes whereby a computer generated design is converted to a three-dimensional (3D) object. This methodology was originally used for the manufacture of models and prototypes (hence the description "Rapid Prototyping") [1] and has becoming an increasingly important tool for designers and manufacturers [2]; indeed in some cases this approach is being extended to the manufacture of small numbers of complex articles. One particularly exciting development is the use of Rapid Manufacturing for the production of biocompatible components for medical use, where the production of a one-off component is a necessary requirement. Examples include the manufacture of dental prostheses [3] and hip joints [4].

Perhaps the best known of this group of rapid manufacturing techniques is Stereolithography. Here, a polymer (usually highly cross-linked) is formed layer by layer *via* photoinduced polymerization; the light source, which is either a highly focused beam or a laser, is used to initiate polymerization through the photodegradation of an initiator to form radicals, cations, or carbene-like species as discussed below. This maximizes dimensional control, with ultimately the resolution being determined largely by the wavelength of the excitation light (see below). A related approach is that of selective laser sintering; in this case, a layer of powder is laid onto a surface and a CO_2 (or similar) laser sinters, i.e. heats and fuses together, the powder. Such systems can be utilized to form objects from polymers, ceramics, and metals [5]. A rather different approach is to deposit the material (also one layer at a time) through a nozzle; this approach includes techniques where conventional/ink-jet printing processes, are converted to 3D systems (once again in a layer-by-layer approach) or systems where a fine-stream of polymer is pushed out on to the surface, a technique known as fused deposition molding [6].

G.R. Mitchell (✉)
Centre for Rapid and Sustainable Product Development,
Polytechnic Institute of Leiria, Leiria, Portugal
e-mail: geoffrey.mitchell@ipleiria.pt

P.J. Bártolo (ed.), *Stereolithography: Materials, Processes and Applications*,
DOI 10.1007/978-0-387-92904-0_5,

Of the many types of materials which may be used, the most common include metals [7], ceramics [8], and polymers [9]. Each of these materials may offer different advantages in terms of factors such as mechanical strength and biocompatibility, but in the case of polymeric systems it is their versatility which makes them particularly attractive; and it is these materials which form the basis of the discussion included in this chapter. In all the typical rapid manufacturing system, the computer converts the design into layers and then the solid object is prepared layer-by-layer; it is intended in this discussion to outline the types of materials available for manufacturing by this approach, their advantages and limitations, and in particular to compare how the properties of articles manufactured in this way differ from those prepared by conventional approaches to polymer manufacturing, such as molding extrusion, and casting.

5.2 Polymeric Materials

5.2.1 Polymerization

In principle, the synthetic chemist can produce an almost infinite variety of organic polymers, by virtue of the versatility and controllability of organic chemistry. The reality is somewhat more restrictive. Commercial enterprises are constrained to those materials manufactured industrially, and cost is a frequent restriction (though this will depend on the application). Thus, commercially, some polymers are classified as commodity plastics; these are cheap to produce in a high volume but their physical properties will be unremarkable (Table 5.1).[1] In contrast Engineering Plastics tend to be more expensive largely because they are produced in lower quantities, but also because they have specific superior properties, examples of which include mechanical strength or thermal stability. Polymers are also classified according to their processibility and may be referred to as thermosetting or thermoplastics; the former are materials which once processed cannot be melted or dissolved to form new objects, this is usually due to extensive cross-linking.[2] Thermoplastics can be melted or solvent processed, a factor which is important in their manufacture and in recycling. The majority of commodity and engineering plastics tend to be thermoplastics [10].

Historically chemists tended to classify synthetic polymers as addition polymers or condensation polymers. The original definition of Carothers [11] required, that for addition polymers, the molecular formula of the monomer should be identical with the structural unit, while for condensation polymers the monomer should

[1] That being said the strength to weight ratio of polyethylene which allows the production of huge (and problematic) numbers of polyethene bags, can hardly be said to be unremarkable, other than in terms of the ubiquitous nature of these articles.

[2] This definition also might apply to rubbers, but rubbers are deformable; thermosetting plastics are generally rigid.

Table 5.1 Some typical polymers, classifications, and uses

Polymer	Repeat unit	Commodity polymer	Engineering polymer	Step growth	Chain growth	Comments
Polyethylene (Low-density)	$–CH_2CH_2–$	✓			✓	Produced in high volume many uses in for example packaging.
Polyethylene (High-density)	$–CH_2CH_2–$	✓			✓	Non branched produced by Ziegler Natta process used for bottles, pipe, cable insulation.
Polypropylene	$–CH_2CH(CH_3)–$	✓			✓	Rope, cordage, car parts, containers.
Poly (vinyl chloride)	$–CH_2CHCl–$	✓			✓	Pipes flooring, wire insulation building materials.
Polystyrene	$–CH_2CHPh–$	✓			✓	Packaging (foam and films) insulation.
Polyethylene Terephthalate	$–OCH_2CH_2–O–(C_6H_4)–$	✓	✓	✓		Unusually considered both as a commodity polymer and an engineering polymer. Used to make soft drink bottles and fibers (especially fleeces).
Polyamide	e.g. $–CONH(CH_2)_5–$		✓	✓		Polyamides are used to form fabrics orientation in fiber formations
Polyimide	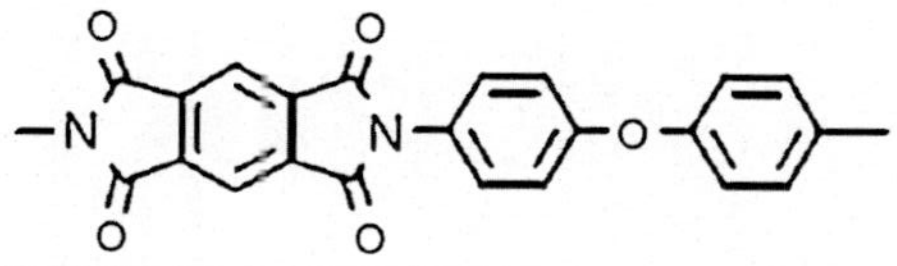		✓	✓		The example shown is Du Ponts Kapton. Resistant to high temperatures used e.g. to insulate wires.

(continued)

Table 5.1 (continued)

Polymer	Repeat unit	Commodity polymer	Engineering polymer	Step growth	Chain growth	Comments
Polycarbonate	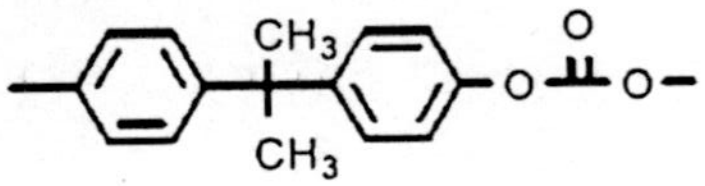		✓	✓		High impact strength- used for safety shields safety helmets etc.
Phenol formaldehyde resin	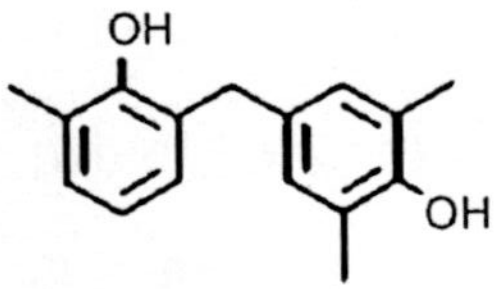			✓		Highly cross-linked system thermosetting polymers. Thermosetting-Plastic
Rubber	$-CH_2C(CH_3){=}CHCH_2-$					Natural polymer – lightly cross-linked to form an elastomer.

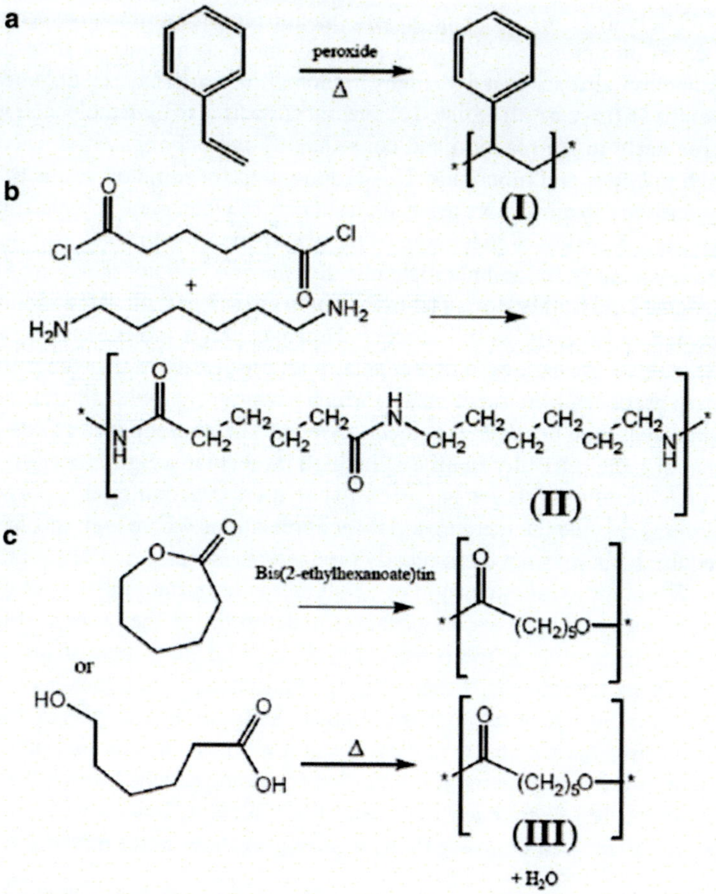

Fig. 5.1 (**a**) Polymerization of styrene via a chain-growth process, (**b**) formation of Nylon 6,6 via a step growth process, and (**c**) alternative chain-growth (*top*) and step-growth (*bottom*) routes to poly(ε-caprolactone)

be different, with the polymer obtained from the monomer by intermolecular condensation. On this basis polystyrene represents a simple addition polymer (Fig. 5.1a), while Nylon 6,6 produced from adipoyl chloride and hexanediamine (Fig. 5.1b) is a typical condensation polymer. However, although these terms remain in usage, they can give rise to confusion, particularly in cases, where a similar polymer can be produced from either mechanism (as in the case of the polyester shown in Fig. 5.1c). As a consequence, polymer chemists tend to classify polymers according to the way in which propagation occurs. Where propagation occurs in a stepwise fashion we have a step-reaction or step-growth system. Where there is addition of a monomer unit onto an activated polymer chain we have a chain-reaction or chain-growth polymer. The importance of this in relation to the usage of polymeric systems in rapid processing applications will become clear from the following discussion.

For chain-growth systems polymerization requires initiation by an activated species such as an anion, a cation, and probably most commonly, a free radical [10]. The initiated chain can react with many monomers to form material of quite high molecular weight. In this case all initiated chains are converted to high molecular weight polymer, quite early in the reaction and consequently the polymerization mixture consists of high polymer and monomer. This requirement for initiator means that chain growth systems are in most cases the systems of choice when photoinitiated polymerization is required. Because of the nature of the polymerization process, step-growth polymers are less readily formed photochemically; thus no initiator is required for step-growth system. This further differs from the chain growth in that all the monomer is partially reacted quite early in the process. However, high molecular-weight polymers are difficult to obtain, and only occur at high conversion; fortunately such systems are often produced from functionality which leaves groups with dipoles, or the potential to form hydrogen bonds in the final polymers. The consequence of this is that not withstanding the difficulty in producing high molecular weight materials, intermolecular forces are often higher in engineering polymers than commodity polymers. Thus such systems may be the ideal material for applications where high mechanical strength is required, but may not be so readily produced in situ as part of the prototyping process.

While the synthetic origin of a polymer is important the way in which a polymer is processed, it is also a major factor in determining the properties of the materials. A full discussion of the possibilities is beyond the scope of this article, but it is important to note that processing, in particular, techniques such as extrusion or spinning impart additional mechanical strength to the polymer system through promoting backbone crystallization and alignment. This is particularly the case for step-growth systems; it is spinning for example which imparts the high strength to polyamide fibers such as Nylon 6 and Kevlar; indeed with careful processing polyethylene can be made to be as strong as these latter materials [12].

5.2.2 *Cross-Linking*

One significant property of polymers not present in low molecular weight compounds is the possibility of forming cross-links between neighboring chains. This cross-linking means that the chains have restricted translational mobility. From the point of view of the applications described here, this cross-linking provides 3D stability; additionally such materials cannot melt (for crystalline polymers) or dissolve. This last point is particularly important in the development of 3D prototypes since it can ensure that the final material can be separated from monomer or prepolymer. Cross-links may be formed post-polymerization by the presence of reactive units in the chain of a polymer,[3] as is the case for the vulcanisation of natural rubber (utilizing the reactivity of carbon–carbon double bonds in the polymer backbone) and it is this approach which is used in many stereolithographic techniques. Alternatively, cross-linked systems may be formed during polymerization if the functionality of the monomer is greater than

[3] It is important to form such systems that the side-groups exhibit different reactivity to the polymerisable units in the monomer prior to polymerization.

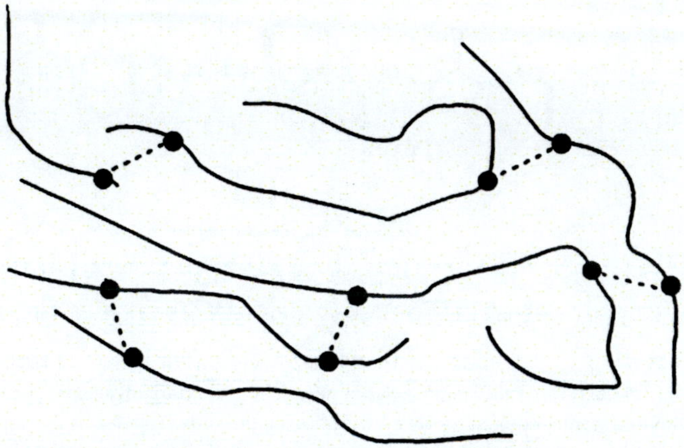

Fig. 5.2 Schematic representation of cross-linking in a polymer (the broken line represents short-length cross-linking chains)

required for a linear polymer (an example of such a case is the phenol formaldehyde resin shown in Table 5.1).[4] Here the final form of the polymer must be determined prior to polymerization; an illustration of such a system is the use of polyurethanes in reaction injection molding [15]. The properties of the cross-link system have a high dependence on the density of cross-links, for example with low cross-link density then rubber elasticity is possible, with the polymer showing reversible deformation; higher cross-link densities give rise to extremely intractable hard resinous materials.

The production of 3D polymeric materials can be achieved in many ways, but a particularly common one is *via* photochemical polymerization or cross-linking. The use of optics, and particularly lasers provides a high level of spatial control and resolution; this can be couple directly to a computer design. The following section details ways in which visible and ultraviolet light can be used to initiate polymerization or crosslinking and describes the properties obtained from various material combinations (Fig. 5.2).

5.3 Photopolymerization

5.3.1 Free-Radical Systems

Although photopolymerization can be induced in a number of ways, the most commonly used systems involve the mediation of free radicals. The interaction of

[4] It is also possible that photoactive units can be added which degrade the polymer chain to form for example free-radicals, which may themselves combine [13, 14]; similarly ionising radiation (e.g. γ-radiation) will induce free-radical formation on the chain; the high energies involved may result in unwanted degradation products such as chain-scission.

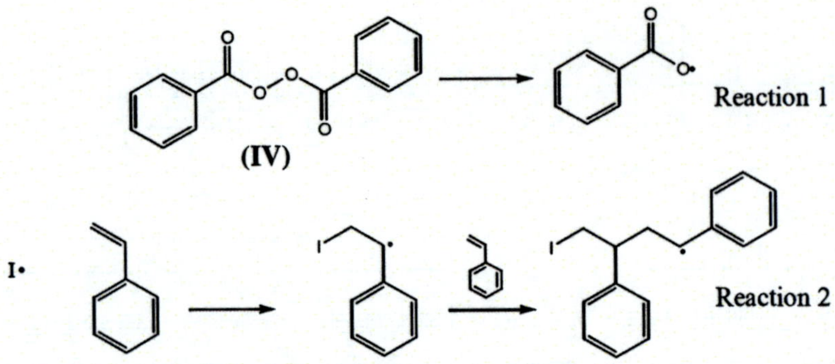

Fig. 5.3 Photolysis of benzoyl peroxide produces predominantly benzoyl radicals; these can add to a vinyl monomer such as styrene to produce a polymeric species

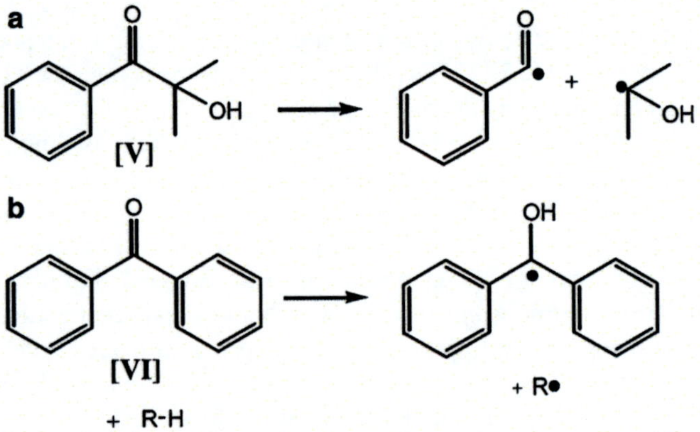

Fig. 5.4 Type I (**a**) and type II (**b**) photodegradation of aromatic ketones; the radicals produced can initiate polymerization

light can produce free radicals via homolysis of weak bonds as in the homolysis of benzoyl peroxide (Fig. 5.3 – Reaction 1); alternatively, the generation of a triplet species from a ketone can be thought of as the formation of a di-radical. This is the case with the type II photodegradation of ketones [14, 16], as shown in Fig. 5.4.

In the Type I case cleavage occurs; Type II systems react by triplet formation followed by hydrogen atom abstraction. The radicals produced add on to vinyl systems such as styrene (Fig. 5.3 Reaction 2) to induce polymerization or to add on to reactive units in the polymer chains. Common free-radical initiators include the peroxides and aromatic ketones described above but others including the azo compound 2,2'-azobis (2-methylpropionitrile). The most common monomer systems are acrylates and methacylates, for example the Monomer (**VII**) is commonly (with $m = n = 1$) is often used in dental applications [17]. Styrene based resins are also used in some applications; this mono-functional monomer alone, however, will produce a polymer

Table 5.2 Factors Influencing Behavior of photopolymer systems [18, 22]

Factor	Consequence
Switch from mono to diacrylate	Rate of polymerization increases and material becomes cross-linked; flexibility decreases; hardness increases, improved solvent resistance
Switch from di- to tri-acrylate	Rate of polymerization increases and material becomes cross-linked; flexibility decreases; hardness increases, improved solvent resistance; more unreacted alkene groups
Use of styrene rather than acrylate	Lower cure speed
Changing the linking groups in a diacrylate	Aliphatic polyurethanes give soft flexible materials; aromatic polyesters and polyethers produce hard scratch resistant coatings
Use of mono-acrylate containing two oxygen atoms in a heterocyclic structure	Greatly enhanced reactivity and speed of cure
Pigmented system	Polymerization lis ess efficient as less light available to interior of sample – switch to specialist initiator required
Change of excitation wavelength to shorter UV	Resolution improves (130 nm for 193 nm excitation wavelength), problems with penetration deep into the sample
Use of siloxane system	Lower glass transition temperature, and more flexible final product
Cationic system	Oxygen less important; reaction may continue in dark In general cationic polymerization is slower than radical process (particularly for epoxy systems) Moisture interferes with cationic systems – not with radicals Shrinkage less of a problem for cationic systems

which is soluble in a range of common solvents. Thus it is combined with an unsaturated polyester (containing for example maleic or fumaric acids) to produce a cross-linked material. Decker [18], however, suggests that this material, while used in some applications due to the low cost of the monomers, is generally inferior to acrylate systems in terms of cure speed and subsequent material properties. The structural advantage of acrylates and methacrylates is demonstrated by (**VII**); the acid functionality of the basic unit allows for an almost infinite structural variation, including the incorporation of both flexible and rigid units, and by having diacrylates (or even triacrylates) ensuring that photopolymerization is accompanied by cross-linking. The consequences of such modifications are indicated in Table 5.2.

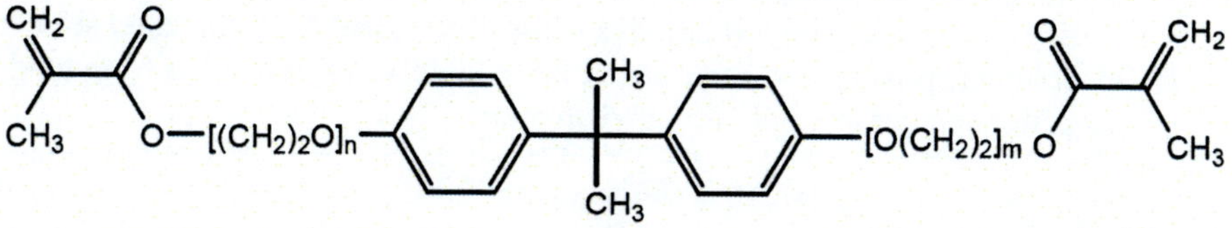

(VII)

5.3.2 Cationic Systems

Photoinitiation of cationic chain-growth polymerization is also relatively common [19], a range of systems are used but commonly these involve triarylsulfonium (**IX**) and diaryliodonium (**VIII**) salts. Photolysis of these compounds results in the formation of aryl cations [Reaction 3], although other species such as cation radicals and radicals may be formed by a competing homolytic process.[5] The cation species may themselves initiate polymerization or they may generate Brönsted acids by subsequent reactions which then react with the monomer. There are a range of monomers which may be polymerized by these initiators (indeed any monomer which can be polymerized cationically might be expected to react); however those most commonly found in the type of application of interest here are vinyl ethers and epoxides. The system is to some extent complimentary to the free-radical systems, these monomers are not generally polymerized by such means. Furthermore, cationic polymerization can proceed in the presence of oxygen and (once ionic species have been produced) the polymerization can occur in the absence of light; the polymerization may be to some extent living. In marked contrast, free-radical polymerizations will discontinue within a fraction of a second on removal of the light source.

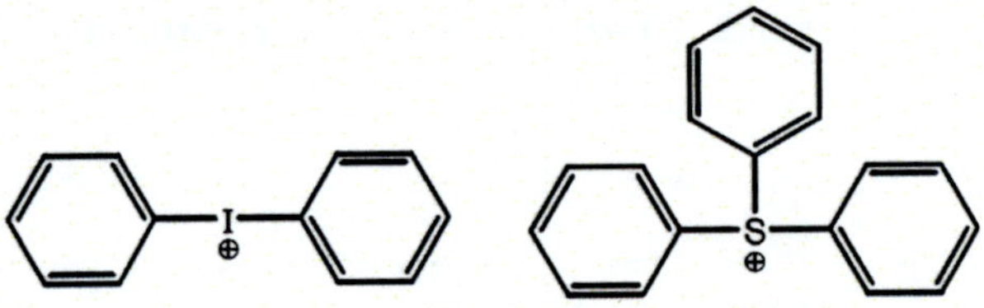

(VIII) (IX)

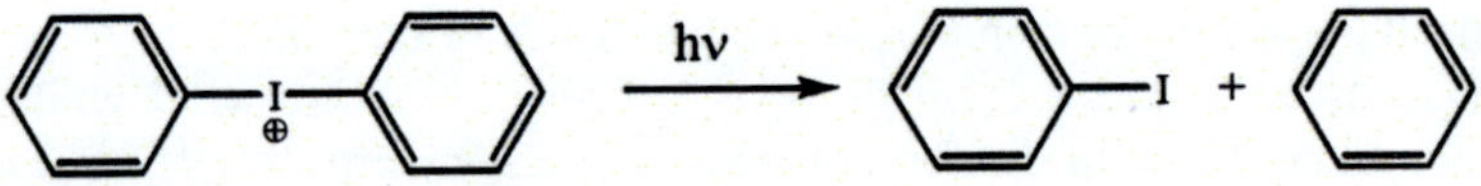

Reaction 3

A particular advantage of cation systems especially epoxides is that they offer the potential of postpolymerization cross-linking; thus they may for example be incorporated as side groups on a vinyl monomer without involvement in the free-radical process, vinyl ethers in contrast, although poorly reacting in many circumstances, may become involved in the free radical process [20]; this can result in premature cross-linking which will of course prevent the shaping of the resin. The choice of epoxide is important, thus while glycidyl ether (**X**) is rather slow to react; cyclohexyl epoxides are cured rather more rapidly, a fact that has been used [21] to prepare photocurable polydimethylsiloxanes [Reaction 4].

[5] Radicals produced may themselves react with the "onium" to form cationic species and a further radical – thus giving unexpectedly high quantum yields for initiation [19].

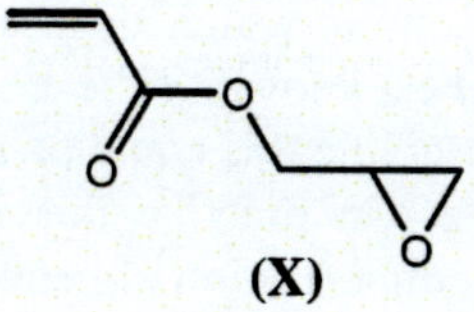

(X)

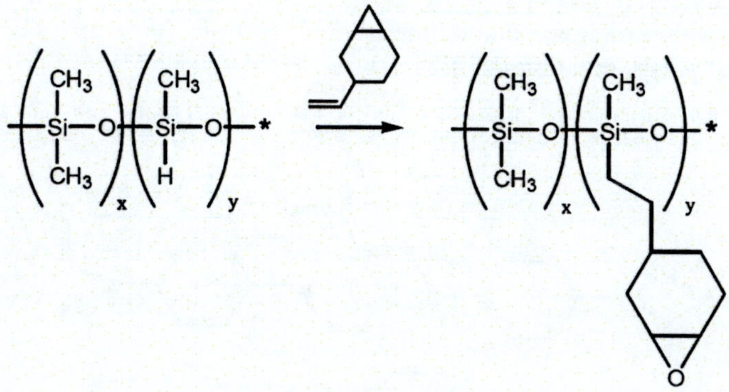

Reaction 4

While the use the photogeneration of free radical species and cations (particularly acids) are common features of photopolymerization, there may be disadvantages; free-radical polymerizations for example are inhibited by oxygen. In this regard the photogeneration of bases represents an attractive alternative. There are, however, substantially fewer anionic initiated photopolymerization systems, although recently some promising systems are being developed [23]. An example of a photodegradable carbamate is shown (**XI**); in this case a primary amine (**XII**) is generated [24]. In general the amines produced can be used to initiate an anionic polymerization; amines such as this have often been used to cross-link epoxy systems [25]; alternatively cyclopentadienyl anions generated from metallocenes have been used to polymerize cyanoacrylates [26].

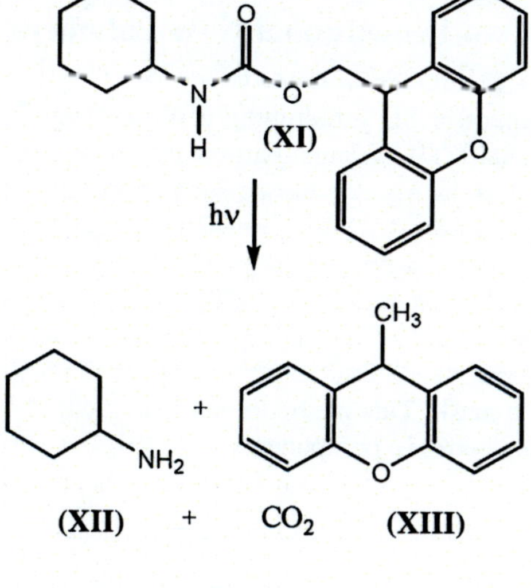

The discussion above has focused on initiators for chain growth processes; however there some photoreactions which do not involve the generation of a reactive initiator. Perhaps the best known of these reactions is the $2\pi + 2\pi$ photocycloaddition of alkenes; in principle this could be used to produce a polymerization (in a step growth manner) in practice this is often used to cross-link a preformed system by the photodimerisation of pendant side groups. In this regard α,β-unsaturated carbonyl compounds are particularly useful [27] and polymers with cinnamic acid [**XIV**] side groups have been successfully used as negative photoresists (see below). Photodimerisation reactions have also been used as topological polymerizations [28].

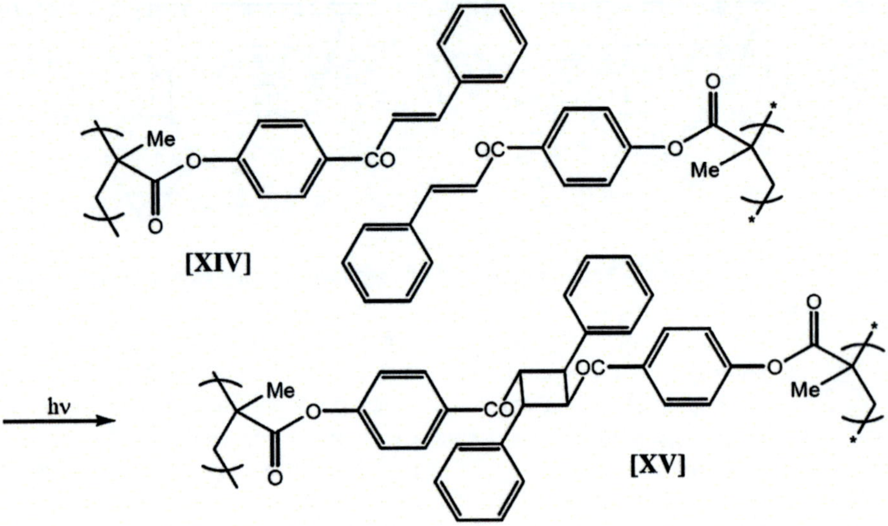

[XIV]

[XV]

5.4 The Two-Dimensional Case Polymers for Photolithography

The development of 3D prototyping could be considered as an extension of the two-dimensional (2D) variants, particularly as a layer-by-layer approach is often used. In this section we look at one such approach namely photolithography; this shares many features with this well-known 3D process, stereolithography, and moreover this technique can be applied to the construction of solid objects itself, particularly at the nanometre level [29] and we discuss this first. Photolithography is a technique which has developed over many years; it involves coating a surface with a light sensitive material, often a polymer, and exposing selected areas to light by means of a mask. This methodology is particularly important for the manufacture of microelectronic components [29, 30], but has been seen as having great potential in other areas such as microfluidic systems [31].

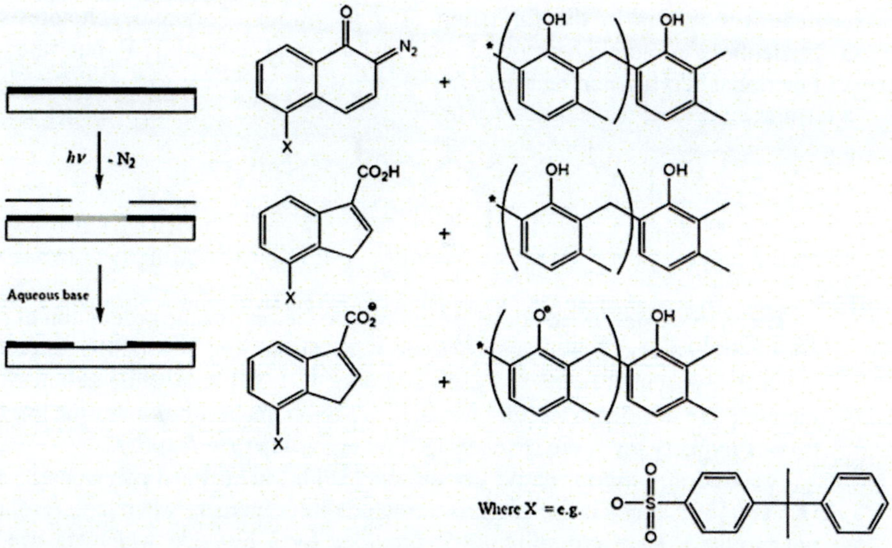

Fig. 5.5 A positive photoresist: In the presence of the ketone the polyphenolic compound is insoluble. Irradiation at the appropriate wavelength results in the formation of the acid shown and the polymer becomes soluble in aqueous base. The irradiated area can be dissolved away

Photolithography methodology is particularly important for the manufacture of microelectronic components. There are two major classifications of this technology: a positive photo resist is one where the area exposed to light becomes soluble and is removed in the process; for a negative photoresist the area exposed remains insoluble. The positive variant is exemplified by the well-known Novolac system [30]; here the presence of the anthroquinone is designed to make the polymeric phenol system insoluble. On irradiation, the photochemical transformation indicated in Fig. 5.5 results in a mixture which is soluble in basic solution; but only in the irradiated regions; thus only the irradiated portion can be removed. This system has a number of advantages including nontoxicity of the developing medium, and superior plasma etch resistance; the best systems of this type have been used to construct integrated circuits with features as small as 350 nm [32].

As stated above negative photoresist become less soluble (or indeed insoluble) in the exposed areas. This may be achieved by altering the polarity of the polymer, by changing the oxidation state within the system, or through crosslinking. The earliest systems were based on poly (vinylcinnamate) [33]; here photo cross-linking is achieved by the $2\pi + 2\pi$ photo cycloaddition of the alkene. This technology, first marketed by Kodak, was first applied to printed circuit boards, nowadays, many other systems are applied; these include Novolac systems and (meth)acrylates. While there are many materials for such applications a typical example is provided by the polyimide precursor (**XVI**) shown below, which is cross-linked photochemically to form an insoluble material; on removal of the unreacted precursor, heat treatment results in the formation of a polyimide. Polyimide photoresists are particularly attractive because of their high thermal stability [34], but other multi-functional acrylics can be used.

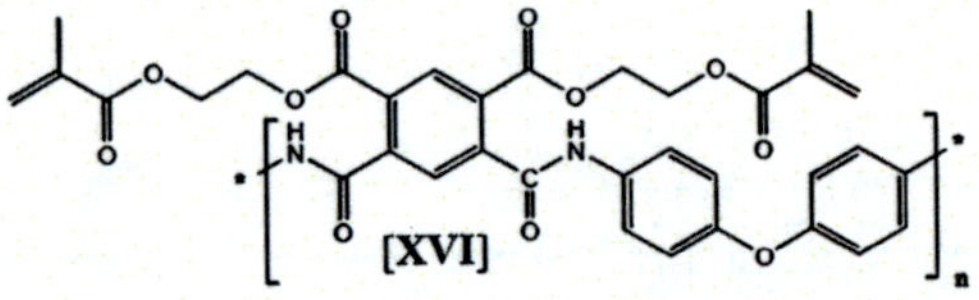

[XVI]

The systems described above are entire satisfactory for the production of printed circuits; however, there are problems with resolution; it would seem that the resolution provided by these systems is poor [30, 35] thus while suitable for the manufacture of printed circuit boards, the requires for resolution for the microelectronics industry are more stringent and other systems are favored.

A commonly-used negative photoresist [36] is the cyclized polyisoprene system (**XVIII**) [30], which is used in the microelectronics industry when high resolution is not required. Here cross-linking is provided by a bisazide which on irradiation produces nitrene units (analogous to a carbene); these insert in the double bonds in the isoprene system leading to a cross-linked structure, as illustrated in Fig. 5.6; such resists are imaged with wavelengths above 436 nm [30]. In fact it may not be necessary to introduce extensive cross-linking into the system since a small proportion of linked chains may render the polymer sufficiently insoluble [28], except that lightly cross-linked systems may swell in the solvent used to remove unreacted material consequently; this may adversely effect the film boundaries. An alternative popular resist material is SU-8, which is an epoxy based system [11]. Here the polymer (based on the Bisphenol A system) is cross-linked on irradiation by a photocationic initiator as described above. A particular feature of this is that the

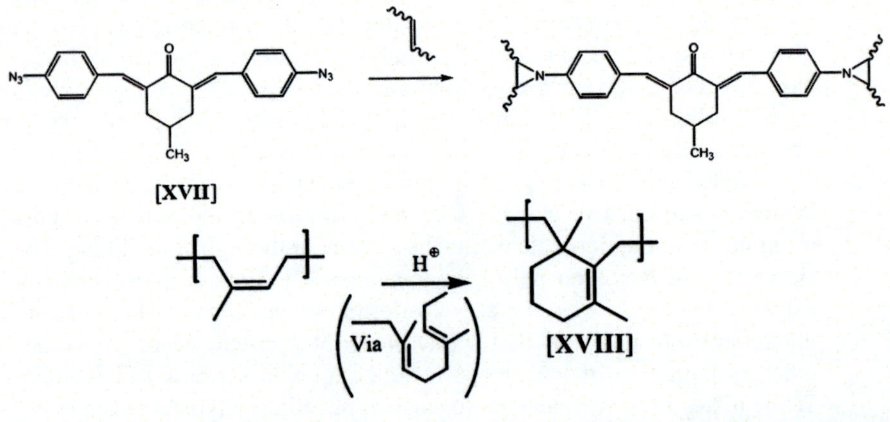

Fig. 5.6 Bisazide [30] used in photo-crosslinking carbon-carbon double bonds in a polymer backbone

material has a relatively low absorbance at the irradiation wavelength (typically 365 nm), which means that light can penetrate significantly through the sample. Thus this is the material of choice for high aspect ration materials [37] and coatings of up to 3 mm may be prepared. The ability to prepare micron- scale-structures with substantial depth has considerable potential in the area of microelectromechanical systems [38]; in such applications the methodology is no longer 2D, but the size of these objects are rather smaller than that we are considering in this chapter.

5.5 The Three-Dimensional Case Polymers for Stereolithography

The potential of resists systems is enormous and based on well established principles; however, with the exception of high aspect ratio photoresists, the technology is largely concerned with 2D imaging. For 3D imaging alternative methodologies are used; however, possibly the most commonly used methodology essentially adopts photochemical development of a surface film of polymer and adapts it to three dimensions; this process is known as stereolithography. Here a solid prototype is constructed via the computer control of a laser, which writes onto a liquid polymer precursor to solidify the material. The process simply writes a particular structure on layer after layer of the polymer until the solid object is produced. Thus, most of the chemistry involved bears close similarities to that used for negative photoresists; in particular the use of cross-linking to solidify an object. Previously the method was found to be expensive and time-consuming, so its use until recent years has been restricted mainly to the construction of prototypes. More recently the method has been developed and commercialized by companies such as 3D Systems of Valencia California and machines such as that shown in Fig. 5.7 are now available [39].

The development of commercialized Stereolithography systems has been accompanied by rapid improvements of the resins available. That being said, most resins rely on the radical polymerization of (meth)acrylates or cationic polymerization of epoxides (or vinyl ethers) and in fact might be formulations containing both. The fundamentals of this chemistry has been discussed above in Sections 5.3 and 5.4; thus compound **VII** used in dental applications may also be used in stereolithography; however, other resins may be based on acrylate-terminated systems for example, polyurethanes, polyesters, polyethers and polysiloxanes, acrylate capped butadienes and acrylamide related systems have also been used. Diacrylates provide cross-linked systems, but more rapid hardening is provided by the use of triacrylates; mono-functionalised acrylates are often put in to reduce the viscosity of the pre-polymer system; it will modify the cure extent and can usefully help to control the properties of the final polymer [40]. Clearly there is substantial scope for modifications, for example the polyurethane system [XXII] shown in Fig. 5.8 is one of a range of similar materials which could be selected. Aliphatic or aromatic diisocyanate might be incorporated; in the example shown the

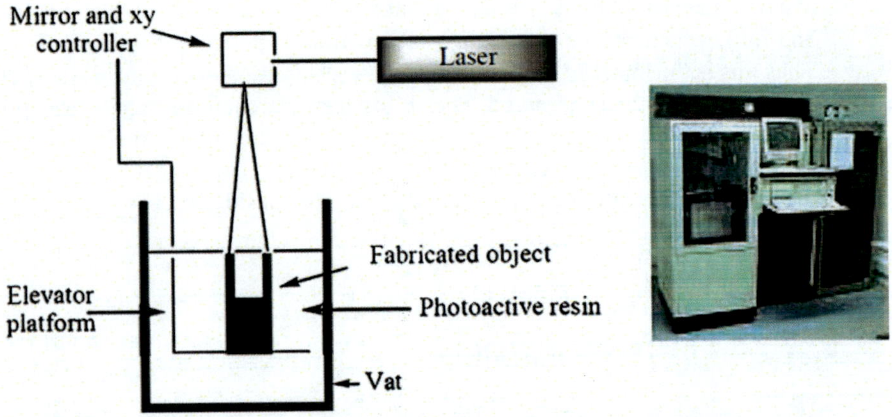

Fig. 5.7 Schematic of the stereolithography process and picture of an actual instrument (by permission of 3D systems Valencena California)

aromatic isocyanate will increase the hardness over that which might be found were an aliphatic isocyanate such as hexyl diisocyanate used. The properties will also depend on the molecular weight of the poly(propylene glycol) used, and for a more rigid system a polyester can replace the central polymer. Similarly, for the polyester based systems (e.g. [XXI]) the nature of the polyester can have a dramatic effect on the reactivity of the resin an also alter the hardness, stability, and the aging properties of the polymer; typically higher molecular weight diacrylates are more viscous, but less reactive and give a more flexible material than the lower molecular weight analogs. There is some interest in the use of liquid crystalline materials in such systems as a way of introducing anisotropic orientation and additional strength [41].

As with the acrylic system there is potentially a diverse range of vinyl ethers and epoxides systems such as those shown in Fig. 5.8, which can be polymerized by a cationic mechanism. Many complex vinyl ethers have been made in laboratory studies, but the industrial application have been to some extent restricted due possibly to the difficulties associated with preparing these monomers particularly as they are usually produced via addition to acetylene [19].[6] Though in fact rather less reactive than vinyl ethers, the epoxides, however are generally more accessible synthetically, but the wide range of monomers available have not been exploited in the same way as for the acrylic systems above. Interestingly, cyclohexyl systems such as [XXVI] are often preferred; this is because the cyclohexyl systems are particularly rapidly photopolymerized (a result of additional strain on

[6] Vinyl alcohol is of course the thermodynamically unfavoured enol tautomer of ethanal and thus the ether is generally formed by addition of ROH across the acetylene triple bond.

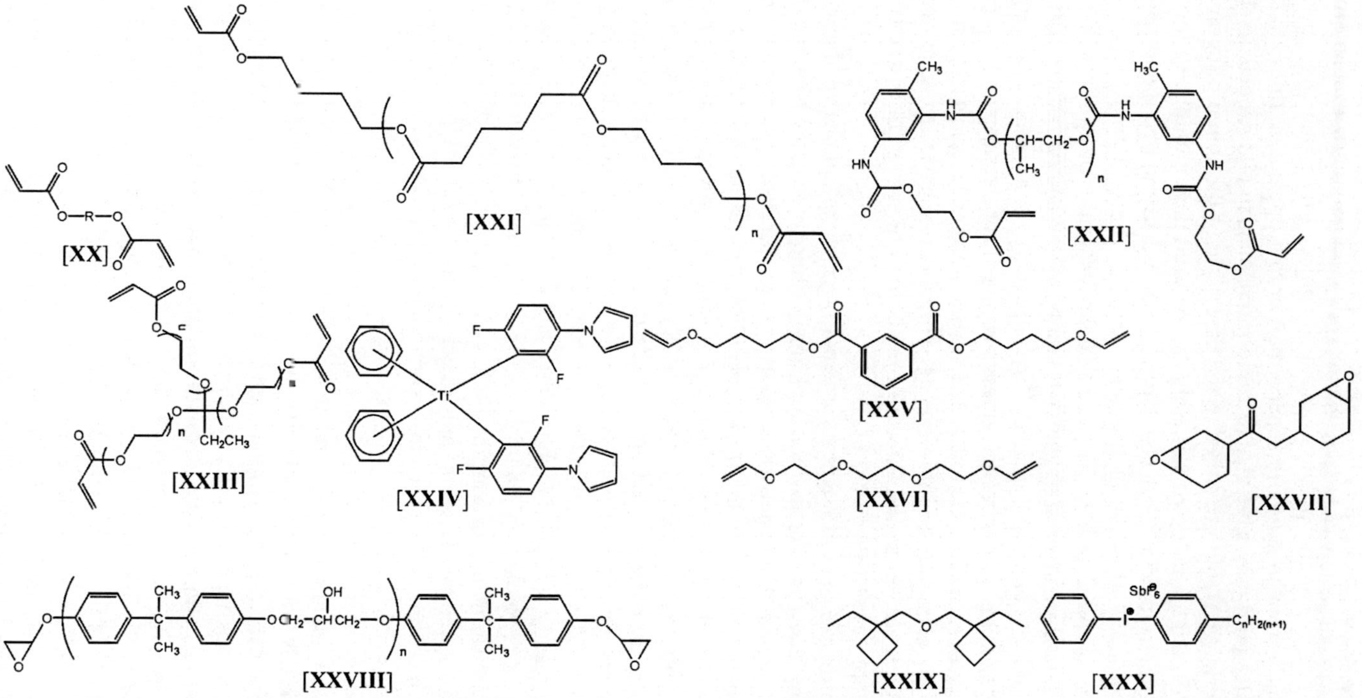

Fig. 5.8 Range of monomers and initiators used for stereolithography; acrylates (and methacrylates such as those shown in structures [XX] to [XXIII] are intiated by free radical initiators, a particularly effective one is Irgacure 784 [XXIV], which sensitive to light with wavelengths of up to 500 nm.; vinyl ethers ([XXV] and [XXVI]), epoxy-based systems [XXVII] and [XXVIII], and the 4-member analog such as [XXIX] can be polymerized by cationic initiators, of which [XXX] is a typical example

the molecule); glycidyl ethers such as [XXVIII] polymerize rather slowly, and it would seem that the polymerization is subject to an induction period arising apparently from an activation energy barrier to propagation; this can be overcome thermally (and may give rise to some interesting spatial aspects to the eventual polymerization), but not suited to the time scale of the stereolithographic process [42]. Although a range of initiators can be used the iodonium system [XXIX] is often preferred; the alkyl chain being tailored to modify the solubility of the initiator and the hexafluroantimonate anion is chosen by virtue of its extraordinarily weak basicity [21]. In an analogous fashion oxetane systems based on the strained four-membered oxygen containing cyclic compound are also photopolymerisable. As with the glicydyl system there is an induction period before polymerization occurs, which is less than ideal for the purpose of stereolithography; however, it has been found that the reaction can be speeded up in a number of ways; for example, copolymerising with more reactive epoxides comonomers [43] and there are some aspects of oxatanes which may be attractive in stereolithography resins (see below).

In principle stereolithography is a more precise technique than the selective laser sintering and fused deposition molding technique described below. The method should provide for precision similar to that observed with the two dimensional systems or particularly the impressive 3D structures made from high aspect ration photoresists. In many ways, however, it is not the polymerization which limits the precision of the technology, but what happens after the light induced polymerization. In most cases post curing is needed to solidify unreacted or partially reacted monomers; this process will for example improve the mechanical properties of the prototype. The step is carried by further treating the model with UV light and heat. Unfortunately, it is often found that the additional polymerization, presumably with increased cross-linking, and the associated thermal effects results in a substantial degree of shrinkage, or warping [44]. It is the control and reduction of this problem that has driven the design of new materials. In this regard the acrylates seem to be rather more prone to problems; in particular shrinkage is substantially more for these systems than for epoxides; the shrinkage arises from a range of effects following the curing process, but it is notable that there may be some trapped radicals remaining in the acrylate samples [45]; furthermore, while acrylates and particularly multi-functional acrylates polymerize extremely quickly under the photolysis the material hardens rapidly leaving a large proportion of the polymerisable groups unreacted. The problems may be controlled to a certain extent by processing conditions, but it would still seem that epoxides provide less shrinkage, even though the polymerization may not be so rapid.

The prevention of shrinkage and the improvement of the mechanical properties of models built by stereolithography have driven a considerable amount of development into resin formulations. In fact some commercial resins appear to be a mixture of epoxides and acrylates and there may be substantial advantages in combining the two types. In addition to this it has been found that for radical systems the inclusion of monomeric acrylates improves the extent of conversion; acrylate esters of oxygen-containing cyclic systems such as [XXXI] and [XXXII] work particularly well in this regard as rather unusually they react rather rapidly [18]. These monomers also enhance the mechanical properties of the prototype; a particular problem with the multi-functional acrylates is that they are brittle; the introduction of these comonomers results in materials with improved flexibility making them more resistant to scratching and shocks. That being said, it would seem that epoxides have the edge in the marketplace [46] and the development of such resins has succeeded to the extent that materials such as DSM SOMOS® ProtoGen™ 18120 (said to be based on oxetane chemistry) may have properties comparable to thermoplastics such as ABS.

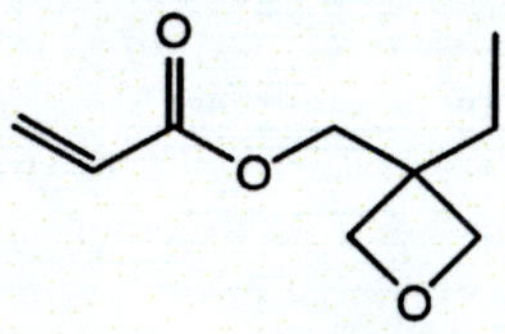

[XXXI]

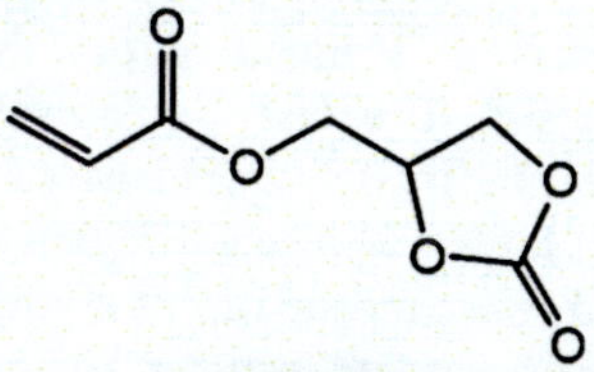

[XXXII]

It is probably fair to say that stereolithography represents the most advanced of the rapid prototyping methods; a natural consequence of the improvement in resin technology is that stereolithography is now being considered much more as a potential tool for the manufacture of parts, particularly when only small numbers are required. While the account above describes attempts to improve the physical properties of the resultant product through the chemical design of the polymers, other approaches have been adopted. The production of ceramic parts is particularly attractive and the use of silica filled resins has been explored in this regard [6, 46]. Problems in light scattering with high loadings of silica can apparently be avoided by the use of nanoparticles; and further improved by refractive index matching which also improves the loadings possible by minimizing Van der Waals interactions [47], although the viscosity of the resulting liquid prepolymer is problematic and requires substantial tailoring of the system. The polymer matrix may be retained, but in some cases it is removed by postmanufacture treatments resulting in a ceramic part [46]. A similar approach that has been explored is the introduction of reinforcing fibers into the photopolymer; this also imparts further strength to the final part, although the process does not seem to readily lend itself to automation [48]. Other attempts have moved away from the layering approach and used parts produced by stereolithography as masters from which silicone molds are produced. These silicone molds can then be used to produce a small number (typically 20 or so) polyurethane materials by vacuum casting, provided relatively low temperatures are used [49]; an alternative, but probably less popular option is to produce the mold itself using stereolithography [50]. That being said despite the accuracy and speed of this process, the polymeric materials my not be ideal and other related methods may be used with alternative materials, in particular, selective laser sintering and fused deposition molding may be used to produce materials from thermoplastics (indeed that the materials will melt is a requirement) and these may offer scope for further applications. Some of the materials used in these applications are discussed below.

5.6 Selective Laser Sintering

Selective laser sintering is a technique which is finding increasing importance as an alternative to stereolithography; the technique can be used for a range of polymers including nylon, polycarbonate ABS, and polystyrene. In addition, there are a range of techniques that involve polymer coated materials. This is a commercial process and involves the selective sintering layer-by-layer of a fine powder; the process is illustrated schematically in Fig. 5.9. The powder (typically 10s of microns in diameter) is spread to form a surface thickness of 75–250 μm. A laser (typically a CO_2 laser) is scanned across the surface at scanning rates between 2 and 100 cm s^{-1}. The laser is controlled by a computer design; where the design indicates the model should be solid, power is applied to the laser and the heat from the laser will fuse the region below the beam together and to the layer below. A useful consequence of this methodology is that the solid produced are supported by the surrounding powder, which eliminates the need to produce supports for overhanging parts [51].[7] The method clearly relies on the presence of material which can be melted, so is not applicable to thermosetting polymers.

The technique of SLS can be applied to both amorphous and semi-crystalline polymers. Examples of polymeric materials include polycarbonate, polyamides, polystyrene and styrene acrylic copolymers. In terms of quality of final product, this depends on a number of factors, for example, studies of polycarbonate have shown that the mechanical properties of the final material depend on the power of the laser – if only weakly sintered then the joins between the particles form places for cracks; at

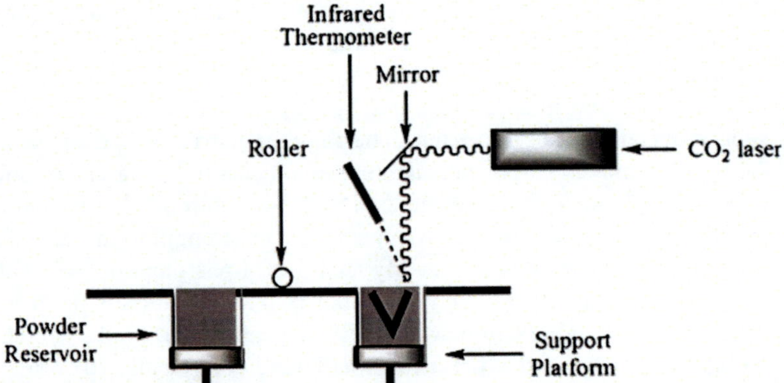

Fig. 5.9 Schematic of the selective laser sintering process

[7] For stereolithography, where the component has overhanging parts the support linking this to the main structure is not formed until after the part; this can only be solved by incorporating additional support struts into the design which are removed when the final model is made.

very high energy densities decomposition may occur [52]; in addition, of course, the heat must not be sufficient to completely melt the polymer, thus for amorphous materials such as polystyrene there is a range of sintering temperatures ranging from a few degrees below Tg to Tg; above Tg the particles all coalesce [53]. The mechanical properties are also determined by the build orientation as there is a tendency for the fracture to occur along the directions of the layers [54]. The selection of materials is important for making molds for techniques such as investment casting, thus for example, high impact polystyrene (polystyrene strengthened by combination with butadiene rubber) is preferable to polystyrene if complex parts are to be made [53].

In addition to polymeric material there is much interest in polymer coated materials; for example a calcium phosphate ceramic material, can be made by coating particles of the inorganic solid with a polymer such as methacrylate. The polymer coated particles are then subjected to the SLS process to produce a prototype which is filled with a solution of calcium phosphate which is then dried to constant weight, and fired to yield porous calcium phosphate, as this process burns off the polymer [55]. In a similar way polymer coated metal powders offer a real approach to the rapid manufacture of metal parts [41].

5.7 Fused Deposition Modeling

The fused deposition modeling process is relatively simple as outlined in Fig. 5.10. In this system the polymer is pushed through a nozzle onto a surface, the polymer solidifies on contact with the surface and in this way a structure is built up layer by layer. The extruder nozzle is controlled by a computer program which guides where material is deposited. The material is molten as it leaves the nozzle and solidifies as it hits the surface. The layer below has to be maintained at a temperature just below the softening point of the polymer to obtain good cohesion. The technique is popular as it is applicable to a variety of thermoplastics. Thus for example materials include polycarbonate, poly (phenylsulphone), ABS, and biocompatible polymers such as poly(ε-caprolactone) (see below). The technique has also been applied to

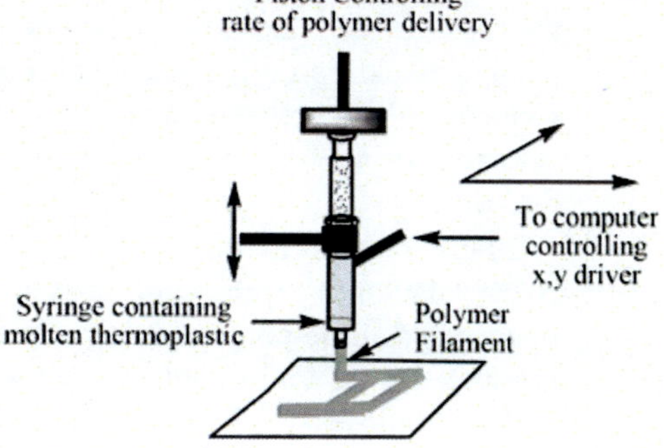

Fig. 5.10 Schematic representation of the Fused Deposition Modelling Process

the formation of a high molecular thermoplastic from a reactive monomer in a process similar to reaction injection molding; a route to forming high strength parts [56]. Polymers filled with solid particles such as iron have also been investigated and this approach is particularly promising for rapid tooling applications [57].

As discussed earlier, biological applications are an area where rapid prototyping and product manufacturing meet; thus a medical application is likely to be highly specific in both terms of size and shape as these are determined solely by the requirements of the patient. This is one area which has developed particularly rapidly over the past few years. A common problem is the development of scaffold to support tissue regeneration. The material requirements here might be quite demanding. Materials may need to be biodegradable, equally they may need to match living tissue in terms of mechanical properties; for example where scaffolds are required to support bone growth, mechanical strength, in particular compressive strength, in contrast for skin replacements may require substantial elasticity. Probably the most important requirement is that the structure produced must support cell growth. One feature of such a porous arrangement is that it does not lend itself well to formation through molding. In addition, as a material will be contained within a biological environment, the polymer must be biocompatible. Thus materials such as gelatin are attractive for scaffolds as are polymers such as polyethylene glycol [XXXIII] and Poly (ε-caprolactone) [XXXIV]. The advantage of fused deposition modeling over the techniques described above is that it offers a mild route to the production of scaffold structures, and that no chemical modification is needed.

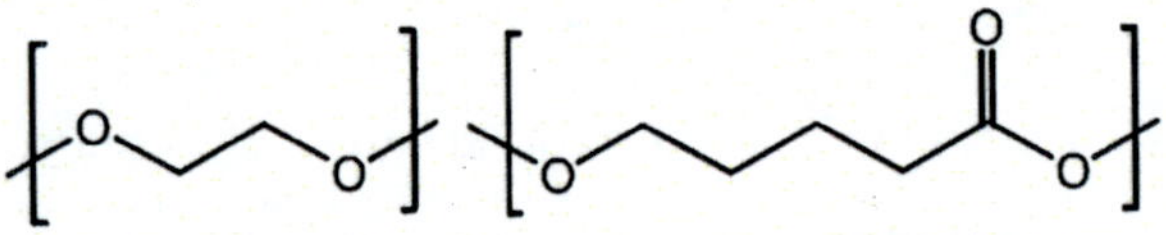

[XXXIII] [XXXIV]

The way in which the rapid manufacturing process operates is likely to have a substantial bearing on the morphology of the final material. For example in the case of stereolithography where photopolymerization or photo cross-linking is involved (particularly with acrylates and styrene-based systems) a common result would be the production of a largely amorphous product with no anisotropy. Although as we have seen photo cross-linkable systems may contain crystallisable materials, in the case of fused deposition modeling, the use of a crystalline polymer is a simple matter; in addition it is released as a single filament (rather than the fused powder formed with selective laser sintering). Furthermore it may be that this level of crystallinity may be enhanced if the polymer is extruded, as in the system we discuss below.

Many polymeric materials such as polyethylene, polypropylene and poly (ε-caprolactone) exhibit semi-crystalline structures and the level and distribution of the crystalline phase is critical to the definition of the properties of the final product. Figure 5.11a contrasts the wide-angle scattering for atactic-polystyrene widely employed in photo-cross-linking resins with that for poly(s-caprolactone).

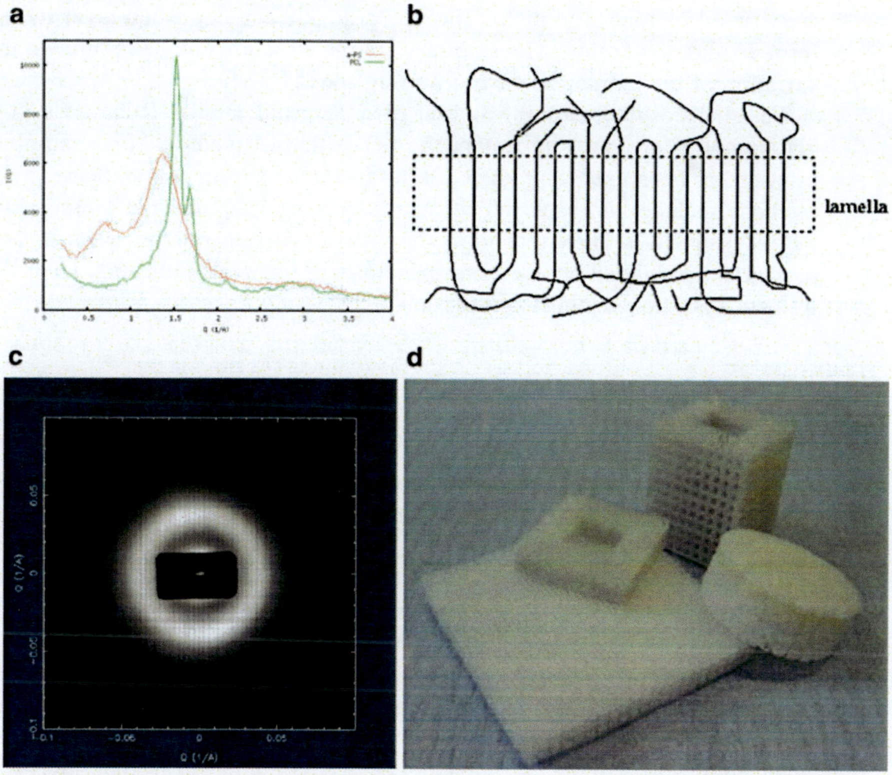

Fig. 5.11 Semicrystalline polymers: (**a**) Wide-angle X-ray scattering patterns for amorphous polystyrene and crystalline poly(ε-caprolactone); (**b**) A end view schematic of a chain folded lamellar; (**c**) Small-angle X-ray scattering of poly(ε-caprolactone) showing the scattering from the chain folded lamellae and (**d**) scaffolds for tissue engineering prepared by a layer addition system involving melt extrusion

The polystyrene curve shows broad peaks typical of an amorphous material. In contrast, the data for poly(ε-caprolactone) show sharp peaks arising from the significant fraction of crystalline material. These crystals are present in the form of thin chain folded lamellae (Fig. 5.11b) typically 10–20 nm in thickness, as confirmed by small-angle X-ray scattering techniques (Fig. 5.11c), but extending over micrometers in the lateral direction. Crystallization can take place in the solid state, but this is very unusual. Normally crystallization takes place either on cooling from the melt phase or precipitation from solution. It is difficult to see how the crystallization processes could be incorporated in to any effective stereolithography process. Hence semicrystalline polymers are excluded from the list of available materials for stereolithography unless added as a particulate. In contrast, the fused deposition system and other melt extrusion processes are particularly suited to semi-crystalline systems. A particular example of this approach is provided by bioextrusion; Fig. 5.11d shows an example of scaffolds produced by the bioextruder at Leira in Portugal [58]. The molten thermoplastic is extruded by means of piston through a nozzle and deposited layer by layer to form a solid on cooling. The semi-crystalline structure can be optimized through variation of the temperatures and writing speed.

Fused deposition molding is one of a range of similar techniques in which 3D structures are built up trough the deposition of material; for example inkjet printing has been used to produce scaffold structures from polyurethane [59]; other methods include laminated object manufacturing [39], and the pressure assisted microsyringe [60]. These techniques have their own material requirements and fit to particular application but the basic principles of material application layer by layer to form a 3-dimensional structure remains.

5.8 Conclusions

Rapid manufacturing offers considerable opportunities; the earliest applications of this technique were to produce prototypes to test designs; however, the methodology, in particular with improvements in computer technology has been extended to small scale production. Two particular applications stand out. The first is in biomedical applications – here the individuality of each situation is such that large scale manufacturing of parts is inappropriate. The second, which has increasing potential, is the situation where small scale manufacture of parts is required. In such situations machining parts may be very expensive; rapid manufacturing offers considerable commercial advantages. The challenge lies in the materials used. In many applications (particularly but not confined to biomedical solutions) properties such as mechanical strength or surface quality are important. The technology is rapidly improving, but considerable challenges remain, for example two-photon photochemical polymerization is being explored as a route to high resolutions and nanoscale device production [61]. As stated above materials are being improved including the production of composites [47, 57], such that the process better resembles manufacturing rather than prototyping, but in addition to this materials are being developed which will allow the prototype to be used to produce a mold, or indeed the technique has been used to make the molds. Many challenges remain but the importance of the approach to industry is such that the advances are likely to be as rapid as those over the past 20 years.

References

1. I. Gibson and D. Shi, Material properties and fabrication parameters in selective sintering process, Rapid Prototyping Journal, 3, 129–136, 1997.
2. R.M. Miranda, G. Lopes, L. Quintino, J.P. Rodrigues, and S. Williams, Rapid prototyping with high power fiber lasers, Materials and Design 29, 2072–2075, 2008.
3. B. Vandenbroucke and J.-P. Kruth in "Bio-Materials and Prototyping Applications in Medicine", Eds P. Bartolo and B. Bidanda, Springer, NY, Pages 109–124, 2008.

4. A.G. Mamalis, J.J. Ramsden, A.I. Grabchenko, L.A. Lytvynov, V.A. Filipenko and S.N. Lavrynenko, A novel concept for the manufacture of individual sapphire-metallic hip joint endoprostheses, Journal of Biological Physics and Chemistry, 6, 113–117, 2006.
5. S. Yang, and J.R.G. Evans, Metering and dispensing of powder; the quest for new solid free-forming techniques, Powder Technology 178, 56–72, 2007.
6. P. Calvert, Freeforming of polymers. Current Opinion in Solid State and Materials Science, 3: 585–588, 1998.
7. M. Agarwala, D. Bourell, J. Beaman, H. Marcus, and J. Barlow, Direct selective laser sintering of metals, Rapid Prototyping Journal 1, 26–36, 1995.
8. W. Höland, V. Rheinberger, E. Apel, C. van't Hoen, Principles and phenomena of bioengineering with glass-ceramics for dental restoration, Journal of the European Ceramic Society, 27 (2007) 1521–1526.
9. A. del Campo and E. Arzt, Fabrication approaches for generating complex micro and nano-patterns on surfaces, Chemical Reviews, 108 (2008), 911–945.
10. M.P. Stevens, Polymer Chemistry an Introduction, 2nd Edition, OUP, New York, 1990.
11. W.H. Carothers, Journal of the American Chemical Society, 51(8), 2548–2559, 1929.
12. S.S. Morye, P.J. Hine, R.A. Duckett, D.J. Carr, I.M. Ward. A comparison of the properties of hot compacted gel-spun polyethylene fibre composites with conventional gel-spun polyethylene fibre composites. Composites: Part A 30, 649–660, 1999.
13. N.S. Allen, J.P. Hurley, G. Pullen, A. Rahman, M. Edge, G.W. Follows, F. Catalina, and I. Weddell, in Current Trends in Polymer Photochemistry, Ed. N.S. Allen, M. Edge, I.R. Bellobono, and E. Selli, Ellis Horwood, Hemel Hempstead, 1995.
14. N.S. Allen, Photoinitiators for UV and visible curing of coatings, mechanisms and properties, Journal of Photochemistry and Photobiology A: Chemistry 100, 101–107, 1996.
15. N.J. Mills, Plastics, Microstructure and Engineering Applications, 2nd Edition, Hodder, London, 123–125, 1993.
16. N. S. Allen, M. C. Marin, M. Edge, D. W. Davies, J. Garrett, F. Jones, S. Navaratnam, and B. J. Parsons, Journal of Photochemistry and Photobiology A: Chemistry, 1999, 126, 135–149.
17. R. E. Kerby, L. A. Knobloch, S. Schricker, and B. Gregg, Dental Materials, 25, 302–313, 2009.
18. C. Decker, Progress in Polymer Science, 21, 593–650, 1996.
19. J. V. Crivello, Journal of Polymer Science: Part A: Polymer Chemistry, 1999, 37, 4241–4254.
20. C. Decker and D. Decker, Polymer, 38(9), 2229–2237, 1997.
21. J. V. Crivello, Nuclear Instruments and Methods in Physics Research B 151, 8–21, 1999.
22. P. Eyerer, B. Wiedemann, K.-H. Duel, B. Keller, Computers in Industry 28, 35–45, 1995.
23. K. Suyama, and M. Shirai, Progress in Polymer Science, 2009, 34, 194–209.
24. H. Du and M.K. Boyd, The 9-xanthenylmethyl group: a novel photocleavable protecting group for amines. Tetrahedron Letters, 42, 6645–6647, 2001.
25. T. Ohba, T. Shimizu, K. Suyama, and M. Shirai,. Photocrosslinking and redissolution properties of oligomers bearing photoamine generating groups and epoxy groups, Journal of Photopolymer Science and Technology, 18, 221–224, 2005.
26. Y. Yamaguchi, B.J. Palmer, C. Kutal, T. Wakamatsu, and D.B. Yang, Ferrocenes as anionic photoinitiators. Macromolecules, 31, 5155–5157, 1998.
27. R. Balajia, D. Grande, and S. Nanjundan, Reactive and Functional Polymers, 56, 45–57, 2003.
28. V. Ramamurthy and K. Venkatesan, Photochemical reactions of organic crystals, Chemical Review 87, 433–481, 1987.
29. Y. Xia and G.M. Whitesides, Soft lithography, Angewandte Chemie (International Ed. in English), 37, 550–575, 1998.
30. J. Rickerby and J. H. G. Steinke, Current trends in patterning with copper, Chemical Review, 102, 1525–1549, 2002.
31. D. Figeys, and D. Pinto, Lab-on-a-chip: a revolution in biological and medical sciences, Analytical Chemistry, 72 (9), 330A–335A, 2000.
32. G. Wallraff and W.D. Hinsberg, Lithographic imaging techniques for the formation of nanoscopic features, Chemical Review, 1999, 99, 1801–1821.

33. M. Chanda, and S. Roy, 2006. Plastics Technology Handbook, 4th ed. CRC Press, Boca Raton, P4–53, 2006.
34. G. Rabilloud, High-Performance Polymers.Vol 3, Polyimides for Electronics, Technip Paris, 2000 P165 – 234.
35. J. M. Shaw, J. D. Gelorme, N. C. LaBianca, W. E. Conley, and S. J. Holmes, Negative photoresists for optical lithography, IBM Journal of Research and Development, 41, 81–94, 1997.
36. A. Reiser; G. Bowes; R. Horne, Photolysis of aromatic azides, Transactions of the Faraday Society, 66, 3194, 1967.
37. H Lorenz, M Despont, N Fahrni, N LaBianca, P Renaud, and P Vettiger. SU-8: a low-cost negative resist for MEMS, Journal of Micromechanics and Microengineering, 7, 121–124, 1997.
38. S.A. Wilson, R.P.J. Jourdain, Q. Zhang, R.A. Dorey, C. R. Bowen, M. Willander, Q. Ul Wahab, M. Willander, S. M. Al-hillie, O. Nur, E. Quandt, C. Johansson, E. Pagounis, M. Kohl, J. Matovic, B. Samel k,10, W. van der Wijngaart, E.W.H. Jager, D. Carlsson, Z. Djinovic, M. Wegener, C. Moldovan, R. Iosub, E. Abad, M. Wendlandt, C. Rusu, and K. Persson, New materials for micro-scale sensors and actuators: An engineering review, Materials Science and Engineering, R 56, 1–129, 2007.
39. X. Yan and P. Gu, A review of rapid prototyping technologies and systems, Computer Aided Design, 26, 307–318, 1996.
40. C. Decker in Radiation Curing in Polymer Science and Technology 3, ed. J. P. Fouassier and J. F. Rabek, Elsevier, Chichester (1993). 1993 Chapter 2 33–64.
41. J.S. Ulletta, T. Benson-Tolle, J.W. Schultz, and R.P. Chartoff, Materials and Design, 20, 91–97, 1999.
42. U. Bulut, and J.V. Crivello, Investigation of the reactivity of epoxide monomers in photoinitiated cationic polymerization, Macromolecules, 2005, 38, 3584–3595.
43. U. Bulut, and J.V. Crivello, Reactivity of oxetane monomers in photoinitiated cationic polymerization, Journal of Polymer Science, Part A, Polymer Chemistry, 2005, 43, 3205–3220.
44. W. L. Wang, C. M. Cheah, J. Y. H. Fuh, and L. Lu, Influence of process parameters on stereolithography part shrinkage, Materials and Design, 17, 205–213, 1996.
45. L. Lecamp, B. Youssef, C. Bunel, and P. Lebaudy, Photoinitiated polymerization of a dimethacrylate oligomer Part 3 Postpolymerization study, Polymer, 40, 6313–6320, 1999.
46. J.-P. Kruth, M.C. Leu, T. Nakagawa, Progress in additive manufacturing and rapid prototyping, Annals of the CIRP, 47, 525–540, 1998.
47. M. Wozniak, T. Graule, Y. de Hazan, D. Kata, and J. Lis, Highly loaded UV curable nanosilica dispersions for rapid prototyping applications, Journal of the European Ceramic Society in Press.
48. D. Karalekas and K. Antoniou, Composite rapid prototyping: overcoming the drawback of poor mechanical properties, Journal of Materials Processing Technology 2004, 153–154, 526–530.
49. A. Rosochowski, A. Matuszak, Rapid tooling: the state of the art, Journal of Materials Processing Technology, 2000, 106, 191–198.
50. V. E. Beal, C. H. Ahrens, and P. A. Wendhausen, Journal of the Brazilian Society of Mechanical Sciences and Engineering, 26, 40–46, 2004.
51. J.C. Nelson, N.K. Vail, J.W. Barlow, J.J. Beaman, D.L. Bourell, and H.L. Marcuss, Selective laser sintering of polymer-coated silicon carbide powders, Industrial and Engineering Chemistry Research, 34, 1641–1651, 1995.
52. H.C.H. Ho, I. Gibson, and W.L. Cheung, Effects of energy density on morphology and properties of selective laser sintered polycarbonate, Journal of Materials Processing Technology, 89–90, 204–210, 1999.
53. J. Yang, Y. Shi, Q. Shen, and C. Yan, Selective laser sintering of HIPS and investment casting technology, Journal of Materials Processing Technology, 209, 1901–1908, 2009.

54. B. Caulfield, P.E. McHugh, S. Lohfeld, Dependence of mechanical properties of polyamide components on build parameters in the SLS process, Journal of Materials Processing Technology, 182, 477–488, 2007.
55. N.K. Vail, L.D. Swain, W.C. Fox, T.B. Aufdlemorte, G. Lee, and J.W. Barlow, Materials for biomedical applications, Materials and Design, 20, 123–132, 1999.
56. J.L. Lombardi and P. Calvert, Extrusion freeforming of Nylon 6 materials, Polymer, 1999, 40, 1775–1779.
57. S.H. Masood and W.Q. Song, Development of new metal/polymer materials for rapid tooling using Fused deposition modelling, Materials and Design 2004, 25, 587–594.
58. J. Mateus, H.A. Almeida, N.M. Ferreira, N.M. Alves, P.J. Bartolo, C.M. Mota, and J.P. de SousaVirtual and Rapid Manufacturing: Advanced Research in Virtual and Rapid Prototyping ed P.J. Bartolo, Taylor and Francis, London, 171–175, 2008.
59. C. Zhang, X. Wen, N.R. Vyavahare, and T. Boland, Synthesis and characterization of biodegradable elastomeric polyurethane scaffolds fabricated by the inkjet technique, Biomaterials, 29, 3781–3791, 2008.
60. G. Vozzi and A. Ahluwalia, Microfabrication for tissue engineering: rethinking the cells-on-a scaffold approach, Journal of Material Chemisty, 17, 1248–1254, 2007.
61. K-S. Lee, R.H. Kim, D-Y. Yang, and S.H. Park, Advances in 3D nano/microfabrication using two-photon initiated polymerization, Progress in Polymer Science, 33, 631–681, 2008.

Chapter 6
Materials for Stereolithography

Serge Corbel, Olivier Dufaud, and Thibault Roques-Carmes

6.1 Introduction

Stereolithography (SL) is a rapid prototyping method for three-dimensional polymer part fabrication [3, 34, 49, 53, 63]. The technique is based on the process of photo-polymerization, in which a liquid resin is converted into a solid polymer under laser irradiation [4, 34]. The models are produced by curing successive layers of the resin material until a three-dimensional object is formed. The advantages of stereolithography are its flexibility in manufacturing parts with different geometries and dimensions, its accuracy and its quickness. The challenge is to extend the stereolithography method to directly fabricate parts with complex shapes and good mechanical properties [30, 47, 58]. Recently, polymer/ceramic composite were successively fabricated by stereolithography [29, 46, 52, 62]. The manufacturing process requires the formulation of a photoreactive medium containing a photocurable resin and powders prior to laser exposure. Once polymerized, the photopolymer constitutes a though matrix around ceramic particles.

For applications suspensions have to satisfy several requirements. It is mandatory to use stable dispersions during the fabrication process (over 1 h). This does call for the knowledge of blending particles and resin of different density. In the same time, the suspension viscosity should be less than 5 Pa s in order to ensure satisfactory layer recoating and reduce the time necessary to obtain a horizontal fresh layer [8]. Generally, the highly loaded ceramic/silica suspensions (nonhollow) are quite viscous [31]. Lately, the suspension must be UV curable with useful depth and resolution. The cured depth must be high enough to avoid an excessive time of fabrication. The adjunction of particles greatly modifies the behavior of the resin under photopolymerization due to the scattering and absorption of the incident light beam by the particles [56, 62]. We also have to manage that low shrinkage occurs during the liquid to solid polymerization process

S. Corbel (✉)
Laboratoire Réactions et Génie des Procédés, UPR 3349 CNRS,
Nancy-Université, 1, rue Grandville, 20451, 54001 Nancy Cedex, France
e-mail: serge.corbel@ensic.inpl-nancy.fr

P.J. Bártolo (ed.), *Stereolithography: Materials, Processes and Applications*,
DOI 10.1007/978-0-387-92904-0_6,

6.2 Stereolithography Principle

In order to create 3D structures, the object is reproduced in a computer-aided-design (CAD) file [34]. Once the object to be built is designed, it is then numerically sliced in layers of a chosen thickness to define the different horizontal cross sections of the object to be manufactured. The 3D shape is broken down into a sequence of cross sections (layers), like the contours of a topographic map. The surface height is adjusted so that it is represented by one layer thickness. Then, the numerical data defining each of the sections are sequentially sent to the machine, in order to build them [34, 35]. Thus, a layer-by-layer coordinates of contour lines or surfaces are obtained. Using CAD information, laser beam radiation (355 nm Nd-YAG laser, Navigator I Vanadate with 20 kHz pulses frequency) is deflected vector by vector on the top of the surface of the liquid (Fig. 6.1).

The UV radiation is absorbed by the photoinitiator which in turn generated free radicals to initiate polymerization of a liquid monomer into a solid polymer. When a solid layer is performed, the deposition of a subsequent layer of photopolymerizable resin on the already polymerized part allowed continuing the manufacturing process. For this purpose, after the exposure step, the supporting platform containing the solid part is largely immersed in the reactive medium and then moved upward such that between the last cured layer and the resin surface there is a new layer of fresh resin. This was made thanks to a motor driving an elevator. The motorized vertical stage accuracy is of 5 μm for the motor controller used in the set-up. During this step of layer recoating, it is necessary to wait until the resin surface becomes completely flat and level in order to have accurate thickness control. This procedure is repeated until the polymer solid part is build. The stacking of the layers leads to a three-dimensional physical part. After a few hours, the platform is raised and the completed part removed. Resolution is clearly related to the film

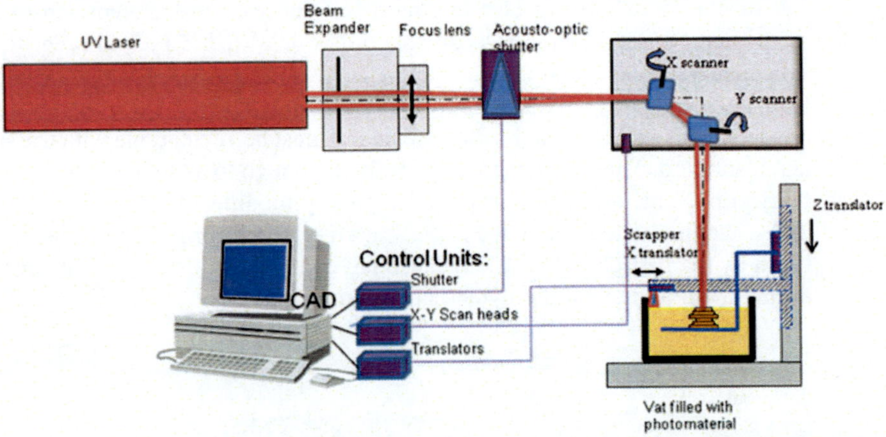

Fig. 6.1 Scheme of the stereolithography process (additive process layer by layer)

thickness and kinetics of photopolymerization and diffusion [53]. A typical layer thickness is 200 μm, line widths are similar. Thinner layers increase the part production time [6], but thicker layers may require longer laser exposure and also longer production times. The maximum part size is really limited by the size of the monomer bath and the dimensions of the photoreactor are convenient to the component's size typically 30 × 30 × 40 cm. Other derived methods, such as liquid-crystal mask-based techniques [46, 52], DMD mask-based techniques [55] or two-photon absorption microfabrication processes have been developed recently [44]. In our case, experiments were carried out by conventional STL since it allows the use of commercial resins optimized for UV curing.

6.2.1 Stereolithography of Polymers

One of the main interests in the STL technique comes from the photopolymerization (curing) in long chain leading to a cross-linked network polymer. A photoinitiator, inside the reactive medium, produces free radicals or cations and induces the polymerization reaction. The irradiated liquid monomer polymerizes by a chain reaction into a solid polymer which does not dissolve in the remaining monomer. In the oversimplified mechanism:

$$\mathrm{I} + \mathrm{Photon} \rightarrow \mathrm{reactive\ species}$$

$$\mathrm{Reactive\ species} + \mathrm{M} \rightarrow \mathrm{insoluble\ polymer}$$

I is a photochemical initiator directly leading, or not, to the reactive species which can be free radicals or ions, and M is a monomer or an oligomer. More complex mechanisms have been used [63], but it is not easy to take all the elementary steps into account and oxygen inhibition on photopolymerization [16]. The power of the lasers used to initiate such photopolymerization process is rather low, typically between 10 mW and 1 W, which makes STL technique economically attractive.

Most materials used are conventional epoxy, acrylate resins, or thermoplastics elastomers. The originals monomers were combinations of multifunctional acrylates which formed cross-linked resins. This composition has several useful properties, such as low viscosity which assists smooth spreading of the liquid material, high photosensitivity, oxygen inhibition, low critical energy, controllable mechanical properties, and relative insensitivity to temperature and humidity changes [6]. The extent of their use in applications has been limited by their poor dimensional stability and by their high volume shrinkage. A major concern is the distortion induced by shrinkage during post-curing. Acrylates showed that the polymerization continued about 2 days after photoinitiation.

Newer photopolymerizable epoxies offer better properties. Thus, in contrast to acrylate resins, they have exceptionally low volume shrinkage and good dimensional stability.

STL has been broadly used as design, working, and master models in the industrial and medical sectors [33] and is currently applied in the medical field. Medical models built represent a new approach for surgical planning and simulation. This technique allows one to reproduce precisely anatomical objects as 3D physical models which can be looked and manipulated by the surgeon [50]. He has a realistic impression of complex structures and can simulate surgical procedures before a real surgical intervention. It is also very attractive in reconstructive surgery and implantation of prostheses, where each "part" has to be adapted to the unique anatomy of the patient and can be made in a short time. This raises the possibility of manufacturing simple prostheses during the course of an operation. For instance, in the reconstructive surgery field, cranioplasty implants have been successfully manufactured using the STL apparatus [21].

The use of STL is well suited for the fabrication of monolithic supports for photocatalysis applications [24]. Photocatalysis using TiO_2 as catalyst is an efficient advanced oxidation process in which the catalyst TiO_2 is deposited on a suitable support and activated by near UV light. It permits to create quickly several geometries which can be modified as necessary. Monolithic supports with complex 3D geometry, have been made in epoxy resin and then have been recoated with TiO_2 (Fig. 6.2b). They were successfully used in photocatalysis for cleaning air contaminated by volatile organic compounds and photocatalytic degradation of organic pollutants in water [22, 23].

STL allows the manufacture of various devices for microfluidic [11] and chemical engineering applications [14]. Microstructures, such as fluid mixers, fluid distributors, and heat exchanger [40], have been elaborated by conventional STL. For instance, Luo et al. examined experimentally the effects of constructal distributors on flow equidistribution in a multichannel heat exchanger [39]. The results showed that the integration of constructal distributors homogenized the fluid flow distribution and consequently leaded to a better thermal performance of the heat exchanger. Micro heat exchangers with internal channel dimensions in

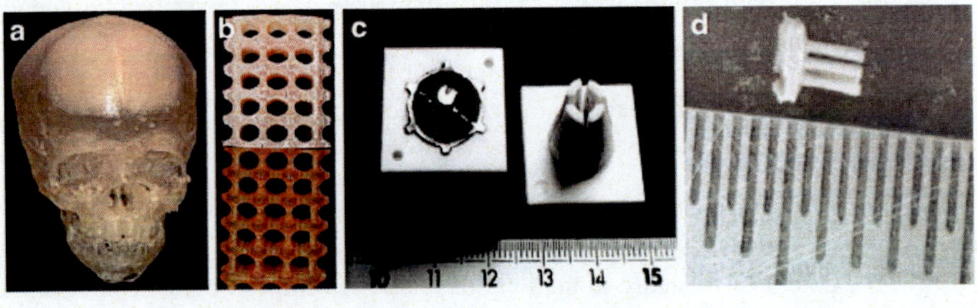

Fig. 6.2 Examples of parts pieces fabricated by stereolithography: (**a**) Skull in Polymer; (**b**) monolithic supports with TiO_2 coating; (**c**) Sintered ceramic mould loaded at 85 wt% alumina; (**d**) Piezoelectric ceramics: Rods loaded at 50 wt% PZT

the millimeter range have been manufactured [14]. The channel geometry was designed in order to produce local mixing of the fluid and to improve the fluid-side heat transfer. Even if the exchanger performance was low in terms of heat transfer due to their polymer walls, these three-dimensional parts allowed the characterization of mass transfer and pressure drop in such structure. Within this framework, research carried out to develop the STL of metallic structures would offer composites with higher thermal and electric conductivities. It also leads to channels and walls with higher mechanical resistance to fluid pressure.

6.2.2 *Fabrication of Ceramic and Metal Components by Stereolithography*

It is difficult to obtain a high level of hardness, high mechanical resistance to fluid pressure, thermal and electrical conductivity when photopolymer are used. In addition, photopolymers have limited applications because of the lack of biodegradability or sufficient mechanical strength. There are efforts to extend the system to other materials, such as ceramics and metals. The methods which are largely used for the fabrication of three-dimensional polymer parts [1, 10, 32, 43, 48, 54, 59–61], has been adapted to the process of three-dimensional ceramics and metallic pieces with final properties (mechanical, thermal, and electrical) closed to those obtained by classical processing techniques. The ultimate aim is the production of component in ceramic and metal material which can be obtained after the debinding and sintering step of the composite object.

The manufacturing process requires the formulation of a photoreactive medium containing a photocurable resin and powders (ceramic or metal particles) prior to laser exposure. Once polymerized, the photopolymer constitutes a though matrix around ceramic particles and confers the cohesion of the green body. This organic phase is subsequently removed by an appropriate thermal treatment, then the sintering of the green part ensures the final properties of the ceramic piece. Various materials have already been successfully shaped in ceramic or metal-based composite materials by inserting different powders (silica, silicon nitride, alumina, hydroxyapatite, (zirconate titanate oxides PZT), aluminum, and copper) in acrylate or epoxy resins [7, 14, 19, 26, 30, 38, 45, 46, 52, 56]. Table 6.1 shows an overview of the various suspensions that have been used in such processes.

A candidate ceramic suspension for the process must satisfy several requirements:

- A high volume loading of particles is required to give dense and high quality ceramic pieces after debinding and sintering. In order to avoid deformation and cracking during binder removal, to reduce the shrinkage during sintering, and to obtain homogeneous and dense ceramic pieces after sintering, it is necessary to minimize the organic concentration in the suspension.
- The suspension viscosity should be less than 5 Pa s, in order to ensure satisfactory layer recoating and reduce the time necessary to obtain a horizontal fresh layer

Table 6.1 Suspension formulation and optical properties for direct fabrication by STL of composites. The values of D_p, E_c, C_d, and Q are calculated using (6.1) and (6.2)

UV curable resin	Powder nature	Particle size (μm)	Particle concentration % V/V	Δn ($n_{part.} - n_{resin}$)	D_p (μm)	E_c (mJ/cm^2)	C_d (μm) $E_0 = 2{,}600$ mJ/cm^2	Q
Epoxy	–	–	–	–	121	13.5	341	–
Acrylate	–	–	–	–	242	38.3	2,750	–
Epoxy	SiO_2	5	50		137	12.7	323	0.0012
Acrylate	SiO_2	3.5	50–60	0.03	140	6.15	800	0.033
Epoxy	Al_2O_3		45		52	32.8	121	0.0037
Acrylate	Al_2O_3	10	53		75	9.3	300	
Acrylate	Al_2O_3	4.4	50–60	0.23	68	4.8	350	0.086
Acrylate	Zircon	4.2	50–60	0.38	50	14.94	200	0.112
Acrylate	Silicon nitride	0.5	30	0.7	20		40	
Acrylate	PZT	5	80 wt%		15	111	25	
Acrylate	Al	6.5	80 wt%		12	1,200	17	
Acrylate	Cu	3	60 wt%		10			

completely flat [8, 9]. The increase of the viscosity of the resin leads to slow resin flows on the top of the build platform, which increases fabrication time and makes a convex mound on the surface of the building part after recoating cycle. If the viscosity is too high, the resin surface never becomes completely flat by gravity alone. Nevertheless, the viscosity of the suspension must be as low as possible to allow a good recoat of the liquid monomer on the polymerized underlayer [25].

- The movement of the scrapper is synchronized with the other operations (Fig. 6.1) and its speed is adapted to the rheology of the liquid. A suspension viscosity of 5 Pa s, is set at the maximum resin viscosity in conventional stereolithography with scrapper to ensure a satisfactory layer recoating. Higher viscosity leads to a paste rather than a liquid which is not possible to process with a conventional STL machine and need special scrapper to spread such pastes.
- The suspension must be UV curable with useful depth and resolution. The cure depth must be high enough to avoid an excessive time of fabrication. The adjunction of particles greatly modifies the behavior of the resin under photopolymerization due to the scattering and absorption of the incident light beam by the particles. The reactivity of an STL resin is characterized by two parameters (D_p and E_c) that define the photosensitivity of UV curable resin during the process. The theoretical expression of the cured depth (C_d, i.e., the thickness of the gelled resin) is derived from the Beer–Lambert law of absorption and can be written as [53]:

$$C_d = D_p \ln\left(\frac{E}{E_c}\right) \tag{6.1}$$

where D_p is the penetration depth of the resin, E and E_c are the laser exposure on the resin surface and critical exposure of the resin at the laser wavelength, respectively. Critical exposure E_c is the laser energy below which the polymerization does not occur. A good resin for STL is characterized by low values of E_c, in order to start the reaction with low energy dose, and by high values of D_p, in order to optimize the cured thickness.

In presence of scattering particles, the penetration depth D_p is expressed as [25]:

$$D_p = \frac{2}{3}\left(\frac{d}{\varphi Q}\right)\left(\frac{n_0}{\Delta n}\right)^2 \tag{6.2}$$

where d is the mean particle size of the powders, ϕ is the volume fraction of ceramics in the suspension, n_0 is the refractive index of the monomer solution, Δn is the refractive index difference between the ceramic and the monomer solution, and Q is the scattering efficiency term. D_p is a function of the volume concentration of powder, the particle diameter and the refractive index difference between the UV curable solution and powder.

The adjunction of particles greatly modifies the behavior of the resin under photopolymerization due to the scattering and absorption of the incident light beam by the particles [56, 62]. It was found that due to light scattering, the fabricated line is wider in width and smaller in depth (low D_p) compared with polymeric fabrication at the same condition (Table 6.1). It was also demonstrated that the higher the refractive indexes difference between the particle and the resin, the stronger the light scattering. The scattering effect can be greatly minimized by choosing the ceramic particles that have the refractive index close to that of the solution. Silica is special in having a refractive index close to that of the resin. Using silica powders characterized by lower values of Δn leads to sufficient depth C_d and D_p values ($C_d = 300$ μm and $D_p = 140$ μm). The alumina suspension shows higher values of Q as compared to silica suspension (Table 6.1). Light scattering by alumina and zircon is much stronger and cure depths are lower ($C_d = 200$ μm and $D_p = 70$ μm). Satisfactory cure depths could be achieved but at the cost of exposure energies about 100 times that of unfilled resins, with a consequent proportional expected increase in build time. Light scattering could be decreased by the use of dyes to tune the refractive index of the resin closer to that of the ceramic [62]. For a higher refractive index difference (silicon nitride), the cure depth is more attenuated ($C_d = 40$ μm and $D_p = 20$ μm) and becomes insufficient for STL applications. The penetration thickness of the laser is the lowest in the case of PZT ($C_d = 25$ μm and $D_p = 15$ μm), aluminum ($C_d = 17$ μm and $D_p = 12$ μm), and copper, due to the predominance of light absorption and scattering [14]. This implies a very delicate processing of piezoelectric ceramic and metal by STL apparatus because of the spreading of thin layers (20 to 10 μm). The STL process which was used for polymer or alumina suspensions shaping is not convenient for the deposition of layers thinner than 50 μm. A new system was developed in the laboratory consisting of injecting nitrogen through a nozzle. The nitrogen flow can spread a layer of a thickness between 6 and 40 μm [14, 45]. The pressure applied by the nitrogen stream on the suspension acted as a substitute for the mechanical pressure of the former scrapper.

Stereolithography has been used to fabricate metal and ceramic parts. After debinding and sintering steps, the resulting ceramic parts have been used either as functional components or as molds for investment casting. The major drawback of the classical fabrication of complex ceramic parts is the cost of the mold, the difficulties to obtain different complicated cross sections and the processing time required to develop the first prototype. STL offers a unique opportunity to fabricate, without using costly molds, complex ceramic parts [5, 7, 14, 15, 19, 29, 30, 36, 46, 62] (Alumina and silica) with final properties (mechanical properties) similar to those obtained by classical processing techniques in a much reduced time and cost (Fig. 6.1) (Piezoelectric ceramics like lead zirconate titanate oxides (Pb(Zr,Ti)03) PZT dedicated to microelectromechanical systems (MEMS), microfluidic applications, and medical imaging applications have been fabricated by STL. The characterization of PZT films, 80–130 μm thick, on silicon substrate showed piezoelectric properties (dielectric permittivity, tangent loss, and remanent polarization) comparable with those of PZT thick films fabricated by other methods [36]. These structures were also integrated in MEMS for advanced piezoelectric microsensors and microactuators with high energy density. Dufaud et al. [15]. have designed and fabricated transducers with 45%vol of PZT similar in shape to those typically obtained with injection molding methods. Microtransducers and micropumps have been elaborated and their contribution in sensing, control, and hydrodynamic fields has been shown [14]. The technique allows the manufacture of various devices for microfluidic applications. A polymer micropump, with a PZT transducer integrated within the upper membrane, was successfully fabricated. PZT/polymer composites have shown excellent aptitudes in the domain of ultrasound medical imaging since the epoxy matrix lowered the acoustic impedance of the composite and made it close to human body impedance [51]. Promising results were also obtained with biocompatible ceramics, such as hydroxyapatite suspensions for medical implant design [12, 27].

Mauzon et al. [45] have demonstrated the possibility of metal components fabrication by direct STL. Aluminum and copper cylinders of 5 mm diameters, microgear, and cube have been realized [38].

6.3 Aluminum Components Fabrication

6.3.1 Raw Materials

Aluminum powders were provided by Goodfellow with an average size d_{50} of 6.5 μm. The 90th percentile particle size (d_{90}) was approximately 12 μm. During this study, aluminum particles are considered as spherical, which was confirmed by optical microscopy. The density of aluminum is 2.7 g/cm^3 and their refractive index is 1.44.

Different resins were used as binders: on the one hand, epoxy or epoxy-acrylates (3,4-epoxycyclohexylmethyl-3,4-epoxycyclohexyl carboxylate trimethylolpropane triacrylate called Somos 6100, Dupont and RPCure 200 AR, RPC) and on the other hand, acrylates (2,2′-bis[4-(methacryloxyethoxy)phenyl]propane called Diacryl 101, Akzo Nobel Chemicals, and 1,6-hexanediol diacrylate called HDDA, UCB) [13].

To improve monomers reactivity, DMPA (2,2′-dimethoxy-2-phenylacetophenone, Aldrich) was used as UV photoinitiator. The addition of a dispersing agent, coating particles surface, was compulsory to ensure dispersions homogeneity and stability. A phosphoric ester (Beycostat c213, CECA) was then chosen as regard as previous experiments [17].

6.3.2 Preparation of Photocurable Suspensions

At first, aluminum powder was diluted in a solvent acetone/ethanol. Then, the metal was coated with 1.5 wt% dispersant with respect to the metal amont (Sonic Dismembrator 550, Fisher Scientifc). The solvent was evaporated at room temperature to allow ester adsorption onto particles surfaces. After drying, powder was deagglomerated with a 50 μm sieve. Granulometric analyses have shown the absence of aggregates larger than 30 μm.

Finally, aluminum was mixed with resin and 1 wt% UV photoinitiator with respect to the metal, during 40 min by ball milling (300 rpm – Retsch PM400, Fisher Scientific).

6.3.3 *Characterization*

Rheological measurements were carried out thanks to a stress-controlled viscosimeter (CSL-100, Carri-Med) in a cone-plane setting on shear rate range $\dot{\gamma}$ of 0.1–300 s^{-1} at various temperatures from 283 to 373 K (10–100°C). Moroever, the stability of aluminum suspensions was checked by sedimentation rate experiments.

The thickness (E_p) and the width (L_p) of a polymerized element has been determined by optical profilometry.

Finally, the features of 3D parts were investigated thanks to a helium pycnometer (Accupyc 1330) for density measurements and to a digital multimeter for conductivity measurements.

6.4 Aluminum Suspensions Properties

6.4.1 *Rheological Behavior*

The influences of binder nature, fillers loading, and temperature on the suspensions rheological behavior was investigated in order to improve the layers recoating by scrappers or gas injection nozzles.

First of all, several resins, standard in stereolithography process, were tested with the stress-controlled viscosimeter. Among these monomers, only HDDA (1,6-hexanediol diacrylate) has revealed a plastic viscosity clearly lower than 1 Pa s at different shear rates and especially for $\dot{\gamma}$ greater than 40 s^{-1}, which is the case with the recoating systems available. Higher viscosities, in particular with high filler loading, will increase the probability of heterogeneous surfaces occurrence or problems in layers recoating due to a viscoelastic behavior. As a consequence, HDDA was chosen as organic binder for aluminum suspensions. It is worth to note that HDDA shows an important shear-thinning behavior, which is useful for dispersions spreading [17].

The rheological behavior of aluminum suspensions was studied for different fillers loading. However, when loading is lower than approximately 70 wt%, parts could not be properly sintered and on the other hand, loading greater than 90 wt% imply dispersions with too high viscosities.

Experiments were thus mainly carried out with 80 wt% aluminum in HDDA monomer. The shear-thinning behavior of the suspensions, (Fig. 6.3), has been fitted by a Sisko model:

$$\tau = \eta_\infty \dot{\gamma} + k \dot{\gamma}^n$$

where k is the consistency coefficient in Pa s and η_∞ is the high-shear Newtonian limit.

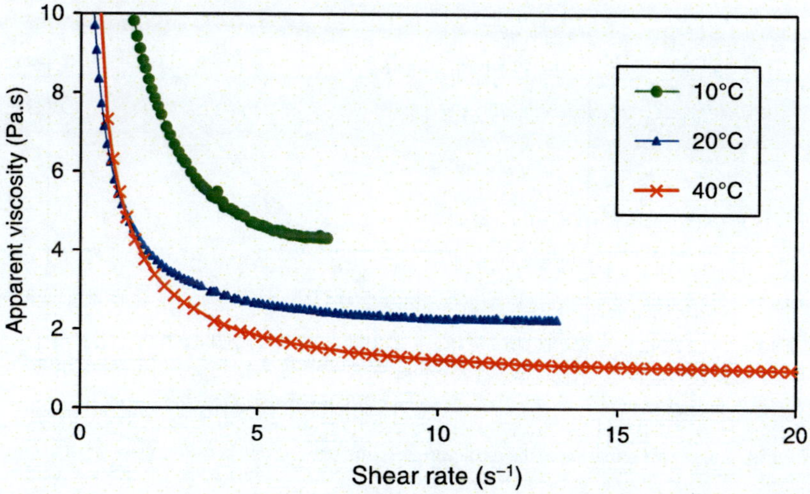

Fig. 6.3 80 wt% aluminum suspensions viscosity versus shear rates at various temperatures [45]

In order to compensate the viscosity increase due to fillers addition, the heating of the suspensions was studied. As shown on Fig. 6.3, when the temperature increases from 293 to 353 K, the apparent viscosity η drops from 2.58 to 0.33 Pa s. This evolution is well represented by an Arrhenius relation:

$$\eta = A \exp\left(\frac{E_a}{RT}\right)$$

where E_a, the activation energy is 34,430 J mol^{-1}, which is close to these of HDDA monomer (32,000 J mol^{-1}), and A, the pre-exponential factor is 2.35.10^{-6} Pa s. Previous experiments carried out on PZT suspensions in HDDA have shown that E_a was also in this range. As a consequence, this parameter probably only depends on binder nature Fig. 6.4.

For further tests, the suspension temperature was limited to 313 K in order to avoid unwanted resin reaction in the presence of photoinitiator.

6.4.2 Dispersions Reactivity

The dispersion reactivity was verified by cured depth (E_p) and width (L_p) measurements. These parameters depend on the density of energy (DE in J/m^2) received by the suspension during a unit of time:

$$\mathrm{DE} = \frac{2P_0}{W_0 v \pi}$$

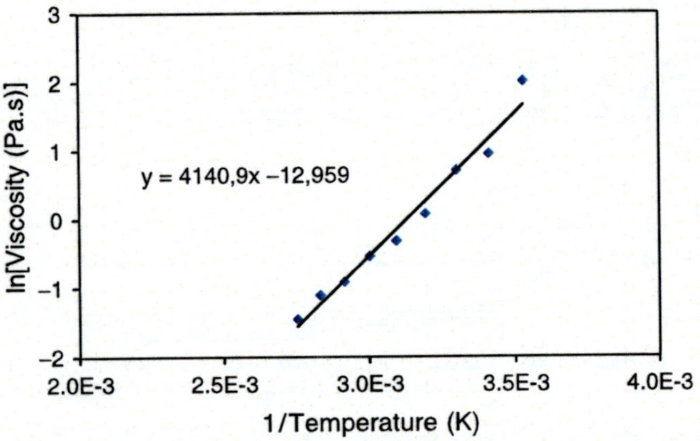

Fig. 6.4 Representation of the Arrhenius relation between suspensions viscosity and temperature [45]

The effects of DE variations on cure width and depth were observed by changing v, the scanning velocity of the laser beam. Laser beam radius w_0 and irradiation power P_0 were kept constant with the respective values:

$$w_0 = 150\ \mu\text{m and } P_0 = 60\ \text{mW}.$$

According to Beer–Lambert's law, the polymerization depth is defined by:

$$E_p = D_p \ln\left(\frac{\text{DE}}{\text{DE}_c}\right)$$

where DE_c is the minimum density of energy necessary to begin the polymerization and D_p, the laser depth of penetration. Similarly, the polymerization width is represented as follows:

$$L_p = \frac{w_0}{\sqrt{2}}\sqrt{\ln\left(\frac{\text{DE}}{\text{DE}'_c}\right)}$$

Increases of polymerized width and depth shown on Fig. 6.5 are consistent with previous observations. Even if there are three polymerization regions for aluminum suspensions, only two of them could be observed on Fig. 6.5 [28]. At very low energy of irradiation, the inhibition of polymerization by oxygen is predominant and prevents the polymerization [18]. When the polymerization starts and below about 3 J/cm², the polymerization rate is not affected by oxygen, the polymerization reaction is fast with a high slope of the curve. At higher energy, the free radicals are likely to react with each other than to form polymer. Then, an increase of the irradiation does not lead to high significant modifications of the depth or width curing and the slope is lower.

Finally, for highly loaded aluminum suspensions, the vertical depth of curing varies from 5 to 20 μm, whereas the lateral width ranges between 250 and 500 μm. Thus, both thick films and 3D components could be fabricated from these dispersions.

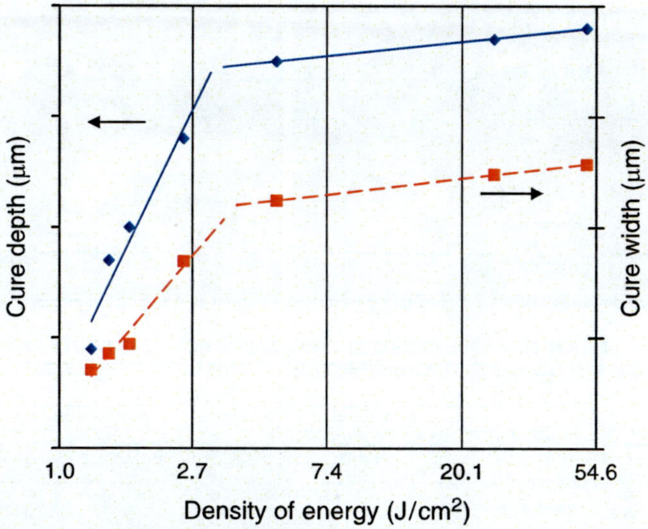

Fig. 6.5 Cure depth and width (*dashed*) versus ln(DE) for a 80 wt% aluminum suspensions [45]

6.5 Three Dimensional Parts Fabrications

6.5.1 Green Parts Fabrications

The aluminum suspension used for green parts fabrication was loaded at 80 wt% in HDDA monomer and kept at 40°C. A new recoating system, based on an oxygen injection nozzle, was developed to allow the deposition of suspensions layers with depth ranging between 6 and 40 μm. In these conditions, the oxygen inhibition on polymerization kinetics and consequently the gain on process accuracy were used to reach optimum lateral and vertical resolutions [13]. Green parts were carefully rinsed with isopropanol.

Two kinds of objects were fabricated by direct stereolithography of aluminum suspensions: on the one hand, metal sheets loaded at 80 wt% aluminum, on the other hand, multilayers components (Fig. 6.6).

Aluminum thick films with an average depth ranging from 4 to 20 μm were laid on glass plates, ceramic plates as well as on copper films. Due to aluminum electrical properties and to layers thickness, these sheets could probably be used as electrodes in MEMS applications.

Three-dimensional structures were also fabricated (Fig. 6.7). The microgear shown on Fig. 6.7 is constituted with 280 layers of 15 μm thickness. Aluminum/polymer composites with cube shape were fabricated in order to be sintered and to be used as demonstrators for conductivity measurements.

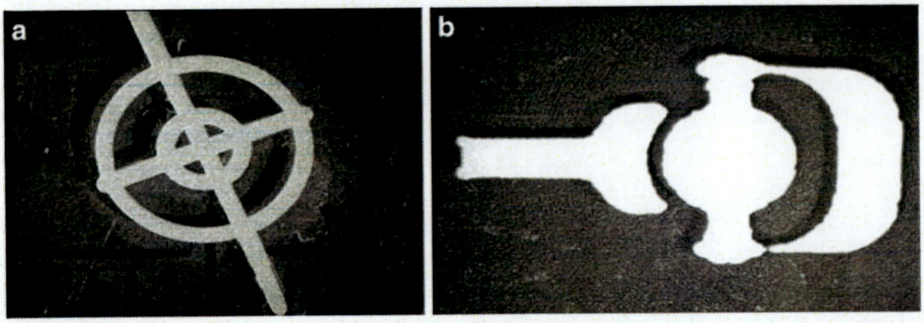

Fig. 6.6 Examples of aluminum sheets loaded at 80 wt% aluminum; (**a**) one-layer target – 15 μm thick; (**b**) first layer of a mechanical articulation [45]

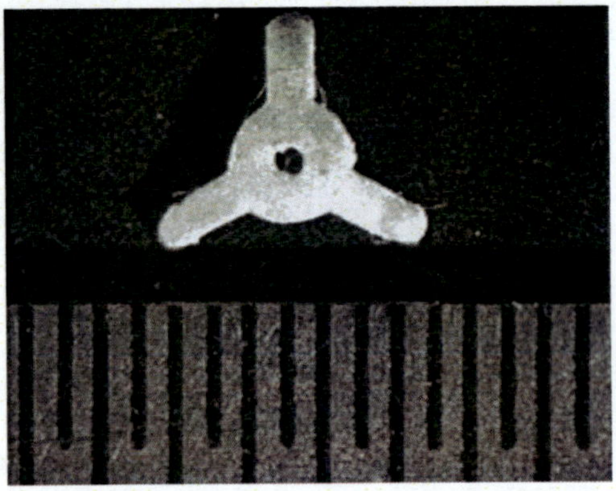

Fig. 6.7 Examples of aluminum components: microgear loaded at 80 wt% aluminum [45]

6.5.2 Sintering

The green parts were debinded and sintered in order to improve their mechanical resistance and to study their electrical, and soon, their thermal properties. Thermogravimetric analysis was carried out to determine the optimum temperature for polymer debinding.

The density of the sintered parts was calculated from analyses performed with a pycnometer (AccuPyc 1330).

As shown on Fig. 6.8, the density of the sintered parts increases with the sintering temperature. At 773 K, the green part is debinded and its density is equal to aluminum one. However, the poor mechanical properties of the parts prevent them to be handling without crumbling. Over 873 K, surfaces are covered with alumina oxide. Indeed, the oxidation of aluminum into alumina begins at 623 K and is complete at 1,313 K. Finally, over 933 K, which is aluminum melting temperature, parts density tends to become the alumina density, with a lack of their electrical properties.

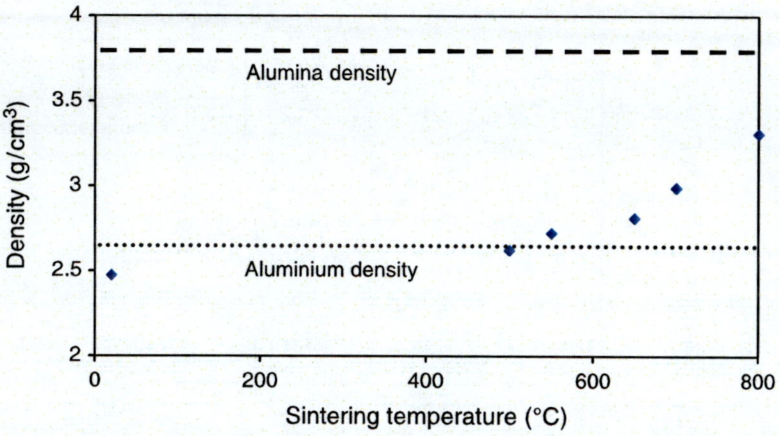

Fig. 6.8 Evolution of the sintered components density versus the sintering temperature [45]

Thus, a temperature of 873 K was chosen in order to avoid the alumina formation. Moreover, the rate of sintering seems to be optimal at about 793 K.

6.5.3 Metal Components Properties

To determine the conductivity of aluminum parts after sintering, their electric resistance was measured between two points distant of 5 mm. Two models were used to exploit these measurements [20]:

1. On the one hand, an empirical relation based on electrical conductivity of electrodeposited nickel foams:

$$\lambda = \frac{1}{4}(1 - \Phi)\lambda_s \tag{6.3}$$

2. On the other hand, based on a theoretical [20] model related to octahedral arrays of metallic wires:

$$\lambda = \frac{(1 - \Phi)}{3\left[1 - 0.121(1 - \Phi)^{1/2}\right]}\lambda_s \tag{6.4}$$

where λ is the electric conductivity, λ_s is the conductivity of the solid and Φ is the average porosity.

The parts porosity was modified by changing the fillers content of metallic suspensions before manufacturing. After stereolithography and sintering steps, measurements show that the volume withdrawal was approximately 10%.

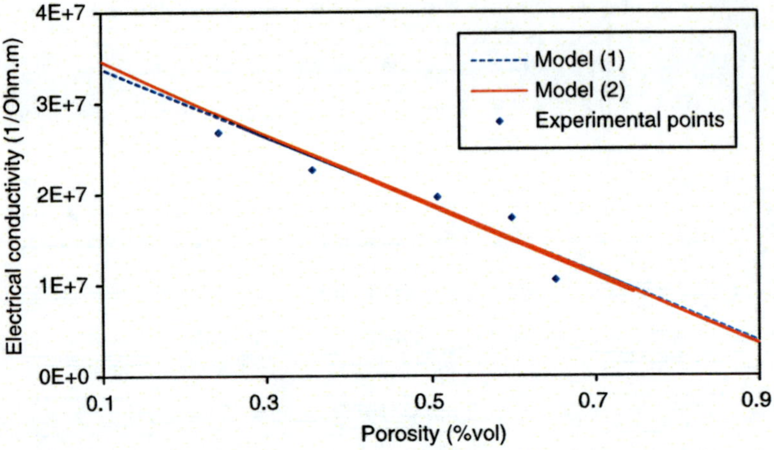

Fig. 6.9 Evolution of aluminum parts electrical conductivity versus porosity [45]

It should be noticed on Fig. 6.9 that both models satisfactorily fit the experimental points. No measurements have been carried out for porosity greater than 65% volume because of the poor mechanical properties of the parts. Moreover, it is highly likeable that no electrical properties would be found below the percolation threshold of aluminum particles.

As a conclusion, electrical conductivities of aluminum parts have been determined and are in accordance with literature models. On top of the advantages due to parts electrical properties, it is also possible to determine their porosity by means of electric resistance measurements.

6.6 Conclusions

Small components or sheets in different material can be made by using direct stereolithography process. From the adequate photocurable suspension of metal or ceramic whose rheological behavior and the dispersion reactivity have been perfectly identified, three-dimensional parts have been built by using a UV laser irradiation with a suitable beam diameter. The technique has been followed by a step of sintering in order to improve the mechanical resistance of the green parts. The determination of the sintering temperature is a crucial point: a too high temperature could alter the quality of the metal and a too low one could be inefficient treatment. Finally, some sintered parts in aluminum were successfully obtained.

This result and the design flexibility, fabrication quickness, and the accuracy of direct stereolithography lead us to think that direct stereolithography is a suitable process to realize small functional parts in alternative with microstereolithography process.

Indeed, the application's field of such fabrications is vast and could be widely used. However, some adjustments remain before exploiting industrially this process.

Acknowledgments The authors would like to acknowledge Jean François Rémy (LSGC) for the granulometric measurements.

References

1. Abe, F., Osakada, K., Shiomi, M., Uematsu, K., Matsumoto M, The manufacturing of hard tools from metallic powders by selective laser melting, Journal of Materials Processing Technology, 111(1–3), (2001), 210–213.
2. Allanic A-L., P. Schaeffer, French Patent n FR2790 418-A1, (1999).
3. André J.C., A. Le Méhauté, O. De Witte, French Patent, 8, (1984), 411, 241.
4. André J.C., S. Corbel, Stéréolithographie Laser, Polytechnica, Paris, 1994, 26.
5. Bertsch A., S. Jiguet, P. Renaud, Journal of Micromechanics and Microengineering 14, (2004), 197–203.
6. Calvert P., R. Crockett, Chemistry of Materials, 9, (1997), 650–663.
7. Chartier T., C. Chaput, F. Doreau, M. Loiseau, Journal of Materials Science 37 (2002), 3141–3147.
8. Chartier T., C. Hinczewski, S. Corbel, Journal of the European Ceramic Society 19 (1999), 67–74.
9. Chartier T., M. Ferrato, J.F. Baumard, G. Coudamy, Acta Ceramica 6 (6), (1994), 17.
10. Choi, D. S., Lee, S. H., Shin, B. S., Wang, K. H., Song, Y. A., Park, S. H., Jee, H. S, Development of a direct metal freeform fabrication technique using CO_2 laser welding and milling technology, Journal of Materials Processing Technology, 113, (2001), 273–279.
11. Cia L., S. Mantell, D. Polla, Sensors Actuators A 94, (2001), 117–125.
12. Dhariwala B., E. Hunt, T. Boland, Tissue Engineering, 10, (2004), 1316.
13. Dufaud O., Corbel S, Oxygen diffusion in ceramics suspensions for stereolithography, The Chemical Engineering Journal, 4082, (2002), 1–8.
14. Dufaud O., H. Le Gall, S. Corbel, Trans. IChemE, Part A, Chemical Engineering and Design 83, (2005), 133–138.
15. Dufaud O., P. Marchal, S. Corbel, Journal of the European Ceramic Society 22 (2002), 2081–2092.
16. Dufaud O., S. Corbel, Chemical Engineering Journal, 92 (1–3), (2003), 55–62.
17. Dufaud, O. (2002), Prototypage rapide de composites céramiques fonctionnels, Ph.D. Thesis, Institut National Polytechnique de Lorraine, 2002.
18. Dufaud, O., Corbel S, Stereolithography of PZT ceramic suspensions, Rapid Prototyping Journal, 8(2), (2002), 83–90.
19. Esposito C. Corcione, A. Greco, F. Montagna, A. Licciulli, A. Maffezzoli, Journal of Materials Science 40 (2005), 4899–4904.
20. Feng, Y., Zheng, H., Zhu, Z., Zu, F, The microstructure and electrical conductivity of aluminum alloy foams, Materials Chemistry and Physics, 78, (2002), 96–201.
21. Forouta M., B. Fallahi, S. Mottavahi, S. Mottavalli, M. Dujovny, Critical Reviews in Neurosurgery, 8, (1998), 203–208.
22. Furman M., S. Corbel, H. le Gall, O. Zahraa, M. Bouchy, Chemical Engineering Science 62, (2007), 5312–5316.
23. Furman M., S. Corbel, H. le Gall, O. Zahraa, M. Bouchy, Virtual Modeling and Rapid Manufacturing, (2005), 589–593.

24. Gerven T.V., G. Mul, J. Moulijn, A. Stankiewicz, Chemical Engineering and Processing 46, (2007), 781–789.
25. Griffith M.L., J.W. Halloran, Journal of the American Ceramic Society, 79 (10) (1996), 2601–2608.
26. Griffith M.L., J.W. Halloran, Manufacturing Science and Engineering, 2, (1994), 529–534.
27. Griffith M.L., T-M Chu, W. Wagner, J.W. Halloran, Solid Freeform Fabrication Conf. (SFF'95), (1995), 31–38.
28. Griffith, M. L. (1995), Stereolithography of ceramics, Ph.D. Thesis, University of Michigan, 1995.
29. Hinczewski C., S. Corbel, T. Chartier, Journal of the European Ceramic Society 18 (1998), 583–590.
30. Hinczewski C., S. Corbel, T. Chartier, Rapid Prototyping Journal, 4(3), (1998), 104.
31. Hinczewski, C., Corbel, S., Chartier, T, Stereolithography for the fabrication of ceramic three-dimensional parts, Rapid Prototyping Journal, 4(3), (1998), 104–111.
32. Ikuta, K., Hirowatari, K, Real three dimensional micro fabrication using stereolithography and metal molding, MEM'93, Proceedings An Investigation of Micro Structures, Sensors, Actuators, Machines and Systems. IEE. (http://dx.doi.org/10.1109/MEMSYS.1993.296949), Proceedings of IEEE Symposium, (1993), 42–47.
33. Im Y.G., S.I. Chung, J.H. Son, Y.D. Jung, J.G. Jo, H.D. Jeong, Journal of Materials Processing Technology 130–131, (2002), 372–377.
34. Jacobs P.F., Rapid Prototyping and Manufacturing: Fundamentals of Stereolithography (Society of Manufactuting Engineers Publishers, Deaborn, 1992).
35. Jacobs, P. F. (1996), Stereolithography and other RP&M technologies, Soc. of Manuf. Engineers, Dearborn (1996)
36. Jiang X.N., C. Sun, X. Zhang, B. Xu, Y.H. Ye, Sensors and Actuators A: 87, (2000), 72–77.
37. Kindernay J., A. Blazkova, J. Ruda, V. Jancovicova, Z. Jakubicova, Journal of Photochemistry and Photobiology A: Chemistry 151, (2002), 229–236.
38. Lee J.W., I.H. Lee, D.W. Cho, Microelectronic Engineering 83, Issues 4–9, (2006), 1830–1866.
39. Luo L., D. Tondeur, H. le Gall, S. Corbel, Applied Thermal Engineering 27, (2007), 1708–1714.
40. Luo L., D. Tondeur, International Journal of Thermal Science 44, (2005), 1131–1141.
41. Luo L., Y. Fan, W. Zhang, X. Yuan, N. Midoux, Chemical Engineering Science 62, (2007), 3605–3619.
42. Luo L., Z. Fan, H. le Gall, X. Zhou, W. Yuan, Chemical Engineering and Processing 47, (2008), 229–236.
43. Male, A. T., Chen, Y. W., Pan, C., Zhang, Y. M, Rapid prototyping of sheet metal components by plasma-jet forming, Journal of Materials Processing Technology, 135, (2003), 340–346.
44. Maruo S., S. Kawata, Journal of Microelectromechanical Systems, 7 (4), (1998), 411–415.
45. Mauzon A., O. Dufaud, H. Le Gall, S. Corbel, 10èmes Assises Européennes de Prototypage Rapide, 2004.
46. Monneret S., C. Provin, H. Le Gall, S. Corbel, Microsystem Technologies 8, (2002), 368–374.
47. Müller, H., Sladojevic, J, Rapid tooling approaches for small lot production of sheet-metal parts, Journal of Materials Processing Technology, 115, (2001), 97–103.
48. Nakagawa, T., Makinouchi, A., Wei, J., Shinizu, T, Application of laser stereolithography in FE sheet-metal forming simulation, Journal of Materials Processing Technology, 50, (1995), 318–323.
49. Nee A.Y.C., J.Y.H. Fuh, T. Miyazawa, Journal of Materials Processing Technology 113, (2001), 262–268.
50. Petzold R., H.F. Zeilhofer, W.A. Kalender, Computerized Medical Imaging and Graphics 23, (1999), 277–284.
51. Pilgrim S.M., A.E. Bailey, M. Massouda, F.C. Poppe, A.P. Ritter, Ferroelectrics 160, (1994) 383–390.
52. Provin C., S. Monneret, IEEE Transactions on Electronics Packaging Manufacturing 25, (2002) 59–63.

53. Schaeffer P., A. Bertsch, S. Corbel, J.Y. Jézéquel, J.C. Andre, Journal of Photochemistry and Photobiology A: Chemistry 107, (1997) 283–290.
54. Shinizu, T., Murakoshi, Y., Wang, Z., Maeda, R., Sano, T, Microfabrication techniques for thick structure of metals and PZT, Symposium on Design, Test and Microfabrication of MEMS and MOEMS, Paris, France, SPIE vol. 3680, (1999), 72–477.
55. Sun C., N. Fang, D.M. Wu, X. Zhang, Sensors and Actuators A (2005).
56. Sun C., X. Zhang, Journal of Applied Physics 92 (8), (2002) 4796–4802.
57. Tondeur D., L. Luo, Chemical Engineering Science 59, (2004) 1799–1813.
58. Voelkner, W, Present and future developments of metal forming: selected examples, Journal of Materials Processing Technology, 106, (2000), 236–242.
59. Wang, J., Wei, X P., Christodoulou, P., Hermanto, H. (2004), Rapid tooling for zinc spin casting using arc metal spray technology, Journal of Materials Processing Technology, 146, (2004), 283–288.
60. Wu, G., Langrana, N. A., Sadanji, R., Danforth, S. (2002), Solid freeform fabrication of metal components using fused deposition of metals, Materials and Design, 23, (2002), 97–105.
61. Yarlagadda, P., Ilyas, I. P., Christodoulou P. (1993), Development of rapid tooling for sheet metal drawing using nickel electroforming and stereolithography processes, Journal of Materials Processing Technology, 111, (2001), 286–294.
62. Zhang X., X.N. Jiang, C. Sun, Sensors and Actuators A: 77, (1999) 149–156.
63. Zissi S., A. Bertsch, J.Y. Jézéquel, S. Corbel, D.J. Lougnot, J.C. Andre, Microsystem Technologies 2, (1996) 97–102.

Chapter 7
Polymerizable Hydrogels for Rapid Prototyping: Chemistry, Photolithography, and Mechanical Properties

Jurgen Stampfl and Robert Liska

7.1 Introduction

7.1.1 Structure of Hydrogels

Hydrogels are cross-linked polymeric structures which are swollen by water [1, 2]. In a more general sense, these polymeric structures can contain solvents other than water, leading to the more general term "gel." Besides the polymer network and solvents, hydrogels can also contain particulate filler materials, typically ceramic particles. The functional and structural properties of hydrogels can be tailored quite easily, as the network density as well as the solvent content can be varied over a large range. The mechanical properties (especially the stiffness) of hydro(gels) are comparable to many biological tissues. Furthermore, the open network in combination with the mobile solvent molecules facilitates the diffusion of nutrients and dissolved gases, which makes hydrogels a widely used material in biomedicine, e.g. for the use in contact lenses [3], wound-healing bioadhesives, scaffolds for tissue engineering [4], and pharmaceutical hydrogel systems. Hydrogels are also used in a number of sensor applications, as the swelling behavior and diffusion coefficient of hydrogels depend on the ambient conditions [5].

Hydrogels can be classified into (1) homopolymer and, (2) copolymer hydrogels as well as (3) interpenetrating polymeric hydrogels. Homopolymer hydrogels are made of a single monomer type which, after crosslinking, makes up the polymeric network. If two (or more) monomers are used, a copolymer hydrogel is formed. Interpenetrating polymeric hydrogels are formed by swelling a first network with a monomer that, after polymerization, forms a network which penetrates the initial network. Hydrogels used for applications in rapid prototyping (RP) are typically either homopolymer or copolymer hydrogels.

J. Stampfl (✉)
Vienna University of Technology, Vienna, Austria
e-mail: jstampfl@pop.tuwien.ac.at

P.J. Bártolo (ed.), *Stereolithography: Materials, Processes and Applications*,
DOI 10.1007/978-0-387-92904-0_7, © Springer Science+Business Media, LLC 2011

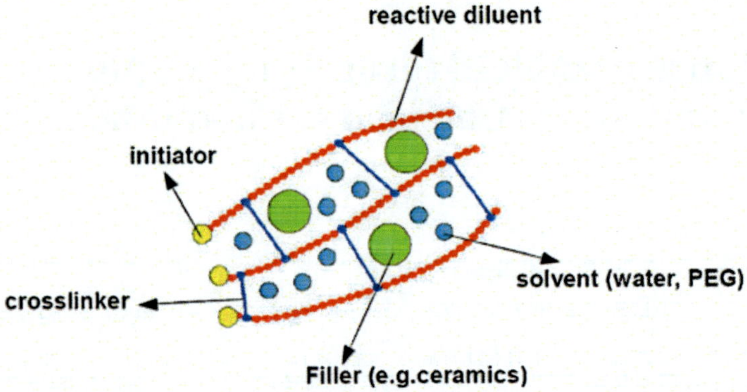

Fig. 7.1 Basic building blocks of photopolymerizable (hydro)gels

Several parameters can also be used to characterize the network structure of a hydrogel [6, 7]: The polymer volume fraction in the swollen state $v_{2,s}$ is one of the key parameters besides the molecular weight of the polymer chain between two adjacent crosslinks ($\overline{M_c}$). The mesh size ξ describes the average distance between two neighboring crosslinks and is a measure for the space which is available between the polymer chains. $v_{2,s}$ is determined by two factors: The chemical potential that drives the solvent into the gel and the elastic retractive forces exhibited by the polymer network which limits solvent absorption.

Artificial hydrogels, in the context of RP, are made of the several components (see Fig. 7.1). A photopolymerizable hydrogel formulation must at least contain a photoinitiator (PI), a multifunctional crosslinker and a solvent. Monofunctional reactive diluents and inorganic fillers might be used too. Reactive diluents allow the formation of copolymer hydrogels and, by adjusting the ratio between reactive diluent and crosslinker, the network density can be tailored quite easily. The viscosity of the monomer formulation, which is a critical process parameter for most lithography-based RP-processes, is also influenced by the amount of (typically low-viscosity) reactive diluents.

The versatility of hydrogels makes them an appealing class of materials for RP applications since:

- It is possible to fabricate parts with mechanical properties (especially elastic modulus) that resemble many biological tissues.
- Due to the availability of a large number monomers, the functional (biocompatibility and biodegradability) as well structural (elastic modulus, strength, and toughness) properties of hydrogels can be tailored over a large range. Especially for tissue engineering applications, hydrogels offer a large potential compared to currently used thermoplastic materials regarding the rate and type (enzymatic or hydrolytic) of biodegradation.
- Lithography based RP processes facilitate the fabrication of cellular structures with defined properties [8] which, in combination with hydrogels, allow the manufacture of parts with targeted mechanical and transport (drug delivery) properties.

7.1.2 Biological Hydrogels: Lessons from Nature

Tissue engineering scaffolds are one of the most promising applications for parts made by RP. Most of the recently done research in this field is focusing on hard tissues (e.g. bone), although the fraction of soft tissue in the human body is significantly larger. Regarding the number of surgeries, applications involving soft tissues (e.g. cartilage, tendon, skin, etc.) are very appealing, and this field of biomedicine could highly benefit from scaffolds with well-defined functional (biodegradeability, biocompatibility, cell adhesion, drug delivery, etc.) and structural properties like elastic modulus, strength, and strain to break. The number of patients in the USA receiving treatment for arthritis-associated or injury-related articular cartilage damage exceeds 600,000 persons per year, leading to costs of more than $15 billion [9].

It makes sense to have a look at the concepts involved in the formation of biological soft tissues, from both a scientific and an economic point of view. Cartilage is one of the tissues being investigated in high detail, thus providing lots of inspiration for the fabrication of artificial tissues.

Human cartilage is a hydrogel made of a crosslinked collagen network with mainly water filling the interstitial space between the collagen fibers. The nutrients required by the cells (chondrocytes) embedded inside the cartilage are provided by the synovial fluid, due to the lack of vascularization. The water content of cartilage is in the range of 65–85 wt%. To a small extent this water is bound inside the chondrocytes, to a larger content it fills the interstitial space between the fibres. Two-thirds of the solid phase of cartilage is made up of collagen. About one dozen different types of collagen were identified, and the collagen composition inside cartilage changes locally, also varying during lifetime. Type II collagen makes up 80–90% of the overall collagen. The individual collagen molecules form fibrils with varying thickness, and crosslinking these fibrils formed the network. Cartilage is mostly loaded in compression, and the impressive longevity of cartilage, even under high loads, can to a large extent be attributed to the proteoglycans [10, 11]. Proteoglycans are cationic macromolecules that inside the cartilage tissue are under high compressive stress. These proteoglycanes take up only 20% of the volume inside cartilage, which they would take up outside the swollen collagen network. The high compressive stress present inside the tissue prevents the initiation and progression of damage under the permanent cyclic loads acting on the cartilage. The sponge-like behavior of cartilage, during loading and unloading, is responsible for providing nutrients to the cells inside the cartilage. During compression synovial fluid leaves the tissue, and is replaced by fresh fluid during unloading. The complex structure of this collagen-proteoglycan-hydrogel in combination with the chondrocytes ability to regenerate damaged tissue are the main ingredients, which enable this soft tissue to survive millions of load cycles during several decades of lifetime.

7.2 Characterisation and Mechanical Testing of Hydrogels

On top of the hydrogel's capability to cover a wide range of functional properties, it is mainly their versatility regarding the mechanical response to outside loads that makes them a very appealing class of materials. In the context of tissue engineering scaffolds, the most interesting mechanical properties are the elastic modulus, which defines the stiffness of the final part, the strength, toughness, and viscoelastic (creep) properties.

The elastic properties (in this case the elastic modulus in unconfined compression E_c) of a gel (PEG-X-DA) can be tailored over a wide range by just changing the content of the solvent (PEG 400), as can be seen from Fig. 7.2. It can also be observed that even at moderate solvent-contents, the elastic modulus of a hydrogel is far lower than the modulus of the solvent-free polymer. Most thermosetting, solvent-free polymers have elastic moduli in the range of 1–3 GPa.

The measured moduli strongly depend on the utilized environmental parameters and the measuring method, due to the complex mechanisms occurring inside the hydrogel during mechanical loading. An example is illustrated in Fig. 7.3, where the compression modulus E_c of a PEG-X-DA (polyethylene glycol-X-diacrylate) hydrogel is measured in dependence of the compressive strain ε. E_c is highly dependent on ε, and a quantitative comparison of different materials is only possible if all relevant experimental parameters are known.

To be able to compare the elastic moduli measured with different methods, a systematic study was performed on thermally cured PEG-X-DA hydrogels. Cylindrical PEG-X-DA hydrogel samples (10 mm diameter and 20 mm height) were prepared and tested using various methods. On a universal testing machine, the

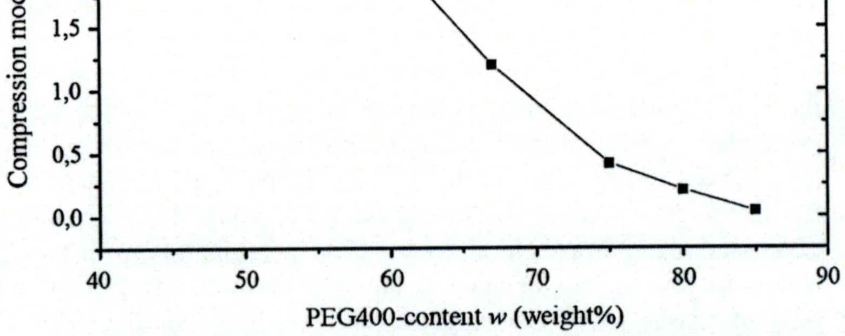

Fig. 7.2 Compression modulus E_c at 7% strain of PEG-X-DA PEG 400-gel [12]

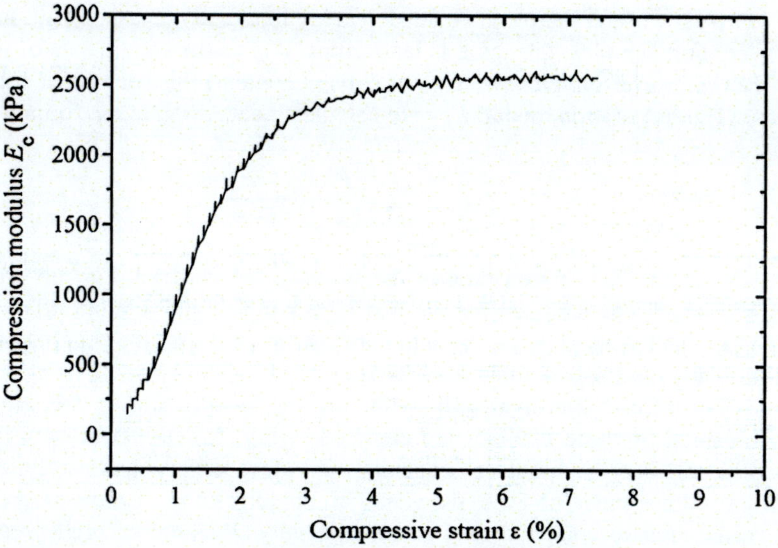

Fig. 7.3 Compression modulus E_c of PEG-X-DA-hydrogel depending on the compressive strain ε [12]

samples were tested in unixial, unconfined compression. On the same machine macroscopic indentation tests were performed using a spherical indenter. The used machine (Zwick Z050) was equipped with a 20 N-load cell of 1 mN-resolution, and the spherical indenter had a diameter of 3 mm.

Nanoindentation was also performed on a NANO Indenter XP-system (MTS Systems) with resolutions of 50 nN in force and of 0.02 μm in distances. The diameter of the utilized flat punch indenter was about 54 μm. Dynamic compression tests were performed on a 2980 Dynamic Mechanical Analyzer from TA Instruments. A static preforce was applied for dynamic compression, which deformed the sample to an average compression of around 7%, before loading the sample dynamically with a 10 μm amplitude at different frequencies.

In the case of static or dynamic uniaxial compression testing, the elastic modulus E_c can be calculated easily from the measured stress–strain-curve if the diameter d and height h of the cylindrical sample is known:

$$E_c = \frac{4h}{\pi \cdot d^2} \cdot \frac{\Delta F}{\Delta S_{cr}} \tag{7.1}$$

ΔS_{cr} is the crosshead travel and ΔF the load increase between two force measurements. $\Delta F/\Delta S_{cr}$ is therefore directly related to the slope of the stress–strain curve.

A critical issue in compression testing is the geometric precision of the sample. If the bottom and top surface of the cylindrical sample are not perfectly parallel, only parts of the sample are loaded, leading to a faulty measurement. In the case of very soft samples, this issue is less critical due to the compliance of the material. But for stiffer samples (with lower solvent content) this becomes a critical issue.

The indentation moduli E_i were calculated according to the formula

$$E_i = \frac{F \cdot (1 - \nu^2)}{2r_{act} \cdot s \cdot \kappa} \tag{7.2}$$

where, ν is the Poisson ratio of the investigated material. For this work, a value of $\nu = 0.4$ was assumed for the cartilage samples and a value of $\nu = 0.5$ for the PEG-DA hydrogel. r_{act} is the actual indentation radius, s the penetration depth, and κ a dimensionless scaling factor [13].

For PEG-X-DA-hydrogels with varying water content the elastic moduli obtained by these methods are depicted in Fig. 7.4. It can be seen that for very low and very high solvent contents the values differ significantly. For medium solvent contents (70–80%) the values are in quite good accordance. At low solvent values, macroscopic indentation and especially DMA deliver higher results than the other methods. At high solvent contents the moduli measured by nanoindentation are significantly above the other values.

The flat curve of the modulus measured by nanoindentation can be attributed to problems either with finding the interface of the specimen or with surface effects due to dehydration. In the case of very soft samples, it is not always clear when the

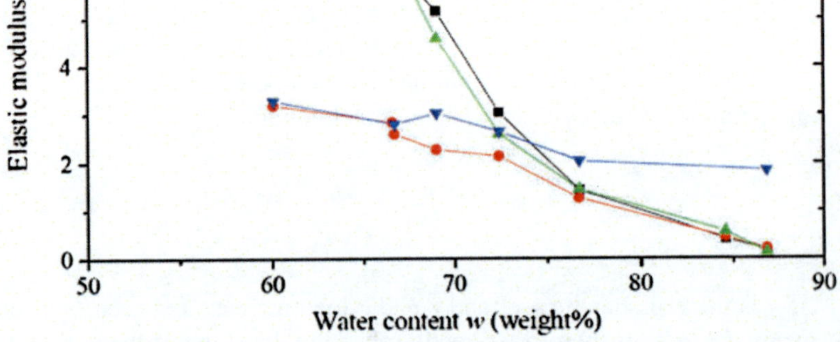

Fig. 7.4 Comparison of elastic moduli E measured by unconfined compression tests, macroscopic indentation using a spherical indenter, dynamic mechanical analysis, and nanoindentation [12]

indenter has hit the surface. Due to evaporating water, the effective water content of hydrogels with a high fraction of water is lowered at the surface, which can lead to an overestimation of the modulus.

The high values of the indentation moduli obtained by macro-indentation can be caused by the inhomogeneous strain inside the indentation zone. The elastic modulus is highly strain dependent, and regions with above-average strain will contribute significantly more to the measured modulus than regions with below-average strain, as shown in Fig. 7.3. A similar effect is probably responsible for the high DMA-moduli measured at low solvent contents. Certain spots are strained more severely than others due to slightly nonparallel top and bottom surfaces, which, once more, lead to an increased average modulus.

Overall it can be concluded that special care has to be taken if the elastic properties of hydrogels are measured and compared with values from data-sheets or literature. If it is possible to prepare specimen with sufficiently accurate geometry, compression testing delivers reliable and reproducible results. Especially in the case of natural hydrogels (e.g. cartilage) the preparation of standard specimen is often not possible. In this case macroindentation in combination with a spherical indenter is the method of choice. The load condition in macro-indentation is comparable to many physiologically relevant situations and the effort for specimen preparation can be kept reasonably small. Macro-indentation enables to obtain both elastic properties and creep data. Nanoindentation and compression DMA have to be performed with special attention, particularly in the case of fairly stiff materials (DMA) and very soft materials (nanoindentation).

In order to assess the suitability of a hydrogel for tissue engineering applications, the determination of the elastic properties will, in many cases, not be sufficient. A critical parameter for the evaluation of cartilage, for instance, is the viscoelastic properties, which determine the longevity of the tissue. The flow of synovial fluid under cyclic pressure is also responsible for providing nutrients to the cells inside the cartilage.

The relaxation time τ under load is defined as shown in Fig. 7.5, where E_{ini} is the initial modulus after loading the sample for the first time and E_{eq} is the long-term modulus of the material. τ is the time after which the modulus decreased to approx. 36.8% ($1/e$) of its initial value. Relaxation times can only be calculated for rather viscoelastic materials with a E_{ini}/E_{eq} of more than 2.7. Long relaxation times indicate that time dependent mechanisms (dissipation processes, fluid loss of the gel, ...) occur fairly slowly, leading to a high resistance against penetration and a fast regeneration after unloading.

In Table 7.1 elastic and viscoelastic properties of artificial (PEG-DA) and natural (human articular cartilage) are summarized. The cartilage samples were obtained in fresh condition and stored in phosphate buffered saline. The elastic moduli were measured using macroscopic indentation tests (spherical indenter). The initial modulus E_{ini} scatters significantly, depending on the sample, as can be seen from Table 7.1. The values of the PEG-DA hydrogel are within the range measured for the cartilage samples. The properties of the cartilage

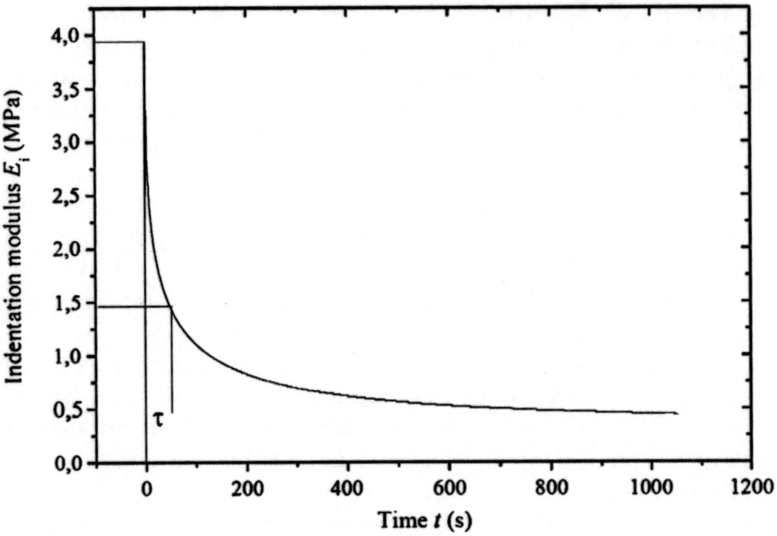

Fig. 7.5 Relaxation experiment (macroscopic indentation with spherical indenter) with human articular cartilage [12]

Table 7.1 Initial elastic modulus E_{ini}, equilibrium (long-term) modulus E_{eq} and relaxation time τ of articular cartilage (2P, 1P, 11) and PEG-DA hydrogel

Sample	Age (days)	E_{ini} (MPa)	E_{eq} (MPa)	τ (s)
2P	0 (53)	5.2 (9.3)	1.3 (0.8)	135 (9.8)
1P	1 (53)	6.8 (7.8)	2.1 (0.6)	45 (14)
11	1 (38)	8.2 (1.8)	0.5 (0.1)	21 (2.1)
PEG-DA 67% H_2O	–	3.7	2.7	>1,800

change after aging for several weeks, due to degradation processes. After the storage time E_{ini} increases in two cases, and drops in one case, it indicates that the determination of elastic constants is not sufficient for the assessment of the cartilage's quality. A more reliable parameter is the relaxation time τ, which significantly decreases for all cartilage samples during the storage time. It can also be seen that the τ values of the cartilage samples are much different from the τ of the PEG-DA-hydrogel, indicating that PEG-DA is probably no suitable material for cartilage replacement due to its viscoelastic properties, despite a similar initial elastic modulus.

7.3 Chemistry of Photopolymerizable Hydrogels

In stereolithography, the selection of the components for the formulation and the photopolymerization process are key factors in the design of hydrogel parts. Liquid resin formulations that contain monomers, solvents, and eventually fillers are cured by the aid of PIs. PIs form a large amount of initiating radicals when irradiated with light of particular wavelength. Those radicals start the free radical polymerization chain reaction and the liquid formulation is converted into the gel material (Fig. 7.6).

In the first step (*initiation*) (1), the PI dissociates into radicals ($Ri^{\bullet}$) due to the absorption of light. Those $Ri^{\bullet}$ add onto monomer molecules to start the chain reaction (2). During the chain growth reaction (*propagation*), the addition of other monomer molecules occurs whereas chain *termination* takes place, because of recombination or disproportionation (3). Oxygen can cause inhibition of the radical polymerization. It can be attached to the reactive ends due to its bi-radical character. As the oxygen centered radical is unable to initiate the polymerization, recombination reactions with growing radical chains terminate the polymerization (Fig. 7.7).

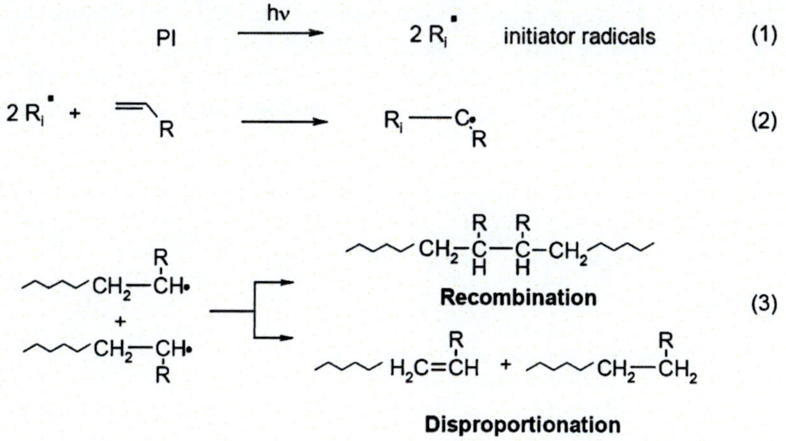

Fig. 7.6 Principle of photopolymerization

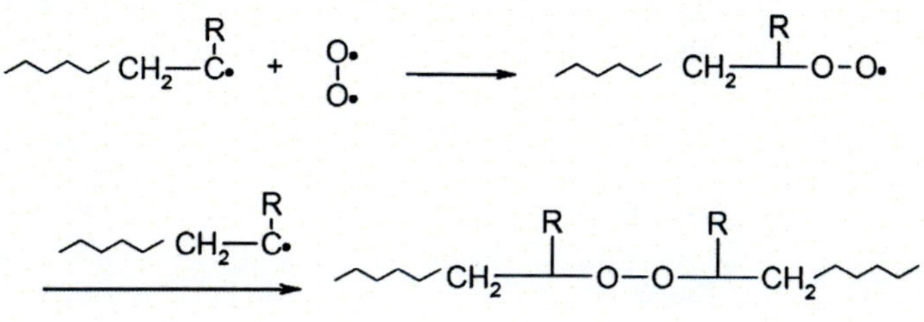

Fig. 7.7 Inhibition due to oxygen

Care must be taken to reduce the amount of oxygen present, since lower double bond conversion can result in migratables unwanted in biomedical applications.

The PI is the key compound of such formulations, because it converts UV–Vis radiation energy into chemical energy by forming radicals that start the polymerization. This can either happen through photo-fragmentation as a result of α-cleavage (Norrish Type I) or through hydrogen abstraction or electron transfer from a donor molecule (Norrish Type II). Typical examples for Type I initiators are benzoin ethers, dialkoxy acetophenones, hydroxyalkyl phenones, benzoylphosphine oxides and morpholino ketones. α-Cleavage of these initiators commonly takes place next to the carbonyl group of a benzoyl chromophore. Figure 7.8 shows the cleavage reaction of a frequently applied water-soluble initiator for hydrogels.

PIs forming radicals through hydrogen abstraction from a *co-initiator* belong to *Type II* initiators. They react in a bimolecular way by an electron–proton transfer from the co-initiator to the excited ketone. Examples of such PIs are benzophenone, thioxanthone, anthraquinone, xanthone, fluorenone, benzil, ketocoumarine, and camphorquinone. Amines, often used as suitable donors, transfer an electron to the excited ketone, forming a radical ion pair. Subsequent proton transfer forms reactive radicals, whereas in most cases only the amine based radical is able to initiate the polymerization. Figure 7.9 shows the mechanism of camphorquinone/ethyl dimethylamino benzoate well known from dental applications and often used for biomaterial applications, due to the low toxicity [15].

Since all the applied monomers are very reactive and polymerize easily owing to thermal stress, it is necessary to use *inhibitors* that scavenge accidentally formed

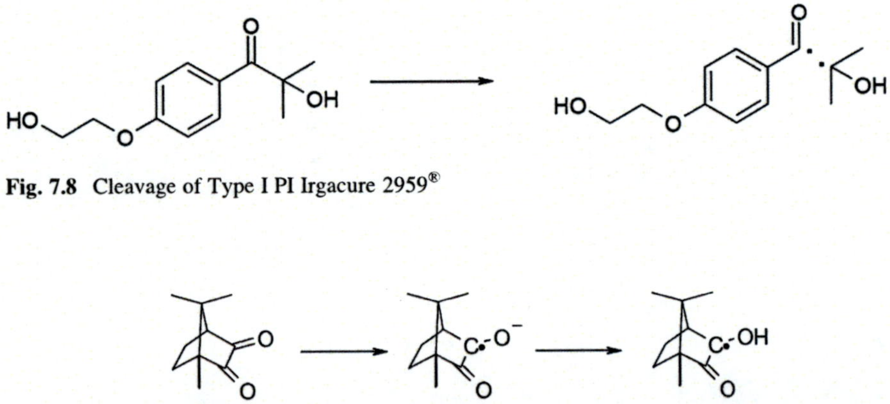

Fig. 7.8 Cleavage of Type I PI Irgacure 2959®

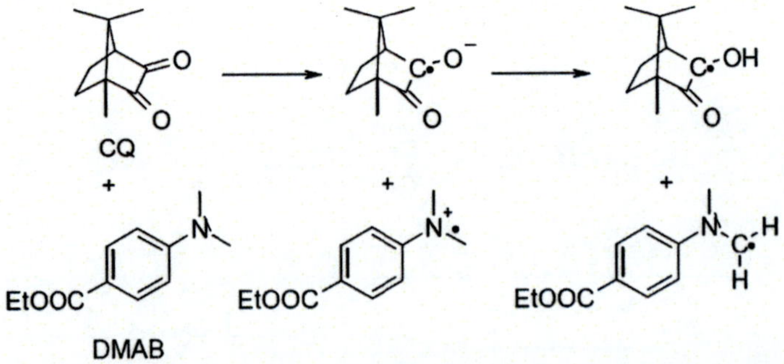

Fig. 7.9 Electron and proton transfer of type II initiators

Fig. 7.10 Function of MEHQ

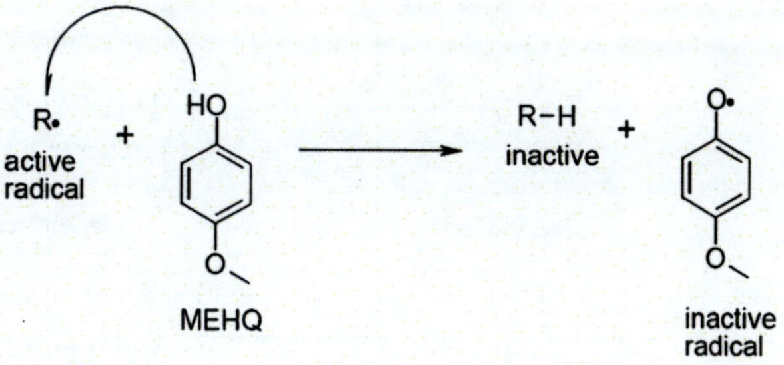

Fig. 7.11 Common monomer for photopolymerization

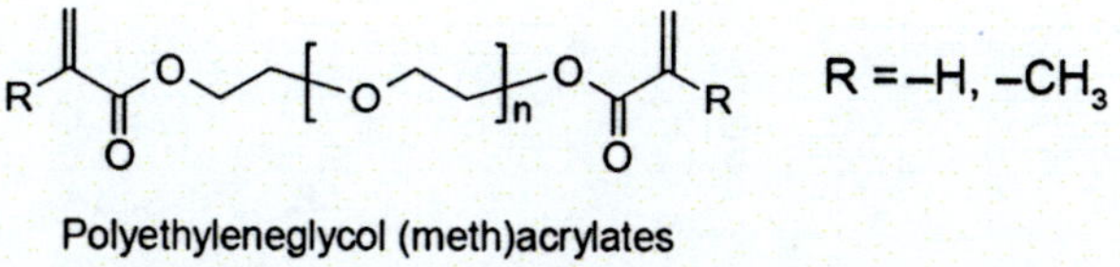

radicals before they start the polymerization. A simplified mechanistic scheme of a commonly used inhibitor hydroquinone monomethylether (MEHQ) is shown in Fig. 7.10.

Common monomers for hydrogels are well known from the coatings industry, and can be divided either into higher molecular weight oligomers with reactive side- or terminal groups or into low molecular reactive diluents with different numbers of reactive groups. The backbone of the reactive oligomers essentially determines the swellability of the gel. Flexible chains with low intramolecular interaction are necessary. Polyethylenegycols with water soluble backbone fulfill these requirements, so are mainly used in photopolymerizable hydrogels. Polyethyleneglycols are prepared by ring opening polymerization of ethylene oxide to give oligomers and polymers with tuneable molecular weight that are commercially available up to 35,000 g/mol. Conversion with (meth)acrylic acid chloride leads to a broad series of commercially available monomers (Fig. 7.11).

Acrylates as reactive group are characterized by a high reactivity towards radical polymerization. Low storage stability and significant irritancy, due to the Michael addition reactions with amino-groups from proteins, are some general drawbacks. Methacrylates circumvent these due to the presence of the methyl-group with its sterical hindrance. Both reactive groups are bound to the oligomer backbone via an ether group that can be considered as degradable. If non degradable systems are desired, acrylamides show significant higher stability towards hydrolysis and present a higher reactivity than acrylates, due to some kind of preorganization by hydrogen bridging. However, the toxicity of the monomer is often an issue. Alternative reactive groups in photopolymerization are maleimides, diacrylamides, and fumarates, but only the latter group was used once in hydrogel systems. Especially maleimides suffer from significant toxicity. Similarly to the field of biocompatible photopolymers, only a limited number of reactive groups exist, and

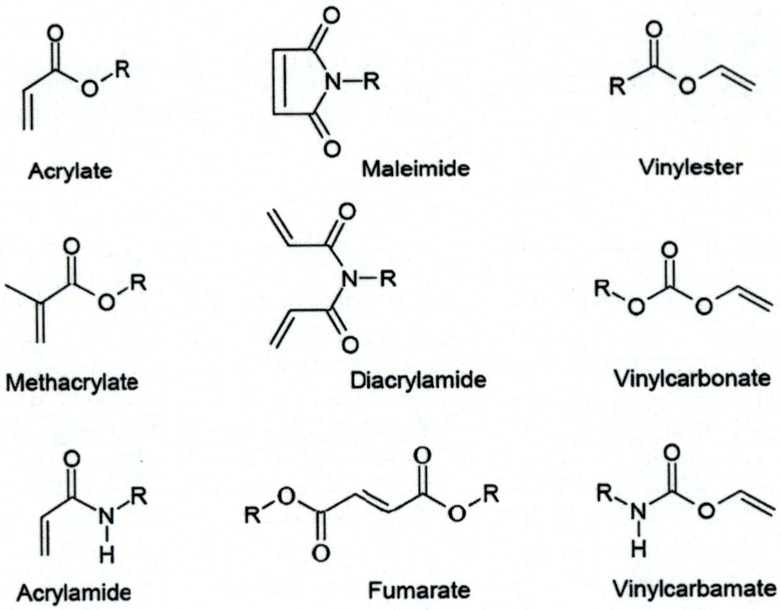

Fig. 7.12 Functional groups in photopolymerization

all of them have their own benefits and drawbacks, alternative functional groups that can be cured by radical polymerization are interesting. For instance, vinyl acetate can easily be polymerized resulting in a polymer that is well known for its low toxicity and usage in the food industry. Unfortunately, only few monomers are commercially available, mainly mono-vinyl esters, due to the high sensitivity of all known synthetic routes towards functional groups. Additionally, it is well known that significantly more chain transfer reactions occur compared to acrylates during polymerization. Interestingly, very little is known on their capability of photopolymerization (Fig. 7.12).

Another disadvantage of crosslinked acrylates regards their degradation products based on polyacrylic acid (Fig. 7.13), which is rather stable towards biodegradation and cannot easily be transported within the human body. This leads to a local decrease of pH and an adverse effect on cells might occur. On the opposite, poly(vinyl esters) form less harmful poly(vinyl alcohol) as degradation product, so it is a widely used product in the food industry. Acidic decomposition products, such as adipic acid, have low molecular weights and can be easily transported. Similarly to fatty acids, these compounds can be metabolized.

Based on these alternative photoreactive monomers for biomedical applications, it has been of interest to gain more readily cleavable groups, as all of these crosslinked systems have generally significantly slower degradation rates than linear polymers. Vinyl carbonates are a promising group to meet these requirements, and they can be synthesized directly from alcohols and vinyl chloroformate. A similar conversion with amines, instead of alcohols, leads to nearly nondegradable

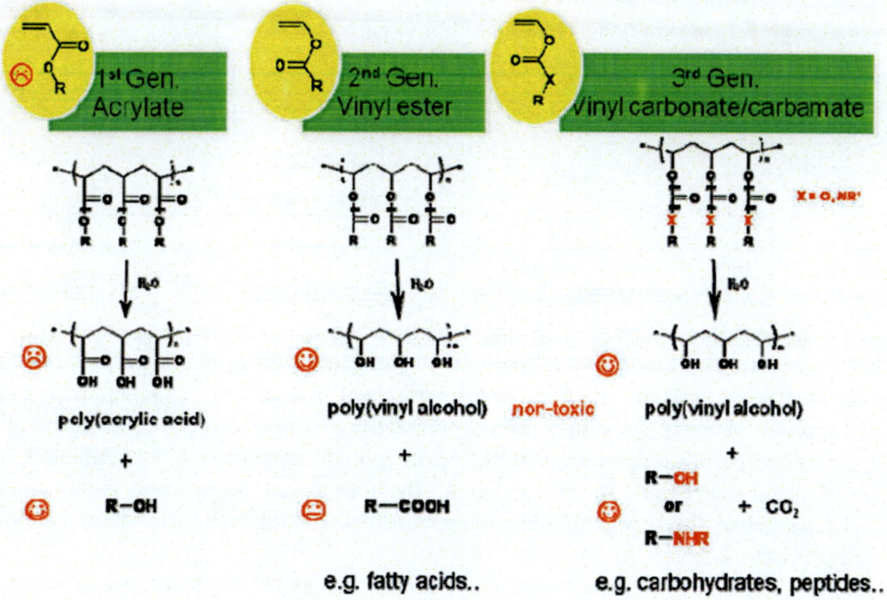

Fig. 7.13 Degradation of polymers

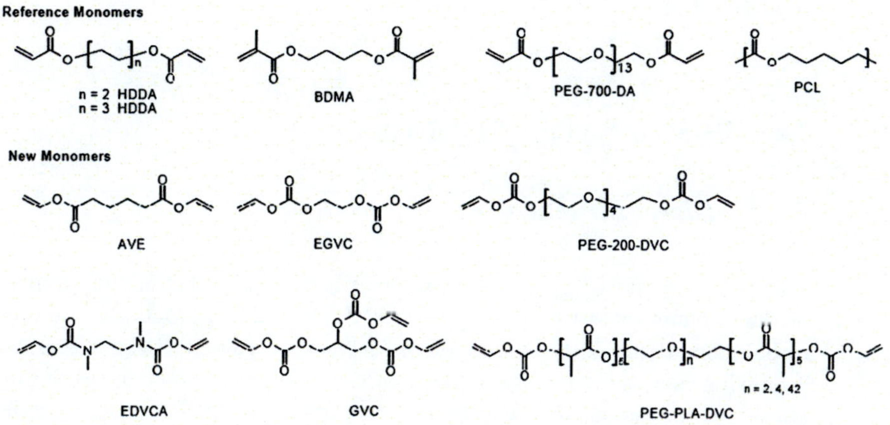

Fig. 7.14 Monomers based on vinyl esters, -carbonates and -carbamates

materials that might show good cell adhesion due to the carbamate group. Figure 7.14 shows a small selection of monomers synthesized very recently.

Recently, it has been shown that the photopolymerization speed lies between acrylates and methacrylates, giving double bond conversion very similar to that of the reference compounds. The mechanical properties (measured by nanoindentation)

Table 7.2 In vitro cytotoxicity with osteoblast and endothelial cells

Monomer	In vitro LC50 (mM)
Acrylate	<0.01
Methacrylate	~1
Vinyl ester	>1
Vinyl carbonate	>10
Vinyl carbamate	>10

are mainly determined by the backbone of the oligomers as expected, and are only influenced to a small extent by the functional group. To determine possible cytotoxic effects of free, nonpolymerized monomeric residues that might be released from the grafts after implantation, the influence of increasing doses of monomeric compounds was tested in human umbilical vein endothelial cell and osteoblast (MG-63) cell cultures. In both cases, significant higher monomer-tolerance (Invitro-LC50) has been observed for the new generation of monomers (Table 7.2).

Especially in the case of vinyl carbonates and carbamates, no cytotoxic effect was found up to the limit of the monomer solubility in the cell culture medium. Recent degradation studies showed that degradation rates of pure cross-linked vinyl carbonates (PEG-200-DVC) present the same order of magnitude as PLA, and can be easily tuned by the addition of low amounts of more stable vinyl esters and vinyl carbamates.

7.4 3D-Structuring of Hydrogels

7.4.1 Direct Structuring

The versatility of hydrogels regarding their functional and structural properties can be further enhanced by shaping the materials, using lithographic or dispensing techniques. Particularly, tissue engineering and guided tissue repair are very promising new areas for the use of hydrogels. Although human tissue contains a considerable amount of water, the major emphasis of the RP-based scaffold fabrication was placed upon melt and powder processing of biodegradeable thermoplastic materials. Recently, photopolymers were also systematically investigated regarding their suitability as substrate for tissue engineering [14, 15]. Hydrogel parts can be manufactured with reasonable effort by incorporating water into photoreactive resins. There are two possibilities of incorporating cells into the scaffolds: (a) seeding of cells onto the surface of the scaffold subsequently to scaffold fabrication and (b) the incorporation of cells into the scaffold fabrication process. Lithography based processes usually rely on the first method, whereas dispensing systems are also capable of directly printing cells into hydrogels. Hydrogels have been recently investigated regarding photocurable formulations for tissue engineering [16–18]. The focus of these studies was to achieve the required microporosity by templating with sacrificial porogens that can be washed out of the final scaffold. Simple three-dimensional shapes could be manufactured using conventional lithographic techniques allowing the fabrication of microfluidic devices. To further increase the achievable shape complexity these photo-curable hydrogels need to be modified to make them suitable for RP-techniques.

7.4.1.1 Lithography Based Direct Structuring

Lithography based RP-processes are an appealing shaping technology for hydrogels owing to several reasons. Using the right PI and monomer system (see Sect. 7.3), a wide variety of hydrogel systems can be structured by lithographic methods. Scaffolds for the cultivation of cells typically require pores in the range of 5–500 μm, so lithography based methods are currently the only RP-techniques capable of achieving such feature resolutions. Several groups used microstereolithography for shaping photopolymerizable hydrogel systems [19–21]. Feature resolutions down to 5 μm with a minimal layer thickness of 10 μm can be achieved using such laser-based SLA-systems [22]. SLA-systems typically expose the resin from above, so air is in permanent contact with the polymerizable formulation. In the case of water-containing resins, this leads to evaporation of water, and consequently to a continuously falling resin level, which can be avoided by using a resin containing a solvent with a high boiling point (e.g. poly (ethylene glycol) – PEG). After the part has been built, it is taken out of the resin tank and the PEG is exchanged by water, leading to a hydrogel structure (Fig. 7.15). The resin formulation for these parts contained cyano ethyl acrylate, PEG-DA, and PEG 300 in a composition of 9:1:10. If PEG is exchanged by water, the originally transparent parts become opaque. If undercuts with good geometric resolution are to be fabricated, absorbers have to be added to the resin. These absorbers limit the penetration depth of the incoming laser light to the targeted layer thickness.

If conventional stereolithography is not sufficient regarding the required feature resolution, two-photon polymerization (2PP) is an interesting alternative. By exchanging the UV sensitive initiator with an appropriate two-photon-initiator [23, 24], hydrogels can be structured using 2PP [25]. On top of the higher achievable feature resolution, 2PP offers another benefit. 2PP is capable of writing inside the photosensitive resin in contrast to all other RP-techniques, which add material layer by layer. This would allow the in vivo manufacturing of functionalized structures in living tissue. In vivo structuring is conceivable by using the low-toxicity-resins described in Sect. 7.3 or photopolymerizable hydrogels, which are used as barrier coatings [18] for the prevention of thrombosis and restenosis after vascular injury.

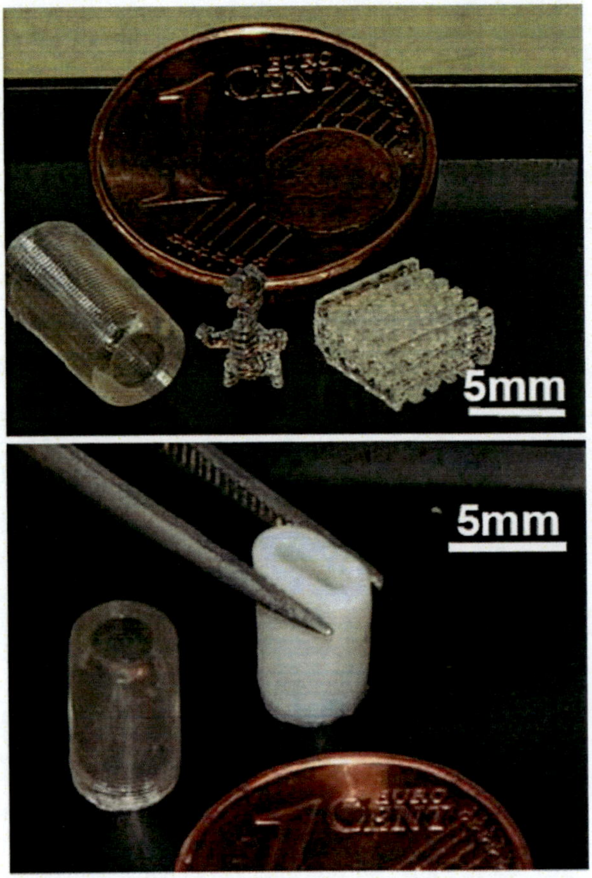

Fig. 7.15 Hydrogel parts made by microstereolithography

7.4.1.2 Bioplotter

Many hydrogel materials, which are used in biomedical applications and for cell culture experiments (e.g. gelatine, agar, agarose, etc.) cannot be structured by lithographic RP-methods without chemical modification. To make these materials accessible [26, 27] to RP techniques, a bioplotter was developed allowing the extrusion of a gel-based plotting material into a liquid plotting medium. Hydrogel formation in such a system can be achieved by chemical reaction of materials, dispensed by either by a two-component dispenser or by plotting one component into a liquid medium containing a co-reactive component. Alternatively, single component thermoreversible hydrogels can be plotted at elevated temperatures into a liquid medium with temperatures below the gelation temperature. Materials suitable for this physical gelation process include agar and gelatine.

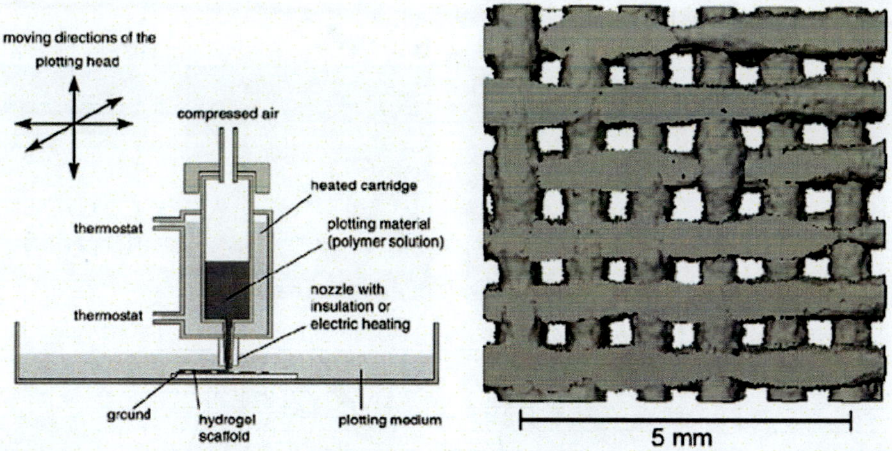

Fig. 7.16 Working principle of a bioplotter (*left*) and visualization of an agarose based scaffold by computer tomography (*right*) [26]

The basic working principle of such a bioplotter is depicted in Fig. 7.16. A heated cartridge contains a liquid polymer solution, which is extruded into a plotting medium by applying pressure inside the cartridge. The cartridge is moved by a X-Y-Z stage through the plotting medium, and during extrusion physical or chemical solidification of the plotting material takes place. A computer-tomography image of a scaffold fabricated by the bioplotter technique is displayed in Fig. 7.16 (right). The achievable feature resolution is not as good as with other RP-techniques. Due to the soft nature of the plotting material, delicate features are hard to obtain. In the context of tissue engineering, the main advantage of the bioplotter can be attributed to the mild conditions under which the solidification of the gels takes place.

7.5 Indirect Structuring: Gelcasting

Gelcasting [28] in combination with RP [30, 31] has been used for the indirect fabrication of ceramic parts. It is possible to fabricate fully dense (>99.5% of the theoretical density) ceramic parts made of a large number of different ceramic powders, including alumina, silicon nitride, silicon carbide, and zirconia, using gelcasting. The gelcasting process is based in an highly filled ceramic slurry, which, in addition to the ceramic powder, contains a solvent (typically water) and dissolved monomers (Fig. 7.17). The slurry contains a thermal initiator and dispersing agents used to further increase the solid loading. A high solid loading will

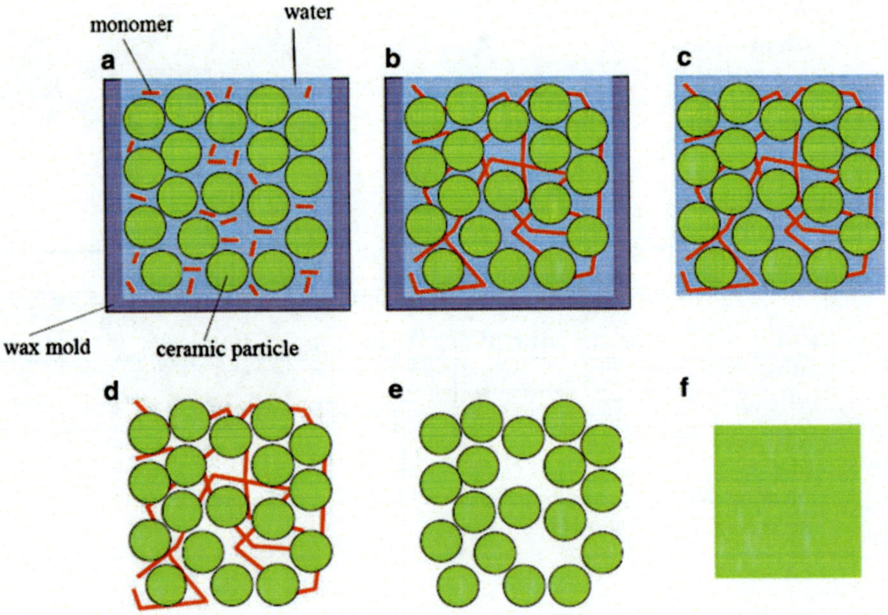

Fig. 7.17 Processing steps for gelcasting

enable to reduce the linear shrinkage after sintering, and depending on the used ceramic particles, it will be possible to achieve values of 40–60 vol%.

The gelcasting process relies on a mold into which liquid slurry is cast. If parts with complex geometries need to be fabricated, sacrificial molds must be used, which can be manufactured by RP processes like wax-ink-jetting (e.g. Solidscape Modelmaker) or MoldSDM [29]. The processing steps, based on sacrificial wax molds made by RP, are depicted in Fig. 7.17. The wax mold is filled with the liquid slurry, with a typical composition of 45 vol% ceramics, 45 vol% water and 10 vol% organics (monomer, dispersing agent, and initiator). After casting (Fig. 7.17a), the slurry is cured at a slightly elevated temperature (50–70°C) until the slurry has solidified, due to the formation of a highly crosslinked polymer-network (Fig. 7.17b), leading to the formation of a ceramic-filled hydrogel. By further increasing the temperature (Fig. 7.17c) to around 100°C, the wax starts to melt of and the green ceramic part is exposed.

By further heating up the green part, the water inside the hydrogel starts to evaporate until the green part is dry (Fig. 7.17d). The evaporated water leaves pores facilitating the debinding of the part (Fig. 7.17a). During the debinding, the temperature is ramped up to around 500°C until all organic constituents have been decomposed, leaving a so-called brown part (Fig. 7.17e), which can be sintered to obtain the final dense ceramic part (Fig. 7.17f).

Drying and debinding are critical steps, as cracks can occur severely degrading the mechanical properties in the final part. Gelcasting is capable of

Table 7.3 Formulation for gelcasting of SiC ceramics [30]

Amount	Substance	Role
10 g	Methacrylic acid amide (MAM)	Monomer
3.33 g	*N*,*N*′-Methylenbisacrylic acid amide (MBAM)	Crosslinker
50 g	Distilled water	Solvent
2 g	Dispersion agent	
300 g	SiC	Ceramic powder
350 μl	10% solution of Ammoniumpersulfate in water	Initiator
35 μl	*N*,*N*,*N*′,*N*′-Tetramethylethylenediamine	Catalyst

fabricating fairly thick-walled parts compared to other ceramic forming techniques like powder injection molding, as the evaporated solvent leaves a porous structure facilitating debinding. The part will shrink during sintering, depending on the solid loading of the original slurry. The linear shrinkage is around 20%, for solid loadings of 45–55 vol% and care has to be taken to avoid sagging and warpage during the sintering step. To facilitate drying and debinding, the amount of reactive organic components should be kept small, so it is important to choose highly reactive monomers in order to have strong and stiff green parts in reasonable time.

A typical formulation for gelcasting is presented in Table 7.3. Depending on the relation of the reactive diluent (monofunctional) and the crosslinker (multifunctional), the stiffness of the final gel can be tailored. If the gel is too soft, the structural integrity of the green part can get lost. If the gel is too stiff, the green part might start to crack during drying or debinding. Since gelcasting is especially suitable for the fabrication of complex thin-walled parts, the viscosity of the slurry is a critical parameter. If the mold has to be completely filled, the viscosity has to be minimized by using appropriate dispersion agents and suitable powders, allowing for a high solid loading at a given viscosity. If the slurry constituents of are properly chosen, the viscosity of the slurry is usually between 2 and 5 Pa s at shear rates of 1 s^{-1}. In this viscosity range, complex wax molds can be filled, if casting is done in a vacuum chamber.

Sample parts made using RP-wax molds are depicted in Fig. 7.18. Thin walled parts (wall thickness down to 0.4 mm) with severe undercuts can be manufactured as illustrated. This clearly indicates that gelcasting in combination with RP widens the available design space for ceramic parts. A further gelcasting benefit is the possibility to process many different ceramic powders, in contrast to lithography-based direct RP techniques, which only work in combination with translucent, non-absorbing powders. Gelcasting presents no problems with interlayer-bonding, which decreases the achievable mechanical properties in many direct RP-methods. Drawbacks of RP are related to the larger number of processing steps, compared to direct methods. Special care must be taken to avoid pores and voids during the casting process, especially if thin-walled parts have to be manufactured.

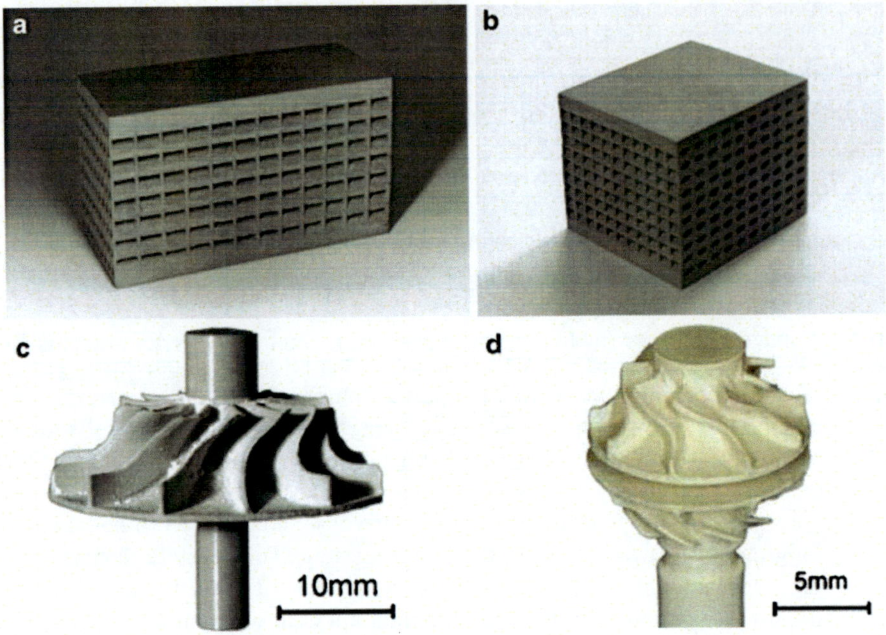

Fig. 7.18 Gelcast parts made of SiC with wall thickness of 0.5 mm (**a**) and 0.4 mm (**b**), from [30]. A sintered Si_3N_4 turbine is shown in (**c**). (**d**) depicts a Si3N4 turbine in the green state, from [31]

References

1. B.D. Ratner, F.J. Schoen, J.E. Lemons, Biomaterials science: An introduction to materials in medicine, Chapter 2 Classes of materials used in medicine, Elsevier Academic Press, New York, 1996.
2. N.A. Peppas, Hydrogels in medicine and pharmacy, CRC Press, Boca Raton, 1987.
3. N.A. Peppas, W.H.M. Yang, Properties-based optimization of the structure of polymers for contact lens applications, *Contact Intraocular Lens Medical Journal* 7, (1981) 300–321.
4. A.J. Engler, S. Sen, H.L. Sweeney, D.E. Discher, Matrix elasticity directs stem cell lineage specification, *Cell* 126 (4), (2006) 677–689.
5. Q.T. Trinh, G. Gerlach, J. Sorber, K.-F. Arndt, Hydrogel-based piezoresistive pH sensors: Design, simulation and output characteristics, *Sensors and Actuators B: Chemical* 117 (1), (2006) 17–26.
6. N.A. Peppas, J.Z. Hilt, A. Khademhosseini, R. Langer, Hydrogels in biology and medicine: From molecular principles to bionanotechnology, *Advanced Materials* 18, (2006) 1345–1360.
7. S.J. Bryant, K.S. Anseth, Photopolymerization of hydrogel scaffolds, in: P.X. Ma and J.Elisseeff (eds.) Scaffolding in tissue engineering, Dekker, New York, 2005.
8. M.H. Luxner, J. Stampfl, H.E. Pettermann, Numerical simulations of 3D open cell structures – Influence of structural irregularities on elasto-plasticity and deformation localization, *International Journal of Solids and Structures* 44, (2007) 2990–3003.
9. K.L. Spiller, S.J. Laurencin, D. Charlton, S.A. Maher, A.M. Lowman, Superporous hydrogels for cartilage repair: Evaluation of the morphological and mechanical properties, *Acta Biomaterialia* 4, (2008) 17–25.

10. C.G. Armstrong, V.C. Mow, Variations in the intrinsic mechanical properties of human articular cartilage with age, degeneration and water content, *The Journal of Bone and Joint Surgery* 64A, (1982) 88–91.
11. G.E. Kempson, H. Muir, S.A. Swanson, M.A.R. Freeman, Correlations between stiffness and chemical constituents of cartilage on the human femoral head, *Biochimica et biophysica acta* 215, (1970) 70–77.
12. S. Gäbler, Viskoelastische eigenschaften von hydrogelen, Diploma thesis, Friedrich Schiller Universität Jena, 2008.
13. W.C. Hayes, L.M. Keer, G. Herrmann, L.F. Mockros, A mathematical analysis for indentation tests of articular cartilage, *Journal of Biomechanics* 5, (1972) 541–551.
14. M. Schuster, C. Turecek, B. Kaiser, J. Stampfl, R. Liska, F. Varga, Evaluation of biocompatible photopolymers I: Photoreactivity and mechanical properties of reactive diluents, *Journal of Macromolecular Science* A44, (2007) 547–557.
15. M. Schuster, C. Turecek, F. Varga, H. Lichtenegger, J. Stampfl, R. Liska, 3D-shaping of biodegradable photopolymers for hard tissue replacement, *Applied Surface Science* 245, (2007) 1131–1134.
16. S.J. Bryant, J.L. Cuy, K.D. Hauch, B.D. Ratner, Photo-patterning of porous hydrogels for tissue engineering, *Biomaterials* 28 (19), (2007) 2978–2986.
17. H.M. Simms, C.M. Bowman, K.S. Anseth, Using living radical polymerization to enable facile incorporation of materials in microfluidic cell culture devices, *Biomaterials* 29 (14), (2008) 2228–2236.
18. K.T. Nguyen, J.L. West, Photopolymerizable hydrogels for tissue engineering applications, *Biomaterials* 23 (22), (2002) 4307–4314.
19. K. Arcaute, L. Ochoa, F. Medina, C. Elkins, B. Mann, R. Wicker, Three-dimensional PEG hydrogel construct fabrication using stereolithography, Mater. Res. Soc. Symp. Proc. Vol. 874 (2005) Materials Research Society L5.5.1.
20. R. Wicker, F. Medina, K. Arcaute, L. Ochoa, C. Elkins, Hydrogel constructs using stereolithography, WO/2006/116180.
21. J. Stampfl, S. Baudis, C. Heller, R. Liska, A. Neumeister, R. Kling, A. Ostendorf, M. Spitzbart, Photopolymers with tunable mechanical properties processed by laser based high-resolution stereolithography, *Journal of Micromechanics and Microengineering* 18, (2008) 125014.
22. A. Neumeister, R. Himmelhuber, C. Materlik, T. Temme, F. Pape, H.-H. Gatzen, A. Ostendorf, Properties of three-dimensional precision objects fabricated by using laser based micro stereolithography, *Journal of Laser Micro/Nanoengineering* 3 (2), (2008) 67–72.
23. C. Heller, N. Pucher, B. Seidl, L. Kuna, V. Satzinger, V. Schmidt, H. Lichtenegger, J. Stampfl, R. Liska, One- and two-photon activity of cross-conjugated photoinitiators with bathochromic shift, *Journal of Polymer Science: Part A: Polymer Chemistry* 45, (2007) 3280–3291.
24. B.H. Cumpston, S.P. Ananthavel, S. Barlow, D.L. Dyer, J.E. Ehrlich, L.L. Erskine, A.A. Heikal, S.M. Kuebler, I.-Y.S. Lee, D. McCord-Maughon, J. Qin, H. Röckel, M. Rumi, X.-L. Wu, S.R. Marder, J.W. Perry, Two-photon polymerization initiators for three-dimensional optical data storage and microfabrication, *Nature* 398, (1999) 51–54.
25. T. Watanabe, M. Akiyama, K. Totani, S.M. Kuebler, F. Stellacci, W. Wenseleers, K. Braun, S.R. Marder, J.W. Perry, Photoresponsive hydrogel microstructure fabricated by two-photon initiated polymerization, *Advanced Functional Materials* 12 (9), (2002) 611–614.
26. R. Landers, U. Hübner, R. Schmelzeisen, R. Mülhaupt, Rapid prototyping of scaffolds derived from thermoreversible hydrogels and tailored for applications in tissue engineering, *Biomaterials* 23, (2002) 4437–4447.
27. N.E. Fedorovich, J.R. De Wijn, A.J. Verbout, J. Alblas, W.J.A. Dhert, Three-dimensional fiber deposition of cell-laden, viable, patterned constructs for bone tissue printing, *Tissue Engineering Part A* 14 (1), (2008) 127–133.
28. M.A. Janney, O.O. Omatete, C. Walls, S.D. Nunn, R.J. Ogle, G. Westmoreland, Development of low-toxicity gelcasting systems, *Journal of the American Ceramic Society* 81 (3), (1998) 581–591.

29. A.G. Cooper, S. Kang, J.W. Kietzman, F.B. Prinz, J.L. Lombardi, L.E. Weiss, Automated fabrication of complex molded parts using mold shape deposition manufacturing, *Materials & Design* 20 (2–3), (1999) 83–89.
30. J. Homa, Direkte und indirekte Strukturierung von keramischen Kreuzkanalfiltern, PhD thesis, TU Wien, 2008.
31. J. Stampfl, H.C. Liu, S.W. Nam, K. Sakamoto, H. Tsuru, S. Kang, A.G. Cooper, A. Nickel, F. B. Prinz, Rapid prototyping and manufacturing by gelcasting of metallic and ceramic slurries, *Materials Science and Engineering* A334, (2001) 187–192.

Chapter 8
Simulation Methods for Stereolithography

Benay Sager and David W. Rosen

8.1 Introduction

Stereolithography (SL) is a layered, additive manufacturing process in which an ultra-violet (UV) laser is used to selectively cure a liquid photopolymer resin in order to physically fabricate a part. Traditional SL systems use a UV laser with galvanometer-driven mirrors to scan a particular cross-section on the build surface. The limits of the resolution, both theoretical and empirical, need to be established so that accuracy and surface finish of SL-fabricated parts can be predicted.

In this chapter, we present two new models of the SL process. The first is an extension of the well known analytical irradiance model for SL that better captures the effects of an angled laser beam that draws finite-length scans. The second models the chemical reactions taking place in SL resins and yields estimates of degree of cure (DOC), polymer concentration, monomer concentration, and temperature distributions over time. An additional type of SL simulation model is discussed briefly that promises higher fidelity through a kinetic Monte Carlo (KMC) approach to modeling SL chemical reactions.

8.2 Analytical SL Irradiance Model

8.2.1 Motivation

For modeling purposes, some general behavior of the laser beam has been widely assumed. Traditional SL laser beam modeling assumes that the laser beam has a constant diameter and irradiance profile, and is always perpendicular to the build surface. While this general behavior is a simplified form of the laser beam

D.W. Rosen (✉)
The Woodruff School of Mechanical Engineering, Georgia Institute of Technology, Atlanta, GA 30332-0405, USA
e-mail: david.rosen@me.gatech.edu

P.J. Bártolo (ed.), *Stereolithography: Materials, Processes and Applications*,
DOI 10.1007/978-0-387-92904-0_8, © Springer Science+Business Media, LLC 2011

Table 8.1 Analytical modeling assumptions for SL laser beam

Parameter/ phenomenon	Standard model	Extended model
Beam diameter	Beam has constant diameter and circular shape	Beam size and shape changes according to location on surface
Beam irradiance	Constant Gaussian irradiance profile	Irradiance profile changes depending on the point of focus
Beam angle	Laser beam is always perpendicular to build surface	Laser beam angle with vertical changes during scanning
Refraction	Refraction does not affect cure profile	Refraction changes size, shape, and location of cure profile

characteristics, it does not reflect the dynamic nature of the SL build process, and hence resulting cured parts. These assumptions and their characteristics will be called the Standard Model in Table 8.1. The standard model was first presented in 1992 by Paul Jacobs and Thomas Pang [1] and has served the SL industry well as evidenced by the longevity of 3D Systems' SL machine product line.

New applications for SL will emerge if resolution and surface finish can be improved, without impacting build time and cost. Hence, it is imperative to know the exact cure profile of individual scan vectors, as well as entire parts. The standard model ignores important phenomena that result in different cured shapes in different parts of the SL vat, such as the angle that the laser beam strikes the vat surface. It is well known that surface finish is highly dependent on the angle of the surface during the SL build (Reeves and Cobb 1997) [2], but also depends on the surface's location in the vat.

Surprisingly little literature exists on the laser–resin interactions in SL. Some researchers extended the standard model to include aspects of laser beam focusing [2]. Others have investigated the chemical curing reactions and their influence on residual stresses, warpage, and other shape errors [3, 4], or their influence on temperature distributions in the vat and cured shapes [5, 6]. The research on curing reactions did not incorporate any laser phenomena beyond the standard model. Wiedemann et al. [7] modeled the differences in UV absorption between liquid and solid resin, and applied the model to the estimation of part distortion.

After summarizing the standard model, we present a new extended analytical model of SL irradiance that incorporates the characteristics presented in the rightmost column in Table 8.1. A simulation model of the SL cure process is developed and demonstrated in Sect. 8.3. A series of experiments is presented in Sect. 8.4 that tests various aspects of the dynamic model. Discrepancies between predictions and experimental results are highlighted and explained.

8.2.2 *Standard Cure Model*

Central to SL exposure and cure modeling is the irradiance profile of the laser beam used in the process. Irradiance is the radiant power of the laser per unit area (mW/mm^2), and is often denoted by $H(x, y, z)$. For the purposes of this study, the laser beam irradiance profile is considered to have Gaussian distribution. A Gaussian laser of power P_L incident on the vat surface has an irradiance profile of

$$H(r,0) = \frac{2P_L}{\pi W_0^2} e^{-2r^2/W_0^2} \tag{8.1}$$

where, r is the radial distance from the laser beam's center and W_0 is the beam waist radius. The coefficient in front of the exponential term, which is the maximum irradiance, is given the symbol H_0.

The exposure at a point in the vat is obtained by integrating the irradiance over time.

$$E(x,y,z) = \int_{t_{start}}^{t_{end}} H[x(t),y(t),z(t)]dt \tag{8.2}$$

where, x and y define the vat surface and z is positive into the vat. For an infinitely long straight scan vector along the x axis with a constant irradiance profile, (8.2) can be integrated to yield the exposure at the distance (y, z) away from the scan line in the vat [1]:

$$E(y,z) = \sqrt{\frac{2}{\pi}} \left(\frac{P_L}{W_0 V} \right) e^{\left(\frac{-2y^2}{W_0^2} \right)} e^{\left(\frac{-z}{D_p} \right)} \tag{8.3}$$

where, V is the scan speed, and D_p is the depth of penetration. The analytical model embodied in (8.3) is the standard model in this chapter.

During the SL process, the exposure value resulting from each laser beam scan is calculated using (8.3). Since exposure is assumed to be an additive quantity, the individual exposure values resulting from several laser beam scans are summed to obtain the final exposure value for a point in the vat. In the SL process, the critical exposure, E_c, is taken as a meaningful threshold above which a point is solidified. Therefore, the points in the vat which have an exposure value equal to or greater than E_c define the cured profile outline. Even though the threshold model has limitations [8], it has been used extensively.

8.2.3 *Laser and Irradiance Modeling*

A Gaussian beam always either diverges from or converges to a point. Due to diffraction, the beam reaches a minimum value, d_0, the beam waist diameter [9], rather than converging to a point. For a Gaussian laser beam, the irradiance has its peak value at the beam waist. The beam waist is dependent on other laser beam variables such as divergence angle and wavelength in the following fashion:

$$d_0 = \frac{4\lambda}{\pi\theta_b} \tag{8.4}$$

It is also useful to characterize the extent of the beam waist region with a parameter called the Rayleigh range, which is the distance from the beam waist where the diameter has increased to $\sqrt{2}d_0$. The collimated region of a Gaussian beam waist is equal to $2z_R$ [9]. The Rayleigh range is calculated as:

$$z_R = \frac{d_0}{\theta_b} = \frac{4\lambda}{\pi\theta_b^2} = \frac{\pi d_0^2}{4\lambda} \tag{8.5}$$

The irradiance at any given point along a laser beam depends on the radial and perpendicular distances, d and z respectively, from the beam waist. The irradiance is a function of the maximum irradiance H_0 that occurs at the beam waist (recall the coefficient in (8.1)). Irradiance is given by:

$$H(d,z) = \frac{H_0 \exp\left(\dfrac{-2(d/d_0)^2}{1+(z/z_R)^2}\right)}{1+(z/z_R)^2} \tag{8.6}$$

8.2.4 Extended Model: SL Spherical Coordinates

What is important for characterization of the laser beam is how irradiance changes across the SL vat. The irradiance is affected by two main factors: distance away from beam waist, and the angle between laser beam and vertical. In order to consider these two factors, it is more convenient to use a spherical coordinate system. Quantities along the laser beam, including these factors, can be modeled directly by the ρ coordinate axis of the spherical coordinate system (ρ, β, γ). The correspondence between the conventional Cartesian vat coordinate system and the spherical coordinate system is shown in Fig. 8.1.

In Fig. 8.1a, the origin of Cartesian coordinates is the center of the vat surface, (0, 0, 0). In the spherical coordinate system, ρ is the spherical distance away from point (0, 0, h). The axis X' makes an angle β with axis X. The angle ψ is between the galvanometer mirror and axis X', and angle $\gamma = 90 - \psi$ from basic geometry. The focus depth f in Fig. 8.1b into the resin is the location of the theoretical beam waist, and is typically set to 3 mm when the laser is at the vat center [8]. The location of the beam waist along the laser beam is constant, and it is at a radial distance $r_f = (h + f)$. The point where the center of the laser beam intersects the vat surface,

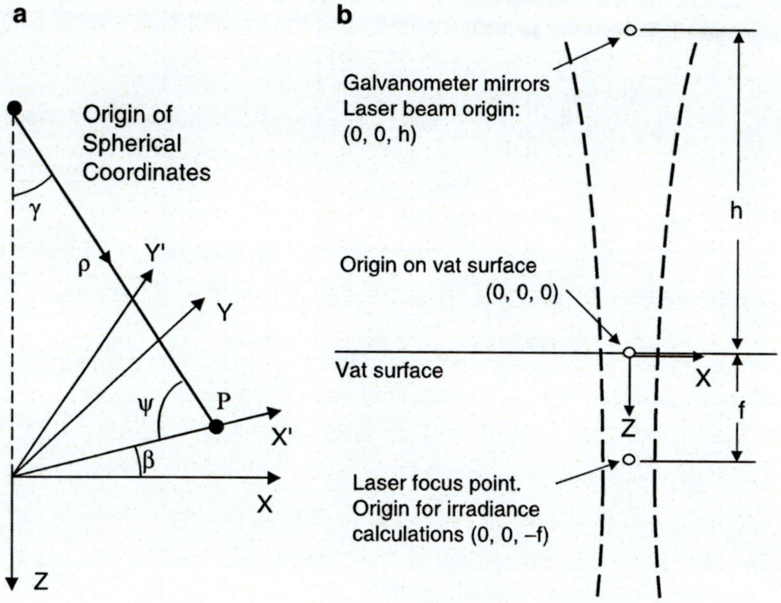

Fig. 8.1 Coordinate systems used (**a**) Origin of systems (**b**) Origin with laser beam

$F(x_F, y_F, 0)$ forms the basis for several irradiance calculations and its spherical coordinates are given as:

$$F(\rho_F, \beta_F, \gamma_F) = \left(\sqrt{x_F^2 + y_F^2 + h^2},\ \tan^{-1}\left(\frac{y_F}{x_F}\right),\ \frac{\pi}{2} - \tan^{-1}\left(\frac{h}{\sqrt{x_F^2 + y_F^2}}\right)\right) \tag{8.7}$$

8.2.5 *Refraction and Absorption Effects*

In the SL process, the path of the laser beam is refracted inside the resin, resulting in a change of the location of the theoretical beam waist. It can be assumed that the maximum angle made with the surface of the resin is not large enough to cause significant laser beam reflection. As is typical, the Beer–Lambert absorption law is assumed to be valid for the SL process [3, 4]. The effects of refraction within the resin are shown in Fig. 8.2.

Using Snell's law for refraction, the relationship between the incident and refractive beam angles can be established. For air, the refractive index is 1. For SL resins, the refractive index is estimated to be 1.5 [10]. It can be assumed that the refractive indices of cured and uncured resins are similar. The beam's refracted path in the resin forms an angle, φ, with the normal, as given in (8.8). The irradiance along this path reaches to the point C, which is the maximum depth of travel achieved within the vat.

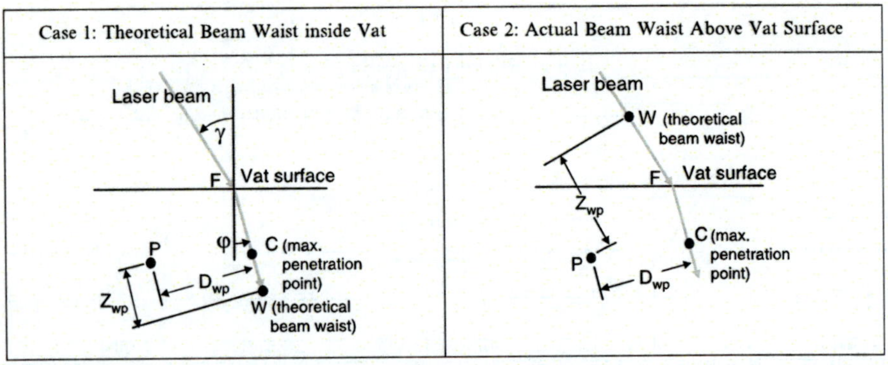

Fig. 8.2 Summary of cases for beam waist location determination

$$\phi = \sin^{-1} \frac{\sin \gamma_F}{1.5} \tag{8.8}$$

In Fig. 8.2b, The Beer–Lambert law for the SL process predicts that the beam irradiance attenuates according to:

$$\text{Attenuation} = \exp\left(\frac{-z_p}{D_p \cos \varphi}\right) \tag{8.9}$$

8.2.6 Irradiance at Arbitrary Point in Vat

Two distinct cases exist for SL irradiance within the vat, as shown in Fig. 8.2. For scans close to the center of the vat, case 1, the *theoretical* beam waist, **W**, is within the vat. The beam waist is called "theoretical" because the focus depth (3 mm) in SL is much larger than the typical depth of penetration (0.25 mm). Therefore, the laser beam never penetrates deep enough into the resin as a result of absorption. For scans that are at the corners of the vat, the *actual* beam waist is either on or above the vat surface, constituting case 2.

To determine which case applies, the distance, $\Delta_F = \rho_f - \rho_F$ between the beam waist location and the point **F** must be found. If the quantity Δ_F is larger than zero, then the beam waist location falls inside the vat (case 1). Otherwise, case 2 occurs. For each case, the location of $\mathbf{W}(x_w, y_w, z_w)$ can be determined relative to the point **F** where the center of the laser beam intersects the build surface. For case 1, since the path of the laser beam will be along the axis X', the location of the beam waist inside the vat is given as:

$$\begin{aligned} W_{\text{case1}}(x_w, y_w, z_w) = &\left(\left(\sqrt{x_F^2 + y_F^2} + \Delta_F \sin \varphi\right) \cos \beta_F,\right. \\ &\left.\left(\sqrt{x_F^2 + y_F^2} + \Delta_F \sin \varphi\right) \sin \beta_F, \Delta_F \cos \varphi\right) \end{aligned} \tag{8.10}$$

In case 2, the path of **W** in spherical coordinates is the same as the path of the focus point **F**. Since the spherical coordinates of **F** are known, the spherical coordinates of the theoretical beam waist can be expressed as ($\rho_F + \Delta_F$, β_F, γ_F). Through some algebraic manipulation, the Cartesian coordinates of **W** for case 2 are obtained as:

$$\mathbf{W}_{\text{case2}}(x_w, y_w, z_w) = (\rho_w \sin\gamma_w \cos\beta_w,\quad \rho_w \sin\gamma_w \sin\beta_w,\quad \rho_w \cos\gamma_w - h) \tag{8.11}$$

To calculate irradiance at any arbitrary point using (8.6), the distances between **P** and **W** that are parallel, Z_{wp}, and perpendicular, D_{wp}, to the beam propagation path must be found. For case 1, the distance D_{wp} refers to the shortest distance between the line **FW** and point **P** (d in (8.6)), whereas for case 2, D_{wp} is the shortest distance between the refracted beam path (line **FC**) and **P**. These parameters are computed as follows [11, 12]:

$$\vec{M} = \vec{F} - \vec{W} \tag{8.12}$$

$$t = \frac{\vec{M} \cdot (\vec{P} - \vec{F})}{\vec{M} \cdot \vec{M}} \tag{8.13}$$

$$\begin{aligned} &\text{case 1: } D_{wp} = ||\vec{P} - (\vec{F} + t\vec{M})||, \\ &\text{case 2: } D_{wp} = ||P - F|| \sin\left(\cos^{-1}\left(\frac{(C - F) \cdot (P - F)}{||P - F||}\right)\right) \end{aligned} \tag{8.14}$$

$$Z_{wp} = (\vec{P} - \vec{W}) \cdot (\vec{F} - \vec{W}) \tag{8.15}$$

Combining (8.14) and (8.15) with (8.6) yields the irradiance equation for point P, where the $\cos\gamma$ term corrects for the increase in area of the laser beam that is incident on the resin surface.

$$H(P) = \frac{H_0 \cos\gamma \exp\left(-\dfrac{2(D_{wp}/d_0)^2}{1 + (Z_{wp}/z_R)^2}\right) \exp\left(-\dfrac{g}{D_p}\right)}{1 + (Z_{wp}/z_R)^2} \tag{8.16}$$

The irradiance model was developed so that it can be used to predict cure profiles. To do so, the exposure at any point in the vat needs to be known by integrating with respect to time, as presented in (8.2). The integration method used here is presented in Sect. 8.3.

8.2.7 *Exposure at Arbitrary Point in the Vat*

To compute the exposure received at a point in the vat, (8.16) must be integrated. However, due to its complicated nature, a numerical integration scheme was

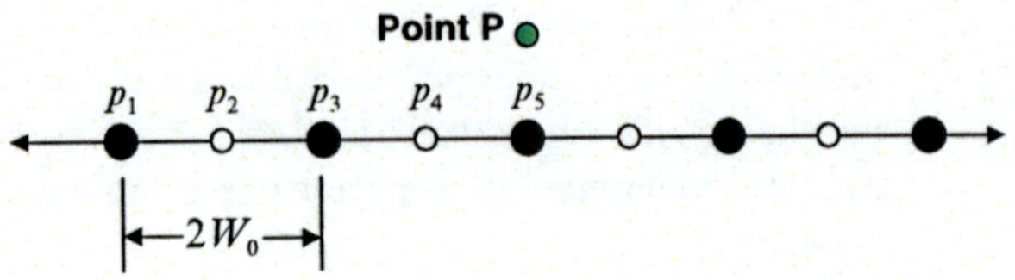

Fig. 8.3 Top view of laser scan points on vat surface

developed. The Simpson's 1/3 Rule integration method was selected due to its reasonably good accuracy and computational ease. To apply Simpson's 1/3 rule, the scan pattern for the SL laser is assumed to consist of four three-point segments that are centered about the point of interest, **P**. At each of these points, the irradiance on the vat surface is calculated using (8.16). The total length of the sampled scan is 4 laser beam diameters, rather than the entire scan vector, since laser beam irradiation attenuates quickly away from the beam center. Since the integration is centered around **P**, the integration is symmetric, so only two segments are integrated over a distance of $4W_0$. The integration pattern is shown schematically in Fig. 8.3.

Constant laser scan velocity is assumed in computing the time, Δt_s, associated with each of the integration points. Applying Simpson's 1/3 Rule to a scan vector results in (8.17), where the irradiance at each point is given by H_{sj} and j denotes the sampling point.

$$E_{n,i}^{\mathrm{P}(x,y,z)} = \frac{1}{3}\Delta t_s[H_{s1} + 4(H_{s2} + H_{s4}) + 2H_{s3} + H_{s5}] \tag{8.17}$$

The exposure term $E_{n,i}^{\mathrm{P}(x,y,z)}$, contains elements that enable a part build simulation, where the part is sliced into layers and each layer is subdivided into scan vectors: n is the layer number and i is the scan number. The exposure at the arbitrary point **P** resulting from a collection of scans in N layers is calculated by:

$$E^{\mathrm{P}(x,y,z)} = \sum_{n=1}^{N}\sum_{i=1}^{I} E_{n,i}^{\mathrm{P}(x,y,z)} \tag{8.18}$$

where, I is the number of scans per layer.

8.3 SL Simulation Method

The SL simulation model is formulated as follows. A part is sliced into N layers using an appropriate slicing scheme (uniform or adaptive); the layer thickness is recorded for each layer. A scan strategy is utilized to generate a set of scan vectors for each layer. Scan velocities are specified for each scan, which can be used as variables in a process planning optimization method [13]. The regions outside and inside the part boundary are sampled with a grid of points. Exposure quantities will be recorded at

```
for each grid point, P
 for each layer n
  if depth(P(z)) < 8*Dp
   for each scan vector i
    if |P(y) - center(scan_i)| < 15*Dp
     for each segment, g, of scan vector
      if Δ_F = ρ_f - ρ_F > 0
         Case 1
       else
         Case 2
       endif
       Compute H_sg(P) using Eqn. 8.16
     end for each segment
     Compute E_{n,i}^{P(x, y, z)} using Eqn. 8.17
     Compute E^{P(x, y, z)} using Eqn. 8.18
    endif
   end for each scan
  endif
 end for each layer
end for each point P
```

Fig. 8.4 SL simulation algorithm

each sample point. Cured part boundaries will be computed by interpolation among sampling points by finding where the computed exposure is equal to E_c.

The SL simulation algorithm is shown in Fig. 8.4. Inputs include layer information, the scan pattern for each layer, and the grid of sampling points. Checks are made within the simulation code to avoid unnecessary computations when **P** is far away from the scan vector. The algorithm was implemented in Matlab as a standalone simulation function. We also have a modified version for use with an optimization-based process planning method.

8.4 Application of Extended Model

The application of the extended model will be presented in two stages. First, the model will be shown to capture phenomena that the standard model cannot, namely the variation of cured shapes caused by angled beams and refraction. Second, the analytical model will be validated by comparing its predictions with measurements from physical parts. This work is more fully reported in [12].

The model embodied by (8.16)–(8.18) was applied to investigate the cured shapes of several types of parts, including a single scan vector and a 10-layer flat part. Part build processes were simulated at the vat center and along vat edges in order to assess the model's capability to capture differences in laser beam angle, focus, and refraction. All simulations and experiments were performed using a SLA250/50 machine using DSM Somos 7110 resin, which has an E_c value of 8.2 mJ/cm^2 and a D_p value of 0.14 mm.

8.4.1 Single Scan Simulation

The cure profile for a 10 mm long scan vector was simulated at two locations in the vat: one at the vat center where the y coordinate is 0 mm, and one at the edge of the vat ($y = 120$ mm). A scan speed of 300 mm/s was used. A comparison of the simulated line shapes is shown in Fig. 8.5.

In Fig. 8.5a, the scan vector profile predicted by our model is shown as the solid blue line, while that predicted by the standard model (8.3) is dashed and black. Discrepancies between the two models can be attributed to the more accurate model of the laser beam in our model. As noted, the conventional cure model assumes a vertical laser beam everywhere in the vat. In contrast, our model accounts for the variation in beam angle throughout the vat and can simulate the effects of this variation, as shown in Fig. 8.5b.

At the vat edge, the laser beam makes an angle of 10.42° with the vertical. With an index of refraction of 1.5 in the resin, the refracted beam angle in the resin is 6.93°, which corresponds very well with the angle of the slanted cured shape (6.95°) in Fig. 8.5b. Note that the cured shape in Fig. 8.5b is wider and shorter than the cured shape at the vat center. Cure depths were 0.265 mm vs. 0.260 mm in Fig. 8.5a, b, respectively. This corresponds with our intuition, since the laser beam energy is spread out in the shape of an ellipse when the beam strikes the vat at a nonperpendicular angle.

8.4.2 Ten-Layer Flat Part

The build process for a ten-layer part was simulated with the new irradiance model to demonstrate the capability of the model to simulate a typical (yet simple) part, and to investigate the extent to which simulated results corresponded to expectations. The part was 10 × 5 × 1 mm in size and was built at the edge of the vat, centered around $y = 120$ mm. Layer thickness of 0.1 mm and hatch spacing of 0.15 mm were used. Scan speeds were 300 mm/s. The scan vectors were repeated along the same paths for each layer. In other words, the x and y scan coordinates were unchanged between layers, in contrast to the typical staggered hatch pattern used in commercial SL machines. This was done to quantify the pronounced effect

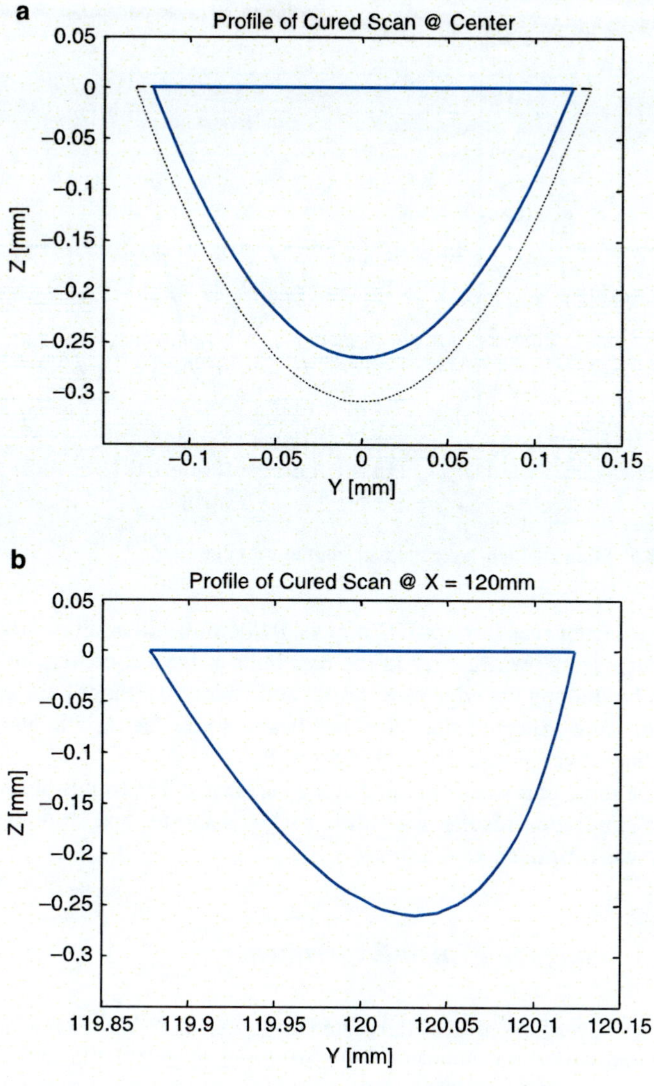

Fig. 8.5 Simulated single scan vectors at (**a**) vat center and (**b**) vat edge

of hatch spacing on the downfacing surface roughness. The output of the MATLAB code was the cure profile and surface roughness average of the cross-section.

Figure 8.6 shows the resulting part profile. Note that the visible cured scan regions along the bottom surface of the part are about 0.006 mm high. Note that the cured scans are spaced 0.15 mm apart, corresponding to the hatch spacing. For this cured profile, a surface roughness of 4.9 μm R_a was determined.

The build process for the same part was simulated at the vat center. With the same scan speed and hatch spacing values, the heights of the cured scan vectors at

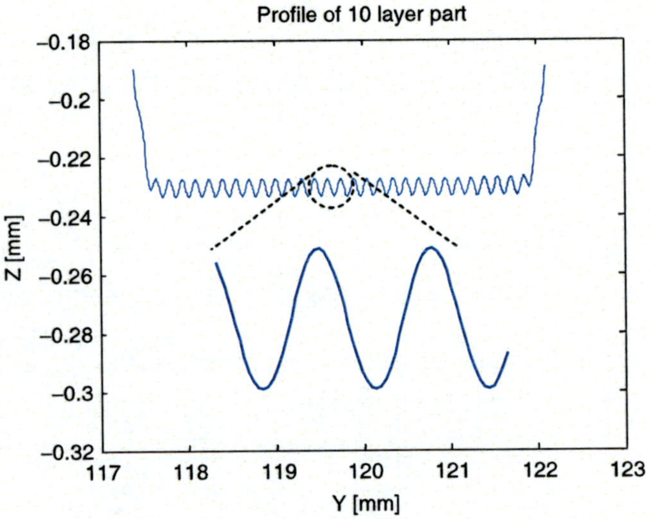

Fig. 8.6 Simulated ten-layer flat part built at vat edge

the vat center was larger (0.02 mm vs. 0.006 mm). In addition, the sizes of the gaps between scans were larger in the part built at the vat center, indicating again that parts built along vat edges are shorter and wider than those built at the center. These results were expected since the laser beam strikes the resin at an angle. The surface roughness was simulated to be 5.8 μm R_a.

The same part was simulated at the vat corner. The trends observed continued for these simulation results, where the part was shorter and smoother. Surface roughness was estimated as 4.3 μm R_a.

8.4.3 Surface Finish Experiments

The ten-layer flat part was fabricated using our SLA-250/50 machine under a variety of conditions. Experiments were performed to assess the repeatability of surface finish results, to quantify surface roughness as a function of hatch spacing and vat location, and to compare measured results with simulation results. All surface roughness measurements were made with a Taylor-Hobson Form Talysurf surface profilometer. Efforts were made to identify and quantify the discrepancies observed between simulated and measured results.

The postprocessing procedure used for all parts, except those presented in Sects. 8.4.4 and 8.4.5, was to rinse parts in 99% industrial grade isopropyl alcohol (IPA) and water for several minutes, followed by 5 min in an ultrasonic bath (Branson Model 5210) in IPA. After washing, the parts were placed in the postcure apparatus (3D Systems) for 30 min, where parts are bathed in UV radiation to ensure that they are fully cured.

Table 8.2 Surface finish at different vat locations

	Vat center, R_a (µm)	X-edge, R_a (µm)	Y-edge, R_a (µm)	Angle of cure profile with vertical at X and Y-edges (°)
Simulated/theoretical	5.8	4.9	4.9	7.33
Measured average (no. of measurements)	10.42 (20)	4.38 (30)	5.67 (30)	8.1 (60)
Standard deviation	1.56	1.91	1.52	1.2

Experimental characterization of the surface roughness at different parts of the vat was carried out in order to determine and quantify the link between laser beam angle and surface roughness. In addition, the correlation between the simulated cure angle and the measured cure angle was documented. To establish this correlation, three parts at the x-edge and three parts at the y-edge of the platform were built. For comparison with the analytical model, two additional parts were built at the center of the vat. All eight parts were built with 150 µm hatch spacing. Ten measurements per part were taken for surface roughness average. The average surface finish and cure profile angle measurements are tabulated in Table 8.2.

Analysis of results in Table 8.2 shows that for the hatch spacing of 150 µm, the experimental results agree well with the simulated results at the edge of the vat. When the cure profile angles are compared, the measured angle of the bullet-shape is 8.1° from the vertical, whereas the simulated value is 7.33°. This is about a 10% difference, which falls within the acceptable range for experimental results. In addition, the R_a values for the x-edge and y-edge parts, which are 4.38 and 5.67 µm respectively, fall within 15% of the simulated value of 4.9 µm.

At the center of the vat, however, the discrepancy between measured and simulated values is 4.6 µm. This is significant and can be attributed to three main factors which will be explored in the next several subsections. However, successful modeling of a significant trend has been displayed in Table 8.2; improvement of surface finish with an angled laser beam. This is seen by comparing the R_a values at the center of the vat and at the edges of the vat for both the simulated and measured results.

8.4.4 Ten-Layer Slanted Surface

The next example demonstrates good correlation between simulated and measured results. A small part was fabricated that was 1.0 mm tall and 7.5 mm long, with a bottom surface that had an angle of 6.1° from the horizontal. Two parts were built using standard processing conditions as determined by the Lightyear 1.3 software from 3D Systems. The parts were hand cleaned. The average surface roughness of the angled bottom surface was 11.6 µm R_a. Figure 8.7 shows that stair-steps are clearly visible on the bottom surface. When the fabrication process of the

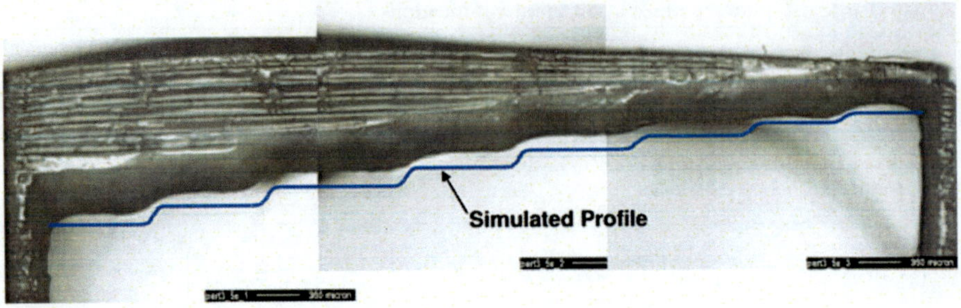

Fig. 8.7 Photo and simulated profile of part with 6.1° surface

part was simulated, the resulting surface roughness estimate was 13.3 μm R_a. As can be seen, the superimposed simulation profile matched well the profile of fabricated part despite a slight bow in the fabricated part (due to the lack of support structures).

8.4.5 *Optimized Process Plan*

The final example is for a part with an optimized process plan. The parameter estimation based method [13] was used to determine the scan speeds for a set of scan vectors to minimize the deviation between the desired surface and the simulated surface profile. The part to be considered is 5 mm high and 10 mm wide, with a down-facing surface of interest at a 45° angle. The 45° surface is 4 mm high. A layer thickness of 0.2 mm is used. Fourteen scan vectors per layer are used with 0.1 mm hatch spacing, except for the two scans nearest the 45° surface, which are spaced 0.05 mm apart. These scans are in the *X* direction; no *Y* direction scans are used. A total of 560 scans were used. Parameters for the SLA-250/50 machine and SOMOS 7110 resin were used. Process planning was performed in 2D (*YZ* plane), not in 3D.

Accuracy and surface finish were controlled by the use of 500 grid points along the 45° surface. Results from process planning were very good. Using these results, the SL build was simulated and resulted in the 45° surface appearing to be very smooth, as shown in the simulated profile in Fig. 8.8. Surface roughness was simulated to be 0.9 μm R_a, and about 0.94 μm RMS. The part was built in our SLA-250/50 machine and the 45° surface does appear to be very smooth. The surface finish was measured at 0.71 μm R_a. This compares very favorably to previously built parts with a 45° angle that were measured at 12.7 μm R_a. Note that the simulated profile is shown offset from the part surface so that both can be seen easily. Solution time was about 55 s on a Dell Inspiron 6000 laptop with a Pentium M processor 1.6 GHz and 500 MB of RAM.

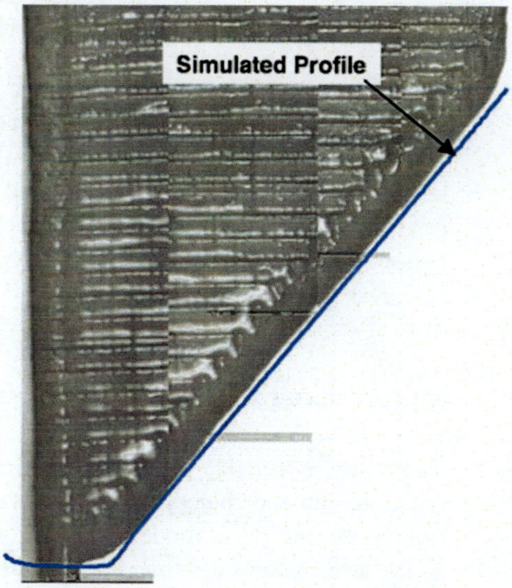

Fig. 8.8 Simulated and fabricated profile of 45° surface part

8.4.6 Sources of Errors

8.4.6.1 Postprocessing Methods

No industry or academia-standard part cleaning procedure exists for SL parts. However, it is known that part qualities are highly dependent upon postprocessing methods, particularly the time of cleaning the part surface. Typically liquid resin is removed from part surfaces by rinsing with IPA and water. For small parts or features, ultrasonic agitation is used sometimes. An agitation time study was conducted using a standard part with 100 μm hatch spacing that involved varying the time, the parts were cleaned in an ultrasonic bath, where the part was placed in a beaker of IPA. The time was varied from 0 to 5 min, where a part with a 0-min ultrasonic bath exposure was hand cleaned. Hand cleaning refers to shaking the part in a beaker filled with IPA using a pair of tweezers. The effect of the ultrasonic bath on SL part surface roughness average was observed clearly: the longer the ultrasonic bath is used, the higher the surface roughness average becomes, while hand-cleaning results in much smoother part surfaces. From this study, it was concluded that ultrasonic cleaning can result in surfaces that are 3–4 μm R_a rougher than those produced by hand cleaning.

8.4.6.2 Resin Curing Reactions

The focus of this study was to use an analytical irradiance model to predict cure shapes in the SLA process. In doing so, the exposure threshold model was applied to determine whether a point is solidified. The threshold model is only a first order approximation of resin behavior. In reality, due to the thermal curing and the dark reaction in the resin, the DOC can vary significantly throughout a part [3, 4, 8]. The extent of polymerization, either spatially or on a molecular weight basis, depends on laser exposure, but also depends on the specific local environment of the reacting monomers [6], which is uncertain. Additionally, regions on the part surface with lesser degrees of cure may be removed by IPA or other solvents, but the solvents do not act linearly or dependably. Both effects (extent of polymerization and solvent actions) introduce uncertainties into the SL process and render accurate predictions of cured shapes problematic.

8.4.6.3 Laser Beam Fluctuations

The actual irradiance profile of the HeCd laser used in the SLA 250/50 machine is not Gaussian and changes over time as much as 10%. The Gaussian beam assumption used in the analytical model is an approximation, and changes in the irradiance profile of the laser beam over time are difficult to model. However, the irradiance profile of the SLA 250/50 laser was recorded for a particular point in time, and used to simulate the resulting cure shape of two overlapping lines. The particular irradiance profile represents a typical variant profile that is commonly observed on our machine.

8.5 SL Cure Simulation Model

A comprehensive SL cure model is developed and validated for simulating part building using radically polymerized resins. In contrast to the semi-empirical E_c–D_p model of resin curing from Sects. 8.2 to 8.4, the comprehensive SL cure model presented here explicitly accounts for the complete photopolymerization process (e.g. photoradical generation, initiation, polymerization, termination) and the simultaneous heat and mass transfer that occur during the polymerization process. In order to address important SL goals that include increasing SL processing speed and improving SL part precision and surface quality, it is necessary to have a quantitative and predictive resin model that can adequately account for changes in resin composition parameters and more subtle SL system parameters such as local resin bath temperature. The first step in developing such a quantitative model is the development of quantitative knowledge of the resin material's photopolymerization kinetics, which is presented in Sect. 8.5.1. This kinetic model is coupled with an SL process model in order to predict the spatial DOC in SL parts. Parts made from a single line scan are compared the predicted DOC to establish a threshold DOC needed to obtain a solid part. Then this threshold DOC criterion is applied to more complex parts to validate the combined kinetic and SL models (Sect. 8.5.2). A DOC threshold model is proposed in Sect. 8.5.3 as an alternative to the E_c threshold model for predicting where the cured boundary of a part lies. Simulation results are presented for a single line scan in Sect. 8.5.4.

8.5.1 Kinetic Model

Ignoring chain transfer reactions, the photopolymerization mechanism for acrylate resin can be briefly described as follows:

$$\left.\begin{aligned} \mathrm{PI} &\overset{h\nu}{\rightarrow} R\bullet \\ \mathrm{M} + \mathrm{R}\bullet &\overset{k_\mathrm{i}}{\rightarrow} P_1\bullet \end{aligned}\right\} \text{ initiation}$$
$$\mathrm{P}_n\bullet + \mathrm{M} \overset{k_\mathrm{p}}{\rightarrow} P_{n+1}\bullet \quad \text{propagation} \tag{8.19}$$
$$\mathrm{P}_n\bullet + \mathrm{P}_m\bullet \overset{k_\mathrm{tc}}{\rightarrow} \mathrm{M}_{n+m}/\mathrm{M}_n + \mathrm{M}_m \quad \text{termination by combination}$$

where, PI, M, In, and Q represent the photoinitiator, monomer, inhibitor, and deactivated radical, respectively; $\mathrm{R}\bullet$ is the primary radical, $\mathrm{P}_n\bullet$ the polymeric radical with a chain length of n monomer units, and M_n the stable polymer molecule with a chain length of n monomer units. The corresponding rates of reaction are described as follows [14]:

$$R_\mathrm{i} = \phi_\mathrm{i} I_\mathrm{a}, \quad R_\mathrm{p} = k_\mathrm{p}[M][P\bullet], \quad R_\mathrm{t} = k_\mathrm{t}[P\bullet]^2 \tag{8.20}$$

where, R_i, R_p, and R_t are the reaction rates of initiation, propagation, and termination, respectively, k_p and k_t are the rate constants for propagation and termination, respectively, ϕ_i is the quantum yield of initiation, I_a the absorbed light intensity in mol/m^3 s, and [M] is the concentration of monomer, and [P•] is the total concentration of chain radicals.

Unsteady kinetics, instead of typical steady state approximation, is adopted here to describe the SL curing process. It is assumed that all monomer radicals are generated during laser drawing, while propagation and termination occur during the dark period [6]. The initial chain radical and monomer concentrations for the dark reaction are thus determined as:

$$[\mathrm{P}\bullet]_\mathrm{i} = 2.3\varphi_\mathrm{i}\varepsilon SE, \quad [\mathrm{M}]_\mathrm{i} = [\mathrm{M}]_0 - 2.3\varphi_\mathrm{i}\varepsilon SE \tag{8.21}$$

where, ε is the photoinitiator absorption coefficient, E the incident exposure, $[\mathrm{M}]_0$ the bulk monomer concentration, and S can be approximated by the initial photoinitiator concentration.

Taking an infinitely small volume of control, in the resin that receives laser irradiance, the following mass and energy balance equations can be derived:

$$\rho C_\mathrm{P}\frac{\partial T}{\partial t} = k\left\{\frac{\partial^2 T}{\partial x^2} + \frac{\partial^2 T}{\partial y^2} + \frac{\partial^2 T}{\partial z^2}\right\} + \Delta H_\mathrm{P} R_\mathrm{P} \tag{8.22}$$

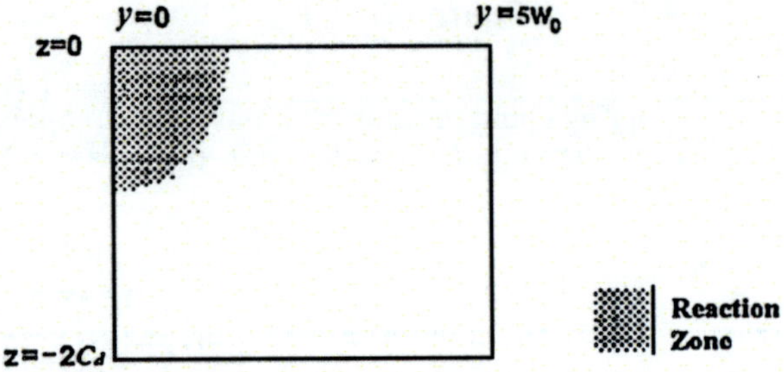

Fig. 8.9 2D domain for single laser drawn line

$$\frac{\partial[\mathrm{M}]}{\partial t} = D_{\mathrm{M}}\left\{\frac{\partial^2[\mathrm{M}]}{\partial x^2} + \frac{\partial^2[\mathrm{M}]}{\partial y^2} + \frac{\partial^2[\mathrm{M}]}{\partial z^2}\right\} + (-R_{\mathrm{P}}) \tag{8.23}$$

$$\frac{\partial[\mathrm{P}\bullet]}{\partial t} = D_{\mathrm{P}\bullet}\left\{\frac{\partial^2[\mathrm{P}\bullet]}{\partial x^2} + \frac{\partial^2[\mathrm{P}\bullet]}{\partial y^2} + \frac{\partial^2[\mathrm{P}\bullet]}{\partial z^2}\right\} + (R_{\mathrm{i}} - R_{\mathrm{t}}) \tag{8.24}$$

where, T is the temperature, D_{M} and $D_{P\bullet}$ are the diffusion coefficients for monomer and chain radicals, and ΔH_{P} is the heat of polymerization which is taken as positive since the reaction is exothermic.

The problem is reduced to a 2D problem considering the repetitive behavior of laser exposure and material curing along a straight line laser scan (x direction). Figure 8.9 shows the domain where the balance equations (8.22)–(8.24) apply [8]. The boundary conditions are as follows:

$$\begin{aligned}
&Q = Q_{\mathrm{i}} \quad \text{at} \quad t = 0,\ 0 \leqslant y \leqslant 5w_0, -2C_{\mathrm{d}} \leqslant z \leqslant 0 && \text{(a)}\\
&\frac{\partial Q}{\partial y} = 0 \quad \text{at} \quad t \geqslant 0,\ y = 0, -2C_{\mathrm{d}} \leqslant z \leqslant 0 && \text{(b)}\\
&Q = Q_0 \quad \text{at} \quad t \geqslant 0,\ y = 5w_0, -2C_{\mathrm{d}} \leqslant z \leqslant 0 && \text{(c)}\\
&\frac{\partial Q}{\partial z} = 0 \quad \text{at} \quad t \geqslant 0,\ 0 \leqslant y \leqslant 5w_0,\ z = 0 && \text{(d)}\\
&Q = Q_0 \quad \text{at} \quad t \geqslant 0,\ 0 \leqslant y \leqslant 5w_0,\ z = -2C_{\mathrm{d}} && \text{(e)}
\end{aligned} \tag{8.25}$$

where, Q represents T, [M], or [P•], and C_{d} is the cure depth which can be obtained by measuring the fabricated part or roughly estimated through a threshold model calculation [1]. Especially for the temperature boundary condition at $z = 0$, heat transfer with the natural air environment is considered, and thus condition (d) in (8.25) becomes:

$$k\frac{\partial T}{\partial z} = h(T_{\mathrm{inf}} - T) \text{ at } t \geqslant 0, \quad 0 \leqslant y \leqslant W_0, \quad z = 0 \tag{8.26}$$

where, the heat transfer coefficient h is taken as 4.18 W/m^2 K [15] for approximation, and T_{inf} is the environmental temperature in the SL chamber.

8.5.2 Simulation Domain and Parameters

FEMLAB, a software package for finite element simulations which provides multidisciplinary application modes (now called COMSOL), was used to obtain the solution to the model established above. Two application modes were employed to complete the model description: diffusion and heat transfer. The boundary conditions and model parameters stated earlier were input into the boundary settings and subdomain settings, respectively. The domain was meshed with triangular, Lagrange-quadratic elements. The partial differential problem established in FEMLAB was solved using a time-dependent-type solver. The time step size is restricted by the absolute and relative tolerance. The solutions were proved to satisfy both h-convergence and p-convergence.

The monomer, Ethoxylated (4) PentaErythritol TetraAcrylate (E4PETeA, SR®494, Sartomer), with the photoinitiator, 2,2-dimethoxy-2-phenylacetophenone (DMPA, Irgacure®651, Ciba), was used as a model compound system to investigate the SL cure process. The kinetics were characterized and the kinetic parameters were evaluated in the work of Tang [16]. The thermal and physical properties of the resin are evaluated for the E4PETeA tetraacrylate with 2 wt% DMPA photoinitiator. They were experimentally determined and compared with literature values and theoretical approximations. The parameters and their corresponding values used in the process model are listed in Table 8.3.

Three types of parts have been built and measured to verify the SL cure process model and assess the predictive capability of the DOC threshold model: single line parts ($V_s = 0.027$ and 0.012 m/s), overlapping cured lines (nine lines, $V_s = 0.46$ m/s,

Table 8.3 Characterized material properties

Material parameters	Symbols	Values	Units
Thermal conductivity	k	0.142	W/m K
Heat of polymerization	ΔH_P	2.85e5	J/mol
Absorptivity (initiator)	ε	19.9	M^3/mol m
Quantum yield of initiation	ϕ_i	0.6	
Coefficient of thermal expansion (monomer)	α_M	0.00177	1/K
Coefficient of thermal expansion (polymer)	α_P	0.00012	1/K
Glass transition temperature (monomer)	$T_{g,M}$	205.65	K
Glass transition temperature (polymer)	$T_{g,P}$	488.35	K
Heat capacity (monomer)	$C_{P,M}$	$5.6 \times T(\mathrm{K}) + 218.6$	J/kg K
Heat capacity (polymer)	$C_{P,P}$	$9.1 \times T(\mathrm{K}) - 1{,}535.5$	J/kg K
Heat capacity (curing system)	C_P	$C_{P,M}(1 - X) + C_{P,M}X$	J/kg K
Density (monomer)	ρ_M	$1{,}128/(1 + \alpha_M(T - 298))$	kg/m^3
Density (polymer)	ρ_P	$1{,}200/(1 + \alpha_P(T - 308))$	kg/m^3
Density (curing system)	ρ	$\rho_M\phi_i + \rho_P(1 - \phi_M)$[a]	kg/m^3

[a] ϕ_M is the volume fraction of monomer

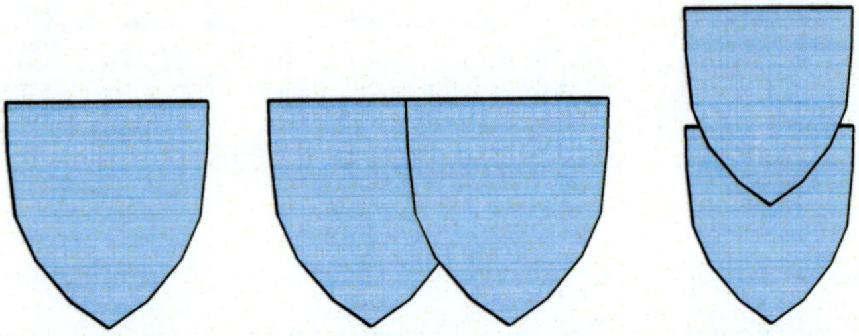

Fig. 8.10 Three basic laser drawing patterns: case I. Single laser drawn line, case II. Overlapping single-layer lines, case III. Stacked single lines

hatching space, $h_s = 12.7\ \mu m$), and stacked single lines (three layers, $V_s = 0.27$ m/s, layer thickness $L_T = 100\ \mu m$). Figure 8.10 shows conceptually the laser scanning patterns of these three types.

8.5.3 Degree of Cure Threshold

The part building process has then been simulated and Fig. 8.11 gives the DOC contour of the built part which corresponds to half of a cured single line scan (case I from Fig. 8.10). For this single line part, the outline is close to the 9% DOC contour. The DOC corresponding to this contour line is defined as the critical DOC, above which the solid part can be formed while below this DOC the resin has not been solidified enough and can be washed away during the postprocessing step. In this sense, the SL cure process model developed in this work can also be referred to as a DOC threshold model. Unlike the exposure threshold model, which only incorporates the exposure, the DOC threshold model takes the reaction and transient intensity effects into account.

The prediction results of all three types of cured lines using the DOC threshold model with the critical DOC range (9–10%) are summarized in Table 8.4. The good agreement between experimental results and simulation results (within 25% error) validates the SL cure process model (i.e. the DOC threshold model). A critical DOC taken outside this range leads to high prediction error.

One source of error is the assumed Gaussian intensity profile of the laser, as discussed earlier. Representing the precise profile should reduce the prediction error. For the stacked line parts the conventional method of recoating after lowering the part in the bath could not be used because the model acrylate parts were made in a small bath inserted in the existing SL bath. Without recoating, less resin may have covered the submerged part, leading to a lower depth of cure experimentally. The simulation results from the three cases show good correspondence with experiments, as well as consistent predictions that a degree-of-cure of 9–10% is sufficient for producing a solid part boundary.

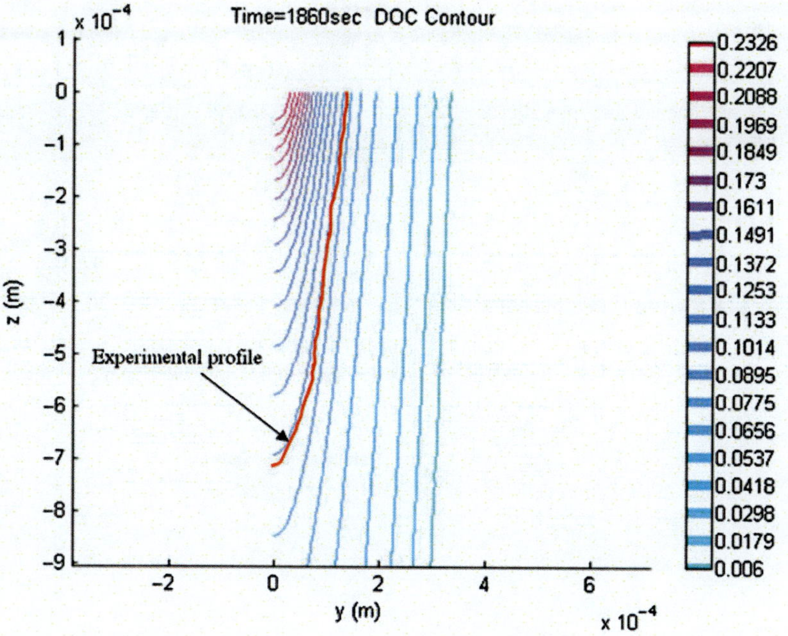

Fig. 8.11 Degree of cure (DOC) contour for parts built at $V_s = 0.027$ m/s

Table 8.4 DOC threshold model prediction: 1. Single line (1) $V_s = 0.027$ m/s (2) $V_s = 0.012$ m/s, 2. Overlapping line, and 3. Stacked line parts

	Depth (μm)				Full width (μm)			
Line type	Experiment (95% C.I.)	Critical DOC	Xc model	Prediction error (%)	Experiment (95% C.I.)	Critical DOC	Xc model	Prediction error (%)
1 (1)	745 ± 10	9%	660	−10	265 ± 4	9%	260	−2
		10%	560	−25		10%	236	−10
1 (2)	957 ± 15	9%	1050	10	322 ± 10	9%	322	0
		10%	940	−2		10%	306	−5
2	1,349 ± 86	9%	1220	−10	468 ± 25	9%	364	−20
		10%	1120	−20		10%	349	−25
3	1,025 ± 52	9%	1285	25	348 ± 8	9%	346	−1
		10%	1150	10		10%	330	−5

Xc model DOC threshold model, *C.I.* confidence interval

8.5.4 Cure Simulation Example

A simple example of a single cured line (case I) is presented here. Additional cases are available [6]. The scan speed is 0.027 m/s along the X axis. The transient response for a point along the center of the line and on the surface $(x,0,0)$ is presented in Figs. 8.12 and 8.13. The laser reaches the selected point after 10 ms have elapsed.

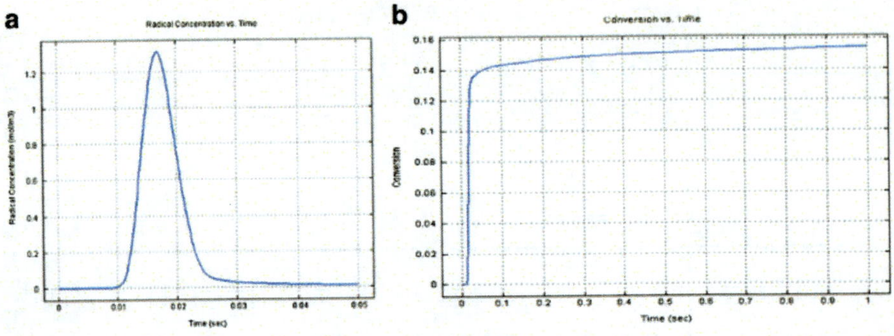

Fig. 8.12 Transients of (**a**) Radical concentration and (**b**) Monomer conversion at a point $(x,0,0)$

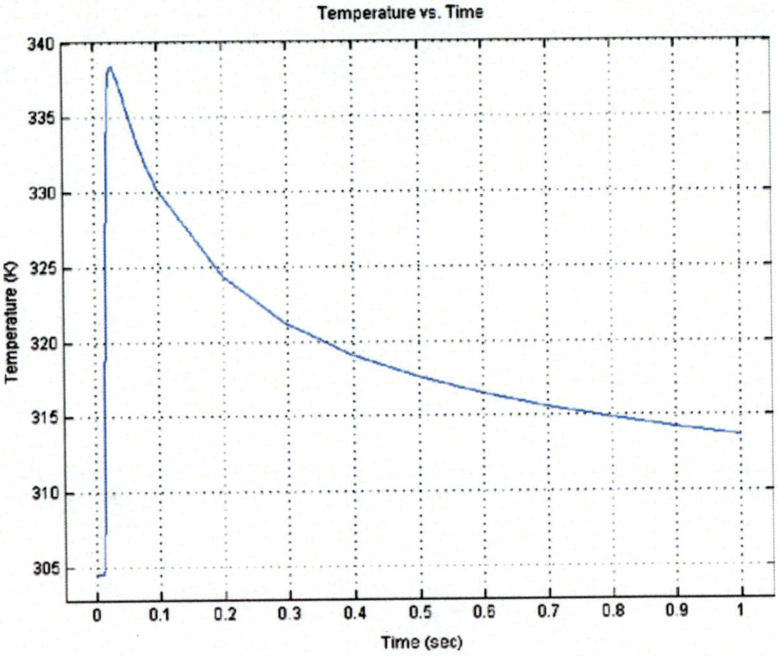

Fig. 8.13 Transient temperature profile for the same point shown in Fig. 8.12

The peak intensity occurs at 17 ms and then the intensity drops to zero at 24 ms. Since this point is along the centerline, this peak intensity also is the maximum intensity of the beam. During this same time period the initiator concentration drops more than 50% and then remains constant. The radical concentration (Fig. 8.12a) is almost symmetric with a peak at 17 ms, corresponding to the laser peak intensity. However, a finite radical concentration is still evident at 50 ms. Almost 10% of the monomer is converted to polymer during the 14 ms exposure time (note the larger time scale in

Fig. 8.12b). Approximately 5% more conversion occurs during the dark period from 24 to 100 ms. After 31 min the conversion at this point reaches 25%. Since the conversion of monomer to solid polymer occurs around 9–10%, the simulation predicts that this particular point solidifies in less than 14 ms. Such rapid solidification is one of the primary benefits of using free radical photopolymerization in SL.

The transient temperature profile for this same point is shown in Fig. 8.13. A temperature rise of more than 20 K occurs during the period of rapid conversion. Since the monomer and polymer have relatively low thermal conductivities it takes a long time for the point to cool down to the starting temperature. For a single line this extended cooling time is not a problem. For multiple parallel lines this temperature rise becomes more significant, leading to faster polymerization and thermal expansion [16]. After cooling there will be residual stresses and dimensional errors due to thermal shrinkage.

8.6 Monte Carlo Simulation Methods

The SL simulation methods presented so far are based on continuous models. However, such models cannot capture the dynamics of interactions among molecules nor the evolution of polymerizing networks. In contrast, KMC models [17] have been developed to simulate molecular interactions, as well as network formation, structure, and evolution, to simulate a wide range of chemical reactions. These models are stochastic in nature. They can be used to study process kinetics and how they change with changes in environmental conditions (e.g. temperature, humidity), polymer structure and properties, such as average molecular weight and degree of branching, and the conditions that favor alternative reaction mechanisms [5].

We have a project underway to extend such models, developed for hyperbranched molecules [18], to SL process with several types of resins. Good correspondence has been achieved for the KMC simulations of monofunctional acrylate and difunctional acrylate materials compared to experiments. Research continues to investigate trifunctional acrylates and hydrogel materials. With KMC methods, we hope to understand the affects of temperature increases during SL processes on reaction rates and polymer network structure. We hope to understand how part surfaces are formed at the molecular level. We want to understand the role of oxygen as a reaction inhibitor and, potentially, how to use it to control the extent of reaction and help to form part surfaces. In process planning, we want to use a DOC model instead of the exposure (E_c) threshold model.

8.7 Conclusions

In this chapter, a series of three SL cure models was presented. The first model extended the standard Jacobs model of SL with a more accurate model of laser–resin interaction. The second was a chemical reaction cure model that enabled a new threshold to be established for predicting part boundaries based on the resin's DOC. The third model was only summarized, as research is still on-going; KMC methods are being applied to develop a new, high fidelity SL simulation model at the molecular level of detail.

Specific conclusions from this work include:

1. The extended model captures expected variations in cured shapes for various hatch spacing values and for different part building locations in the SL machine.
2. Coupled with an optimization-based process planning method, the extended model enables unprecedented smooth surface finishes on down-facing surfaces.
3. The SL cure model enabled the investigation of monomer consumption, polymer conversion, and temperature changes as a function of time, providing important insights into the SL process.
4. By correlating simulated DOC (polymer conversion) with experimental measurements, a DOC threshold can be established for each material. In this work, we found that a DOC of 9–10% yielded solid material and, hence, defined part boundaries.
5. KMC models are a promising avenue for even higher fidelity simulations of SL processes, yielding unprecedented insights into the fabrication of SL parts, their properties, and how their properties vary with reaction conditions.

Acknowledgments The authors gratefully acknowledge financial support from Rapid Prototyping & Manufacturing Institute (RPMI) member companies, CIBA Vision, and the Georgia Research Alliance. Sect. 8.5 was based largely on the Ph.D. dissertation of Dr. Yanyan Tang (now at Intel, Corp.); the authors thank her and her advisors (Drs. Cliff Henderson and John Muzzy at Georgia Tech) for their good work.

References

1. Jacobs, P.F. (1992) *Rapid Prototyping & Manufacturing: Fundamentals of Stereolithography*. Dearborn, MI: Society of Manufacturing Engineers.
2. Reeves, P.E., Cobb, R.C. (1997) "Reducing the Surface Deviation of Stereolithography using In-Process Techniques," *Rapid Prototyping Journal*, **3**(1):20–31.
3. Lu, L., Fuh, J.Y.H., Wong, Y.S. (2001) *Laser-Induced Materials and Processes for Rapid Prototyping*. Boston, MA: Kluwer Academic Publishers.
4. Flach, L., Chartoff, R.P. (1995) "A Process Model for Nonisothermal Photopolymerization with a Laser Light Source I: Basic Model Development." *Polymer Engineering and Science*, **35**(6):483–492.
5. Flach, L., Chartoff, R.P. (1995) "A Process Model for Nonisothermal Photopolymerization with a Laser Light Source II: Behavior in the Vicinity of a Moving Exposed Region." *Polymer Engineering and Science*. **35**(6):493–498.
6. Somvarsky, J., Dusek, K. (1994), "Kinetic Monte-Carlo Simulation of Network Formation: 1. Simulation Method," *Polymer Bulletin*, **33**:369–376.
7. Tang, Y. (2005). "Stereolithography Cure Process Modeling," Chemical Engineering, Ph.D. Dissertation, Atlanta, GA: Georgia Institute of Technology.
8. Wiedemann, B., Dusel, K-H., Eschl, J. (1995) "Investigation into the Influence of Material and Process on Part Distortion," *Rapid Prototyping Journal*, **1**(3):17–22.

9. Tang, Y, Henderson, C, Muzzy, J., Rosen, D.W. (2004) "Stereolithography cure modeling and simulation," *International Journal of Materials and Product Technology*, **21**(4):255–272.
10. O'Shea, D.C. (1985) *Elements of Modern Optical Design*. New York: John Wiley and Sons.
11. Narahara, H., Saito, K. (1994) "Fundamental Analysis of Single Layer Created by Three Dimensional Photofabrication." International Conference on Rapid Prototyping. Dayton, OH: University of Dayton.
12. Edwards, C.H., Penney, D.E. (1994) *Calculus with Analytic Geometry*. 4th ed., Englewood Cliffs, NJ: Prentice-Hall Inc.
13. Sager, B. (2006) "SLA Characterization for Surface Finish Improvement: Inverse Design Methods for Process Planning," Ph.D. Dissertation, Atlanta, GA: Georgia Institute of Technology.
14. Sager, B., Rosen, D.W. (2008) "Use of Parameter Estimation for Stereolithography Surface Finish Improvement," *Rapid Prototyping Journal*, **14**(4):213–220.
15. Fouassier J. (1995) Photoinitiation, Photopolymerization, and Photocuring – Fundamentals and Applications, New York: Hanser Publishers.
16. Pananakis D. and Watts D.C. (2000) "Incorporation of the Heating Effect of the Light Source in a Non-isothermal Model of a Visible-Light-Cured Resin Composites", *Journal of Material Science*, **35**:4589–4600.
17. Goodner, M.D., Bowman, C.N. (2002) "Development of a Comprehensive Free Radical Photopolymerization Model Incorporating Heat and Mass Transfer Effects in Thick Films", *Chemical Engineering Science*, **57**:887.
18. Gillespie, D.T. (1976) "General Method for Numerically Simulating Stochastic Time Evolution of Coupled Chemical-Reactions," *Journal Computational Physics*, **22**(4):403–434.
19. Oguz, C., Unal, S., Long, T.E., Gallivan, M.A. (2007) "Interpretation of Molecular Structure and Kinetics in Melt Condensation of A2 Oligomers, B3 Monomers, and Monofunctional Reagents," *Macromolecules*, **40**:6529–6534.

Chapter 9
Theoretical and Modeling Aspects of Curing Reactions

Paulo Jorge Bártolo

9.1 Introduction

Predicting changes in temperature over the entire sample, due to both the chemical reaction and the laser irradiation effect, is crucial for design optimization of the stereolithographic process. Differences between the temperature within the irradiation volume and that at the surrounding resin can produce stress gradients of considerable magnitude [1]. This is a disadvantage of stereolithography, since such stresses may lead to deficiencies, such as cracking, crazing, and delamination [1]. Moreover, to produce high quality products, it is vital to ensure a fairly high degree of solidification to avoid any post-cure operation, which may lead to distortion and warping of models [2]. Thus, the amount of solid material formed, at a specific time, during the course of the chemical reaction, represents an important parameter in the study and optimization of stereolithographic processes.

Simulations of the behavior of thermosetting resins require analytical models, which can accurately describe the cure kinetics. Moreover, the cure reaction of thermosetting materials is exothermic, and as polymers exhibit low thermal conductivity [3], the temperature and rate of reaction can vary considerably within the curing mass. Thus, the amount of cured material is a function of time and position, as determined by the balance of internal heat generation from the curing reaction, conduction, and heat exchange with the surroundings.

In this chapter, a thermal-kinetic model comprising both heat transfer and reaction kinetics is proposed and implemented by a computational code, using the finite element method. This code helps to understand the mechanism of cure and the influence of different parameters such as the operating conditions, the material properties, and the temperature distribution in the sample. It also enables the modeling of heat flow while curing takes place, assuming that by irradiating

P.J. Bártolo (✉)
Centre for Rapid and Sustainable Product, Polytechnic Institute of Leiria, Leiria, Portugal
e-mail: pbartolo@estg.iplei.pt

P.J. Bártolo (ed.), *Stereolithography: Materials, Processes and Applications*,
DOI 10.1007/978-0-387-92904-0_9, © Springer Science+Business Media, LLC 2011

the material, the affected volume absorbs energy, causing a phase change in the material (*liquid* → *solid* transformation).

9.2 Modeling Approaches to Study and Simulate Cure Reactions

9.2.1 *Relation Between the Glass Transition Temperature and Conversion*

The glass transition temperature, (T_g) "a material characterizing parameter" [4], increases with the fractional conversion, α, due to mobility restrictions over the chain segments, which are associated with a decrease in the free volume (or free space to accommodate the chain movements) during the cure process [5]. As suggested by Fox and Flory [6] during cure reactions, free volume is consumed by addition reactions and by cross-linking reactions that restrict the movement of the chains participating in the network formation. As the cross-linking increases the restriction of chain movements, it leads to an increase of T_g [5] and consequently, non-linear relations between T_g and α can be defined [5–10].

The knowledge of the relationship between T_g and the fractional conversion has been used by several authors [5–10] to study and understand the cure process. These T_g–α relationships have been established by fitting the data, through a polynomial equation [11] or by using the DiBenedetto equation [7, 12] that relates the shift in T_g to the degree of conversion through the following equation:

$$\frac{T_g - T_{g0}}{T_{g0}} = \frac{\left(\frac{E_\infty}{E_0} - \frac{F_\infty}{F_0}\right)\alpha}{1 - \left(1 - \frac{F_\infty}{F_0}\right)\alpha} \tag{9.1}$$

where, T_{g0} is the glass temperature of the uncured polymer, E is the lattice energy, F is the segmental mobility, and the sub-indexes 0 and ∞ refer to the uncross-linked and fully cross-linked polymers, respectively. The ratio of the lattice energies may be approximated by [13]:

$$\frac{E_\infty}{E_0} = \frac{\rho_\infty}{\rho_0}\frac{(M_0)_\infty}{(M_0)_0}\left[\frac{\delta_\infty}{\delta_0}\right]^2 \tag{9.2}$$

where, ρ represents density, M_0 is the molecular weight and δ is a solubility parameter.

Nielson [14] estimated that E_∞/E_0 should be approximately equal to 1.0, while F_∞/F_0 should be zero for highly cross-linked materials. Adabbo and Williams [12] reported values of $0[F_\infty/F_0]$ 1 and $E_\infty/E_0 = 1$, while Enns and Gillham [15] showed that by only taking E_∞/E_0 and F_∞/F_0 as adjustable parameters, a good fit of several T_g–α relationships [16, 17] were obtained.

Other authors [7, 11, 18] have used a model developed by Couchman [17] based on entropic considerations:

$$\frac{T_g - T_{g0}}{T_{g\infty} - T_{g0}} = \frac{\lambda\alpha}{1 - (1 - \lambda)\alpha} \tag{9.3}$$

with

$$\lambda = \frac{\Delta C_{p\infty}}{\Delta C_{p0}} \tag{9.4}$$

where, $T_{g\infty}$ is the T_g of the fully cured polymer and ΔC_{p0} and $\Delta C_{p\infty}$ are, respectively, the heat capacity changes in the glass transition region of the uncured and fully-cured polymer.

Venditti and Gillham [19] modified Couchman's equation and proposed the following model:

$$\ln(T_g) = \frac{(1 - \alpha)\ln(T_{g0}) + \lambda\alpha\ln(T_{g\infty})}{(1 - \alpha) + \lambda\alpha} \tag{9.5}$$

with λ defined by (9.4).

The T_g–α relationships, previously described, were successfully used in different thermosetting systems. Georjon and coauthors [20] used the DiBenedetto equation to describe the relation between T_g and the fractional conversion for an uncatalyzed dicyanate ester monomer. Nevertheless, they obtained better results by using (9.3). Ramis and Salla [7], using (9.1) and (9.4) for an unsaturated polyester resin system, also obtained a good representation for the relationship T_g–α. However, (9.1) does not provide a good description in at least two cases: for highly cross-linked multifunctional epoxy-Novolac systems [21], and for epoxy-rich amine-epoxy systems [9]. In these two cases, (9.3) and (9.5) represent a better alternative. Moreover, when λ is known, (9.3) and (9.5) may be employed without the need for any fitting parameter as in the case of the DiBenedetto equation. In addition, (9.3) and (9.5) assume a knowledge of parameters of the fully cured polymer ($T_{g\infty}$ and $\Delta C_{p\infty}$), which is a difficult task to perform due to the incomplete conversions usually observed during the isothermal cure of thermosetting resins [9]. However, as Hale and coauthors demonstrated [21], instead of the data corresponding to the fully cured polymer, those equations could use the data corresponding to the highest measured fractional conversion (α_M). Therefore, (9.3) and (9.5) can be, respectively, rearranged as:

$$\frac{T_g - T_{g0}}{T_{gM} - T_{g0}} = \frac{\lambda'\alpha}{1 - (1 - \lambda')\alpha'} \tag{9.6}$$

$$\ln(T_g) = \frac{(1-\alpha')\ln(T_{g0}) + \lambda'\alpha'\ln(T_{gM})}{(1-\alpha') + \lambda'\alpha'} \qquad (9.7)$$

where, $\alpha' = \alpha/\alpha_M$ and $\lambda' = \Delta C_{pM}/\Delta C_{p0}$. T_{gM} and ΔC_{pM} are, respectively, the glass transition temperature and the heat capacity corresponding to the maximum achieved value of the fractional conversion.

Hale and coauthors [21] developed a completely different model, with respect to the previous indicated models, based on the work of DiMarzio [22]. To develop such a model, they assumed that T_g increases throughout the cure reaction, due to [21]:

- A decrease in the chain-end concentration
- The formation of elastically effective cross-links
- A decrease in the configurational entropy because the cross-link attached to the network has fewer degrees of freedom. In thermodynamics, T_g means the temperature at which the configurational entropy is zero [23].

The model including the above considerations is represented by the following equation:

$$T_g = \frac{\left(\dfrac{1}{\dfrac{1}{T_{g0}} - \mu\alpha}\right)}{1 - \dfrac{\Phi X}{1 - \Psi X^2}} \qquad (9.8)$$

where, X is a measure of the cross-link density, defined as the number of moles of chains per mole of segments, μ and Φ are constants incorporating the effects of chain ends and cross-links, and Ψ embodies the nonideality effects such as the non-Gaussian behavior and the steric effects on the chain configurations at high cross-link density [21]. The constant μ may be evaluated from $1/T_g$ vs. α in the pregel region of the reaction [21]. The cross-link density X can be elicited by using statistical methods such as the recursive technique developed by Miller and Macosko [24]. Finally, the parameters Φ and Ψ can be fitted using nonlinear regression analysis [21].

Equation (9.8) gives a good description for the relation T_g–α of a multifunctional epoxy-Novolac system [21] and an epoxy-rich amine-epoxy system [9]. However, this model requires a complex and laborious scheme to calculate all the parameters involved.

9.2.2 Theoretical Predictions of the Network Formation

Three classes of network formation are commonly modeled [25, 26]:

- Step growth
- Vulcanisation of chains
- Chain growth

Although different models have been proposed, two main groups can be identified [27–29]:

- Statistical models
- Percolation models

Statistical models are based on the tree-like models [25, 30], where the trees are either generated by linking monomer units or by linking primary chains through a random combination of cross-linked units [25]. Flory [31–33] and Stockmayer [34, 35] developed the first statistical method to study network formation processes. However, the Flory-Stockmayer theory of cross-linking [35] or the theory of branching [34], which neglected intramolecular reactions and assumed that the reactivity of functional groups was independent of the molecular size, if applied to vinyl–divinyl copolymerization reactions, yields reasonable results only if the content of the divinyl component is very low [36]. Significant deviations between experimental data and theoretical values predicted by the Flory–Stockmayer theory have also been found by many authors [37–41]. Three factors have been proposed to explain these deviations:

- Intramolecular cyclization reactions [27, 42]
- Reduced reactivity of pendant vinyl groups [27, 43]
- Molecular shielding effect [27]

Miller and Macosko [24, 44] developed a recursive model. They assumed unequal reactivity of the functional groups, no intramolecular reactions and no molecular shielding [24, 44]. To implement this model, Miller and Macosko used Markovian statistics and the law of total probability of expectation [24, 27, 44]. Due to the complexity of the associated mathematics, this model can only be applied to very simple systems [27].

Gordon and coauthors [45, 46] enlarged the applicability of the Flory–Stockmayer theory by developing the cascade theory. Cascade theory was developed by considering unequal reactivity and substitution effects within monomer units [45, 46], and was implemented using a mean-field treatment of very restricted cyclization [27, 45, 46]. Nevertheless, mean-field models do not deal with radical isolation or shielding of functional groups and they also assume unrestricted chain flexibility [26].

Percolation models [26, 27, 47–49] have also been developed to describe network formation. These models are usually associated with a two or three-dimensional lattice, where each side of the lattice represents a monomer molecule with all its reactive functional groups [26, 27, 47–50]. Percolation models assume intra- and inter-molecular reactions, unequal reactivities, and molecular shielding effects [26, 27, 47–50]. However, the major limitations of percolation models are the impossibility of accurately representing the sizes of monomers and initiators, and the complex computational work that is required [27].

9.3 Modeling the Cure Kinetics

The kinetics of curing reactions has been the subject of many works [10, 51–61]. Kinetic models result in the evolution of reactive species concentration as a function of time by means of a reaction rate expression. Generally, kinetic models are mechanistic or phenomenological [56, 57].

Mechanistic models are based on both the concept of free radical polymerization and the mechanism of reactions with diffusion [10, 56, 57, 62–68]. This approach requires several assumptions and approximations to simplify the complexity of the curing reaction. The model equations often include many parameters which must be determined with aid of numerical optimization schemes [64]. Recently, semimechanistic models have been developed for stereolithography, relating beam exposure to the cure part shape, using a simple threshold approach [69, 70]. These models ignore a variety of transient, thermal, and chemical effects.

Phenomenological models have been developed assuming that only one reaction can represent the whole cure process [10, 51–58], and they are given by the following equation [10]:

$$\frac{d\alpha}{dt} = k_c(T) f(\alpha) \tag{9.9}$$

where, $d\alpha/dt$ is the reaction rate, $f(\alpha)$ is a function of conversion, and $k_c(T)$, the chemical-controlled rate constant, is a function of temperature.

For non-isothermal experiments (9.9) is usually replaced by:

$$\frac{d\alpha}{dT} = \frac{1}{\varphi} f(\alpha) k_c(T) \tag{9.10}$$

where, ϕ is the heating rate.

The simplest and most common analytical form of $f(\alpha)$ is given by:

$$f(\alpha) = (1 - \alpha)^n \tag{9.11}$$

where, n is a constant that corresponds to the reaction order. Substituting (9.11) into (9.9), the following equation is obtained [71]:

$$\frac{d\alpha}{dt} = k_c(T) \cdot (1 - \alpha)^n \tag{9.12}$$

which corresponds to the "so-called" nth order kinetic model [10, 51–58].

The rate constant is supposed to observe an Arrhenius law [10, 56, 57], and therefore can be expressed by:

$$k_c(T) = k_0 \exp\left(\frac{-E}{R(\Delta + T)}\right) \tag{9.13}$$

where, k_0 is a pre-exponential factor or frequency factor, E is the activation energy, R is the gas constant, T is the temperature in °C, $T + \Delta$ is the absolute temperature ($\Delta = 273°C$) and will be indicated as T_{abs}.

Considering the rate constant expression, (9.12) becomes:

$$\frac{d\alpha}{dt} = k_0 \exp\left(-\frac{E}{RT_{abs}}\right) \cdot (1-\alpha)^n \qquad (9.14)$$

For an isothermal reaction, the nth order kinetic expression, given by (9.14), predicts a maximum of the reaction rate at time $t = 0$ [10, 60]. However, if an isothermal process is characterized by a thermogram, showing a maximum value of the reaction rate at any point, rather than the reaction starting point, the nth order kinetic model cannot be applied [10, 60]. In these cases, the nth order models are usually replaced by the so-called autocatalytic models [10, 55, 60]:

$$\frac{d\alpha}{dt} = (k_{c1}(T) + k_{c2}(T)\alpha^m)(1-\alpha)^n \qquad (9.15)$$

where, $k_{c1}(T)$ and $k_{c2}(T)$ are rate constants expressed according to (9.13), m and n are constants and the sum of m and n is the overall reaction order [10, 56–58]. The order of the reaction, according to its definition, indicates the number of atoms, molecules, or reactive groups whose concentration determines the reaction rate.

In isothermal curing processes the degree of conversion (α) can be expressed as [51]:

$$\alpha = H_t/H \qquad (9.16)$$

with $H = H_r + H_{iso}$, where H_t is the heat of reaction at an arbitrary time t, H_r is the residual heat of reaction, H_{iso} is the total heat of reaction during isothermal cure and H is the total heat of reaction. This formula embodies the experimental observation that the peak in the exotherm occurs at $t > 0$ and consequently $k_{c1}(T) = 0$ [10, 72]. Equation (9.15) then becomes:

$$\frac{d\alpha}{dt} = k_c(T)\alpha^m(1-\alpha)^n \qquad (9.17)$$

This expression has been used to describe some non-autocatalytic processes, such as the complex isothermal polymerization reaction of unsaturated polyester resins [10, 55–58]. However, autocatalytic models do not explicitly include the effects of initiator concentrations on the rate of cure and, as a consequence, the kinetic parameters must be recalculated after each change in the resin formulation. Another limitation of these models is the inability to predict the diffusion-controlled effects after vitrification (see Chap. 1).

To characterize the entire curing reaction, Chen and Macosko [73] proposed a two-step plateau kinetic model composed of two simple nth order models:

$$\begin{aligned}\frac{d\alpha}{dt} &= k_1(1-\alpha)^{n_1}(\alpha < \alpha_{onset})\\ \frac{d\alpha}{dt} &= k_2(\alpha_{max}-\alpha)^{n_2}(\alpha \geqslant \alpha_{onset})\end{aligned} \tag{9.18}$$

where, k_1 and k_2 are the rate constants in the chemical controlled and diffusion controlled regions, respectively, n_1 and n_2 are reaction orders, α_{max} is the maximum conversion and α_{onset} is the onset conversion where the reaction shifts from chemical to diffusion controlled.

For photo-curing reactions, Corcione et al. [53] correlated the change of maximum extent of reaction, α_{max}, with light intensity, I, through the following equation:

$$\alpha_{max} = \alpha_0 \times I^c \tag{9.19}$$

where, C is a dimensionless parameter. The rate constant at each temperature is given by:

$$k = k_0(T) \times I^b \tag{9.20}$$

where, b is a dimensionless parameter and $k_0(T)$ is a temperature dependent kinetic constant. A value between 0.5 and 1 for the parameter b may be observed if termination occurs through both unimolecular and bimolecular pathways [10].

Bártolo [10, 56–58] modified (9.17) for stereolithographic processes, incorporating the effects of radiation over the curing process (see Sect. 9.4.1). In Bártolo's model, the kinetics parameters are defined as functions of temperature, resin composition, light intensity, and fractional conversion. Diffusion-controlled effects characterizing the vitrification phenomena are also considered.

9.4 Thermal-Kinetic Model for Photo-Curing Processes

The thermal-kinetic model described here is based on the understanding of the fundamental physical and chemical phenomena governing the behavior of a thermosetting material in photoinitiated curing applications [10]. A number of assumptions were made for its development:

- A simplified isotropic/anisotropic material, with density ρ, specific heat C, thermal conductivities k_r and k_z, and total heat release H
- A well mixed and homogeneous polymeric system
- Any optical scattering effects and the flow of material due to convection or diffusion are both considered negligible
- The internal heat generation is only due to the heat of polymerization

- The absorption of UV radiation is defined by the Beer-Lambert law

According to the model, the temperature field in the exposed region is described by the two-dimensional heat conduction equation in cylindrical co-ordinates (r,z):

$$\rho C\frac{\partial T}{\partial t} = \frac{1}{r}\frac{\partial}{\partial r}\left(rk_{\mathrm{r}}\frac{\partial T}{\partial r}\right) + \frac{\partial}{\partial z}\left(k_{\mathrm{z}}\frac{\partial T}{\partial z}\right) + \rho H\frac{\mathrm{d}\alpha}{\mathrm{d}t} \tag{9.21}$$

transient = conduction + generation

where, T is the temperature, t is the time, α is the fractional conversion and $\mathrm{d}\alpha/\mathrm{d}t$ corresponds to the kinetic model.

The solution of the (9.21) requires both knowledge of the initial temperature, T_{i}, and the initial value of the fractional conversion, α_{i}, in the domain Ω being studied:

$$T(\upsilon, 0) = T_{\mathrm{i}} \tag{9.22}$$

$$\alpha(\upsilon, 0) = \alpha_{\mathrm{i}} \tag{9.23}$$

where, υ represents a generic point in the space. Additionally, the solution must satisfy the following boundary conditions (Fig. 9.1):

- Specified temperature

$$T_{\mathrm{s}} = T(\upsilon, t) \quad \text{at } \Gamma_1 \tag{9.24}$$

- Specified light intensity

$$k_{\mathrm{n}}\frac{\partial T}{\partial n} - I(\upsilon,t) = 0 \quad \text{at } \Gamma_2 \tag{9.25}$$

- Convection boundary condition

$$k_{\mathrm{n}}\frac{\partial T}{\partial n} + h(T(\upsilon,t) - T_{\infty}) = 0 \quad \text{at } \Gamma_3 \tag{9.26}$$

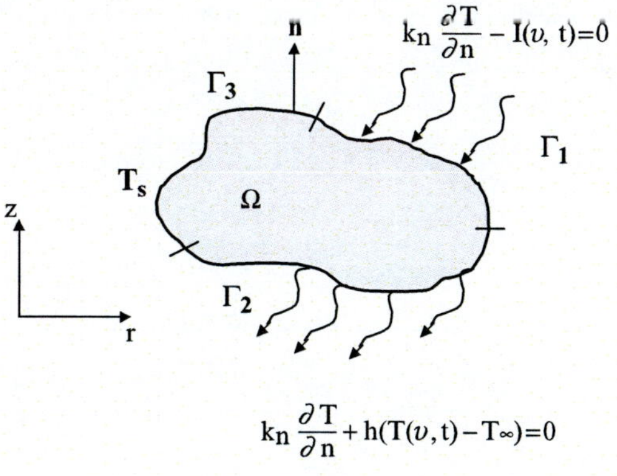

Fig. 9.1 Thermal-kinetic model boundary conditions representation [9, 57, 58]

where, T_s is the specified temperature, I is the light intensity, $\partial T/\partial n$ is the derivative of temperature in the direction normal to the surface, h is the coefficient of heat transfer, and T_∞ is the temperature of the surrounding space. As a final remark, it is worth noting that the term $k_n \partial T/\partial n$ in the r−z geometry, is given by:

$$k_n \frac{\partial T}{\partial n} = k_r \frac{\partial T}{\partial r} \cos(n_r) + k_z \frac{\partial T}{\partial z} \cos(n_z) \tag{9.27}$$

Light intensity values at the resin surface I_s are defined by assuming a Gaussian intensity distribution:

$$I_s(\upsilon, t) = I_0 \exp\left[-2\left(\frac{s(\upsilon, t)}{w_0}\right)^2\right] \tag{9.28}$$

with I_0 being the peak light intensity, $s(v,t)$ representing the position in time of a point under irradiation, and w_0 corresponding to the beam radius. The decrease in light intensity with depth is assumed to obey the Beer–Lambert's law according to the following equation:

$$I_d(\upsilon, t) = I_s(\upsilon, t) \exp[-\varepsilon[\beta]z] \tag{9.29}$$

where, I_d represents the variation of the light intensity along the thickness of the resin layer, ε is the absorptivity of the photoinitiator, $[\beta]$ is the photoinitiator concentration, and z represents the penetration depth.

The Fresnel diffraction theory is applied to evaluate the effect of diffraction for irradiation problems through masks without a collimated light source. According to the Fresnel diffraction theory, the light intensity at the surface of the resin is given by:

$$\begin{aligned} I_P = \frac{I_0}{2} &\left\{[\Im(u_2) - \Im(u_1)]^2 + [\wp(u_2) - \wp(u_2)]^2\right\} \\ &\times \left\{[\Im(v_2) - \Im(v_1)]^2 + [\wp(v_2) - \wp(v_1)]^2\right\} \end{aligned} \tag{9.30}$$

with

$$\Im(u) = \int_0^u \cos\left[(\pi u^2)/2\right] \mathrm{d}u \tag{9.31}$$

$$\wp(u) = \int_0^u \sin[(\pi u^2)/2] \mathrm{d}u \tag{9.32}$$

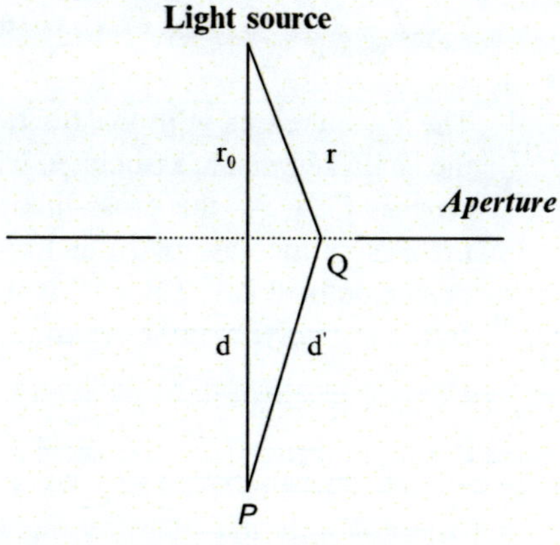

Fig. 9.2 Fresnel diffraction when an aperture is illuminated [57]

being the Fresnel integrals and the two dimensionless variables, u and v (Fig. 9.2), represented by the following equations:

$$u = x\left[\frac{2(r_o+d)}{\lambda r_o d}\right]^{1/2} \qquad v = y\left[\frac{2(r_o+d)}{\lambda r_o d}\right]^{1/2} \tag{9.33}$$

λ is the wavelength.

9.4.1 Kinetic Model

The model used represents a more realistic approach to describe the major phenomena (vitrification, diffusion limitations, incomplete conversion, radical trapping, etc.) occurring during photoinitiated curing reactions [10]. This model is represented by the following equation:

$$\frac{d\alpha}{dt} = \frac{1}{1+\exp\left[\xi(\alpha-\alpha_d)\right]}\phi I^p \exp\left(\frac{-E}{RT_{abs}}\right)[\beta]^q \alpha^m (1-\alpha)^n \tag{9.34}$$

where, ξ is the diffusion constant, α_d is the critical value of the fractional conversion corresponding to the onset of diffusion-controlled effects over the curing reaction, φ is the pre-exponential factor of the rate constant, I is the light intensity, E is the activation energy, R is the gas constant, T_{abs} is the absolute temperature, p and q are constants and the exponents m and n represent the reaction orders, with the sum $(m + n)$ being the overall reaction order. The kinetic parameters, ξ, α_d, m, n and E, are not constants for this equation. Experimentally, it was found that they vary in a nonlinear way with temperature, light intensity, and initiator concentration [10]. To develop the kinetic model, several photoinitiated curing reactions were performed, at different light intensities and temperatures, for resin samples containing different amounts of photoinitiator.

9.4.2 The Glass Transition Model

The T_g, increases with the fractional conversion, due to mobility restrictions over the chain segments, associated with a decrease in the free volume during the curing process [22]. As the cross-linking increases the restriction of chain movements, there is an increase of T_g and consequently, nonlinear relations between T_g and α can be defined [21, 22]. In this research work, the T_g–α relationship is given by the following polynomial equation:

$$T_g = T_{g0} - T_{g0}\alpha + T_{g\infty}\alpha^3 \tag{9.35}$$

where, T_{g0} is the glass temperature of the uncured polymer and $T_{g\infty}$ the glass transition temperature of the fully cured polymer.

9.4.3 Shrinkage Model

The overall volumetric changes of a thermosetting resin during curing processes can be considered a combination of thermal expansion/contraction and polymerization shrinkage. In this case we consider only the polymerization shrinkage described through the following model:

$$\left(\frac{1}{V_0}\frac{\mathrm{d}V}{\mathrm{d}t}\right) = \phi\frac{\mathrm{d}\alpha}{\mathrm{d}t} \tag{9.36}$$

where, V is the volume, V_0 is the initial volume and φ is a shrinkage constant.

9.4.4 Finite Element Discretization

The numerical solution of (9.21), subjected to the initial conditions, (9.22) and (9.23), and to boundary conditions, (9.24)–(9.26), involves two stages of approximation:

- Spatial approximation using the Galerkin method [74, 75]
- Temporal approximation using the Crank–Nicolson method [76]

Fractional conversions, predicted by the kinetic model, are obtained with a fourth-order Runge-Kutta procedure, which enables to integrate numerically (9.30). Figure 9.3 briefly describes the flow of information necessary for the implementation of the thermal-kinetic model.

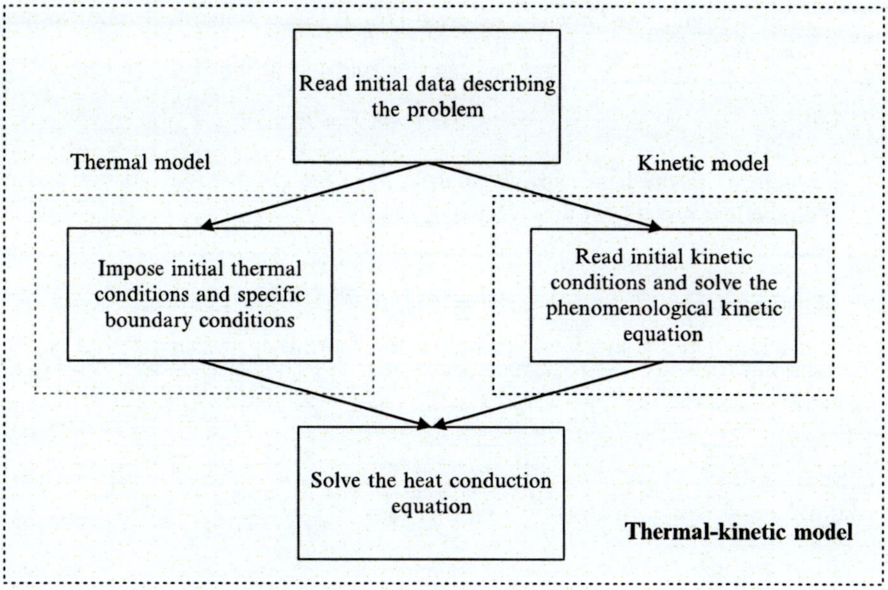

Fig. 9.3 Flow information in the thermal-kinetic model [10, 58]

9.4.4.1 Spatial Discretization

The discretization of (9.21) in space is accomplished by the discretization of the generic domain Ω into an appropriate collection of finite elements Ω_e connected at the nodal points:

$$\Omega = \sum_{i=1}^{k} \Omega_{ei} \tag{9.37}$$

with k being the number of elements and Ω_e representing each element domain.

The interpolation functions N_i are used to interpolate the temperature [74–76]:

$$T(\upsilon, t) = \sum_{i=1}^{j} N_i(\upsilon) T_i(t) \tag{9.38}$$

where, j represents the number of nodes of the considered element and T_i are the nodal temperatures.

Using the Galerkin method the (9.21) can be rewritten at the element level in the following form:

$$\int_{\Omega e} N_i \left\{ \left[-\frac{1}{r}\frac{\partial}{\partial r}\left(r k_r \frac{\partial}{\partial r}\right) - \frac{\partial}{\partial z}\left(k_z \frac{\partial}{\partial z}\right) \right] N_j T_j - \rho H \frac{d\alpha}{dt} + \rho C \frac{\partial}{\partial t} N_j T_j \right\} d\Omega = 0 \tag{9.39}$$

After applying the Greens's theorem [76, 77], and some mathematical manipulations, the following relation is obtained:

$$C\dot{T} + KT = F \tag{9.40}$$

where, C is the heat capacity matrix, K is the conductivity matrix and F is the equivalent nodal heat flow vector, given by:

$$C_{ij} = \int_{\Omega e} N_i \rho C N_j \, d\Omega \tag{9.41}$$

$$K_{ij} = \int_{\Omega e} \left(\frac{\partial N_i}{\partial r} k_r \frac{\partial N_j}{\partial r} + \frac{\partial N_i}{\partial z} k_z \frac{\partial N_j}{\partial z} \right) d\Omega + \oint_{\Gamma e} N_i h N_j \, d\Gamma \tag{9.42}$$

$$F_i = \int_{\Omega e} N_i \rho H \frac{d\alpha}{dt} \, d\Omega + \oint_{\Gamma e} N_i I \, d\Gamma + \oint_{\Gamma e} N_i h T_\infty \, d\Gamma \tag{9.43}$$

9.4.4.2 Temporal Discretisation

The two-time-level Cranck–Nicolson algorithm method of unconditional stability [76, 77] was used to integrate (9.40) with respect to time. According to this algorithm, the unknown values of temperature at time point t_{n+1} are determined through the known temperatures at time point t_n, considering the following temporal approximation:

$$T_{n+1} = T_n + \frac{1}{2} \Delta t \left(\dot{T}_{n+1} + \dot{T}_n \right) \tag{9.44}$$

which can be applied to the matrix (9.40) that becomes:

$$\left(C + \frac{1}{2} \Delta t \, K \right) T_{n+1} = \left[C - \frac{1}{2} \Delta t \, K \right] T_n + \frac{1}{2} \Delta t [F_{n+1} + F_n] \tag{9.45}$$

where the subscript denotes the time at which the corresponding term must be evaluated.

9.4.4.3 Element Formulation

Linear rectangular elements were considered and the concept of isoparametric formulation was used. Isoparametric formulation is fully discussed elsewhere [10, 74–77] and will not be described here.

Table 9.1 The values for the gauss points and weighting factors

$k \times k$	ξ_i	η_j	W_i	W_j
2 × 2	−0.57735	−0.57735	1	1
	−0.57735	0.57735		
	0.57735	−0.57735		
	0.57735	0.57735		

9.4.4.4 Numerical Integration

Numerical integration is used to evaluate the (9.41)–(9.43). The Gauss-Legendre quadrature rules [10, 74–78] were used to perform such a numerical integration. Numerical quadrature formulas [78] in quadrilateral elements have the following form:

$$\int_{\Omega e} F(\xi,\eta)\,\mathrm{d}\xi\mathrm{d}\eta = \int_{-1}^{1}\int_{-1}^{1} F(\xi,\eta)\,\mathrm{d}\xi\mathrm{d}\eta \approx \sum_{i=1}^{k}\sum_{j=1}^{k} F(\xi_i,\eta_j)\,W_i W_j \qquad (9.46)$$

where, W_i and W_j are the weighting gauss factors and k is the number of integration or Gauss points. A 2 × 2 integration rule for linear rectangular elements [10, 74–78] is used to numerically integrate (9.41)–(9.43). The corresponding values for quadrature points and weighting factors [78] are indicated in Table 9.1.

9.4.5 Computer Implementation

Figure 9.4 briefly describes the flow of information necessary for the implementation of the thermal-kinetic model previously described. The formulation earlier presented was implemented through computer software named STLG-FEM (Fig. 9.5), developed for the stereo-thermal-lithographic process (see Chap. 1 for additional information), organized in two levels:

- Main level developed in Visual Basic
- Routine level developed in Fortran 77

The main level acts as a gateway to introduce the program. Through the main level, all the necessary input data parameters are defined. These data are stored in different internal files, managed by the computer during the computational phase.

The routine level, which is the computational level of the software, reads all the input data from the files generated by the main level, on top of performing all computations. It is organized into three basic parts:

- Preprocessor
- Processor
- Postprocessor

and involves several subprograms or routines (Table 9.2).

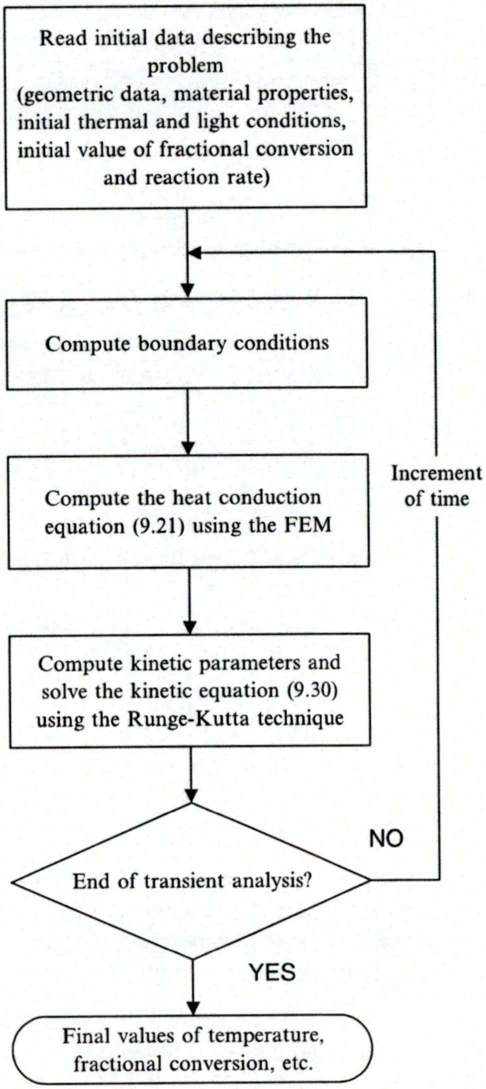

Fig. 9.4 Numerical implementation scheme

In the preprocessor part, all the input data are read. On the other hand, processing involves the generation of the element matrices using numerical integration, the assembly of all the element matrices to form the global ones, the imposition of the boundary conditions and the computation of nodal temperatures, fractional conversions, rates of gel formation and glass transition temperatures. Finally, the post-processor part processes the output data according to a specified format. A flowchart of the code can be observed in Figs. 9.6 and 9.7.

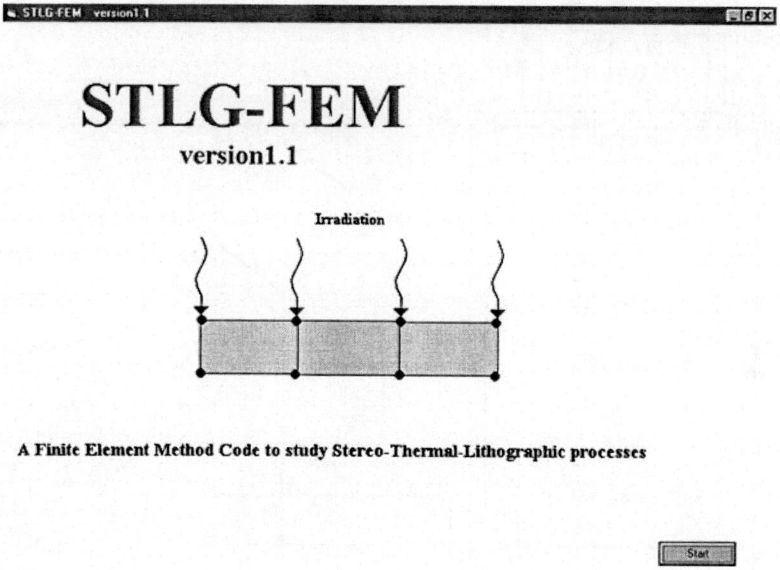

Fig. 9.5 First window-screen of the STLG-FEM software

Table 9.2 Subroutines of the STLG-FEM code

Subroutine name	Description
BOUNDRY	Subroutine to impose temperature boundary conditions
CATAL	Subroutine to compute the kinetic parameters
CONVEC	Subroutine to impose convection effects
EMATRIX	Subroutine to compute [C], [K] and {F}
GENER	Subroutine to define the geometric parameters for mesh definition
GRAD	Subroutine to compute heat fluxes
KINETIC	Subroutine to compute the kinetic equation
LIGHT	Subroutine to impose UV radiation conditions
MESH	Subroutine to generate the mesh
MODEL	Subroutine to define the coordinate system, type of cure process and diffusion effects
MPROP	Subroutine that specified the material properties
RUNGEKUTTA	Subroutine that implements the Runge–Kutta algorithm
SHAPE	Subroutine to evaluate interpolation functions and its derivatives
SOLVER	Subroutine to solve the system of algebraic equations
TEMP	Subroutine to define the parameters for subroutine BOUNDRY
THERMAL	Subroutine to impose thermal or IR radiation conditions
TIMER	Subroutine to compute the matrices and vectors for the time-dependent problem
TRANSI	Subroutine to define the parameters for the time-dependent problem

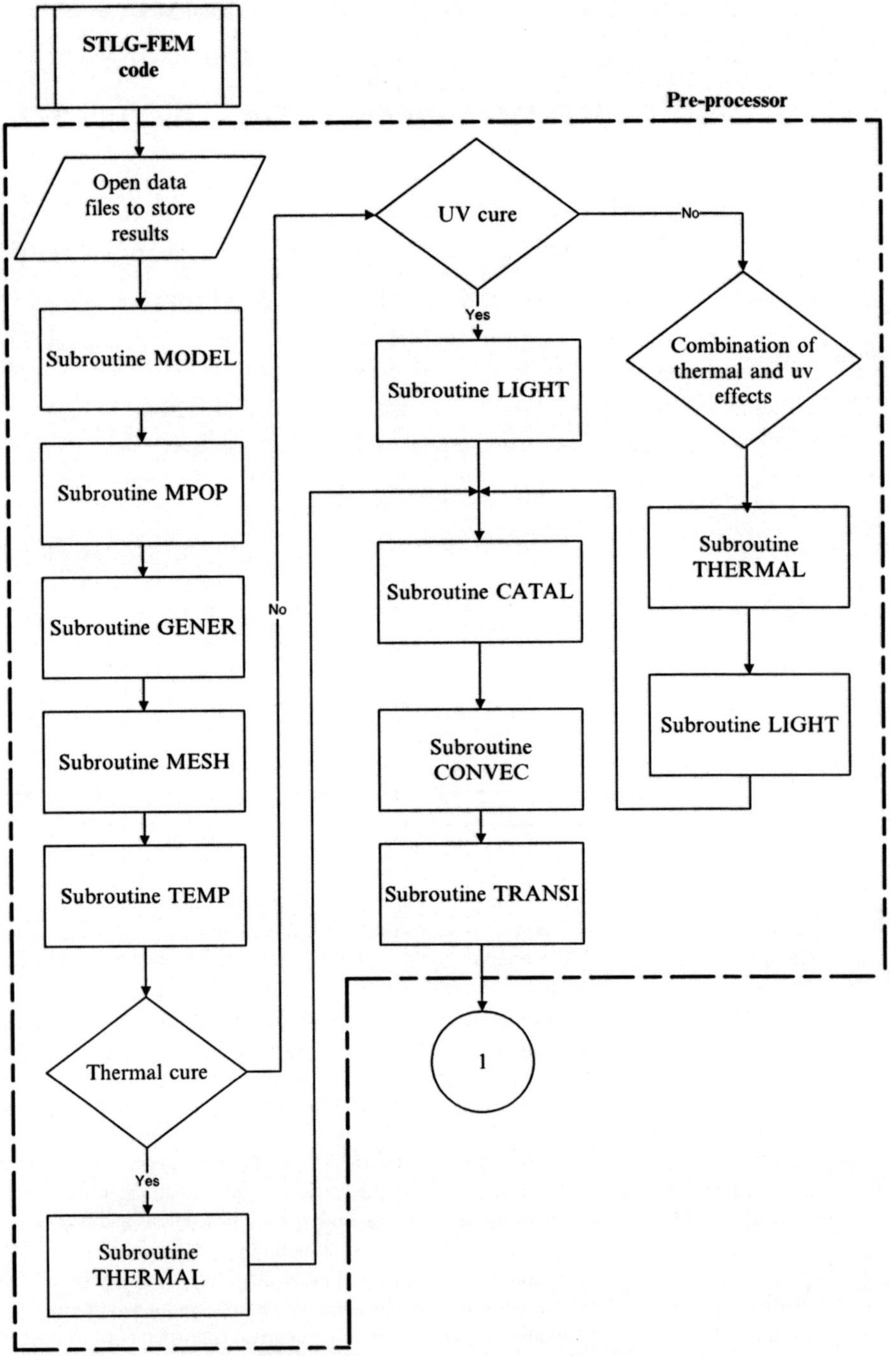

Fig. 9.6 The flowchart of the STLG-FEM code – preprocessor unit

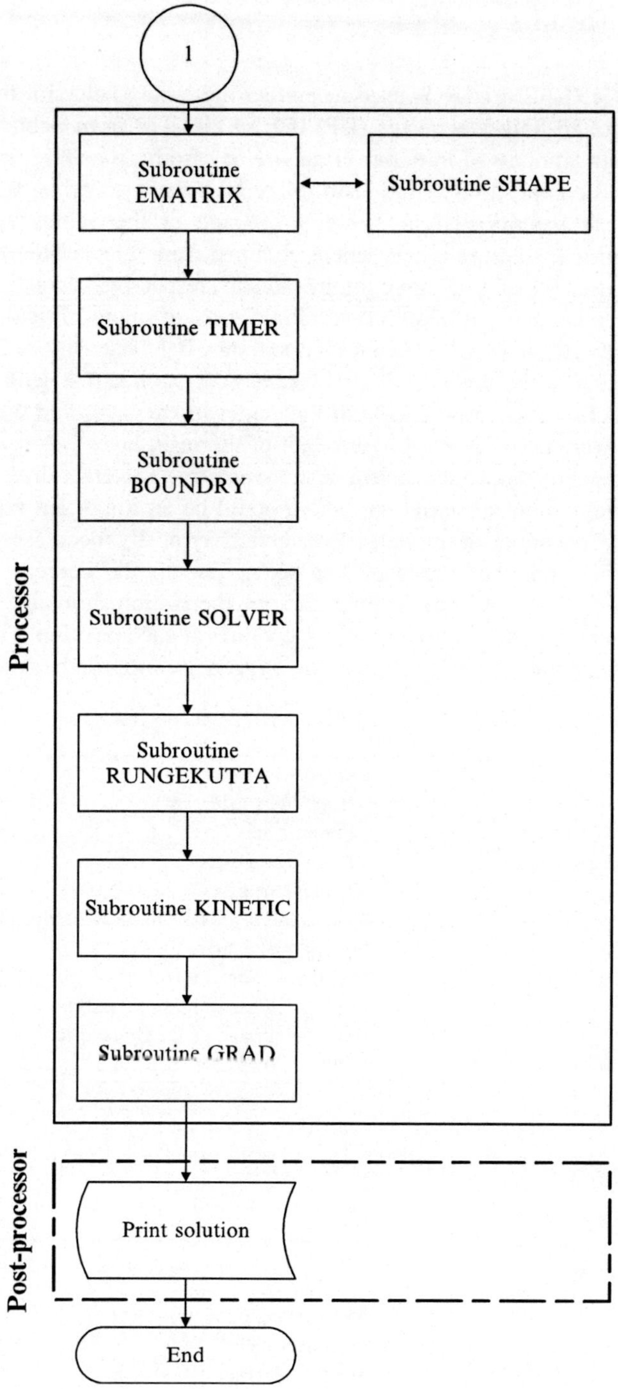

Fig. 9.7 The flowchart of the FEM code – processor and postprocessor units

9.5 Results

The STLG-FEM code is used to predict the curing behavior of a general-purpose unsaturated polyester resins (UP) (CRYSTIC 272 from Scott Badder) containing different amounts of initiators (Irgacure 651 from Ciba-Geigy). Table 9.3 gives the detailed composition of the resin. Material parameter values used in these simulations are summarized in Table 9.4. Some of the values were experimentally determined, whereas others where obtained from relevant literature sources [3].

Figures 9.8 and 9.9 are contour plots of, respectively, fractional conversion and T_g for a layer of UP resin containing 5 wt% of photo-initiator irradiated at room temperature, with a UV beam of maximum light intensity of 200 mW/cm^2 at the center. A resin layer with a thickness 0.1 mm and a length of 0.1 mm was considered. Assuming that both the center of the layer and the center of the light beam were coincident, only one half of the resin layer was considered too, due to the geometry and light distribution symmetry. The fractional conversion contour indicates a conical variation, which could be an important factor in terms of the quality of obtained models. Conversion typically decreases by increasing the distance from the center of the beam, due to the decrease of light intensity. Figure 9.10 shows both the cure and polymerization shrinkage profiles. The values represented in this figure are average values at each position. The results show that, after an induction period (due to the oxygen inhibition), the conversion rate and the

Table 9.3 Composition of the UP resin

Composition (mol%):	
Isophthalic acid	24
Fumaric acid	22
Propylene glycol	25
Diethylene glycol	29
Molar ratio of glycols to dicarboxylic acids	1.2
Styrene content (wt%)	37
Number-average molecular weight (g/mol)	3,111
Unit molecular weight/mol polyester C=C (g/mol)	435
Average number of C=C/polyester	7
Molar ratio of C=C bonds for styrene–polyester	2.5

Table 9.4 Material properties used in the simulations

Property	Value
Density	1.1 g/cm^3
Specific heat	1.674 J/g°C [22]
Thermal conductivity	0.0017 W/cm°C [22]
Heat transfer	0.002 W/cm^2°C [23]
Heat of reaction	335 J/g
Absorptivity of photoinitiator	5 (wt% of PI cm)$^{-1}$ [24]

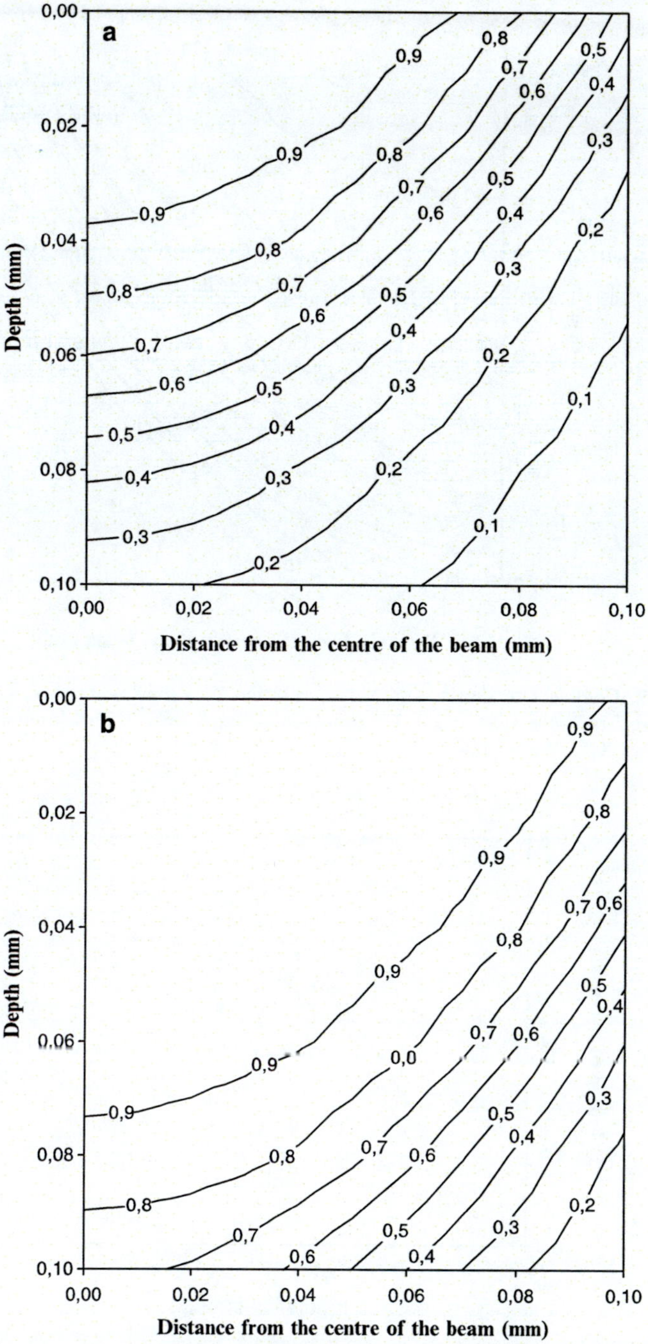

Fig. 9.8 Contour plot showing the variation of fractional conversion as a function of position for two different irradiation times: (a) 0.6 s; (b) 1.2 s. UP resin containing 5 wt% of photoinitiator. Maximum light intensity: 200 mW/cm^2 [56]

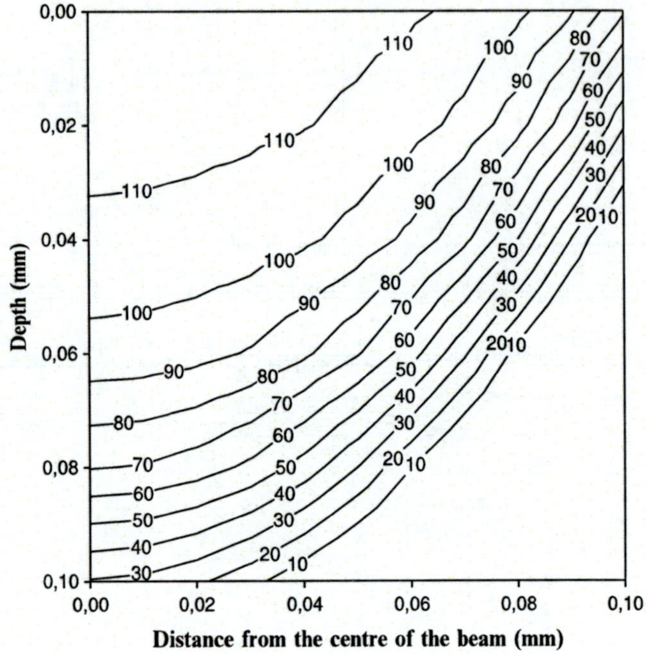

Fig. 9.9 Contour plot showing the variation of the glass transition temperature as a function of position. Irradiation time: 1.2 s. UP resin containing 5 wt% of photo-initiator. Maximum light intensity: 200 mW/cm^2 [56]

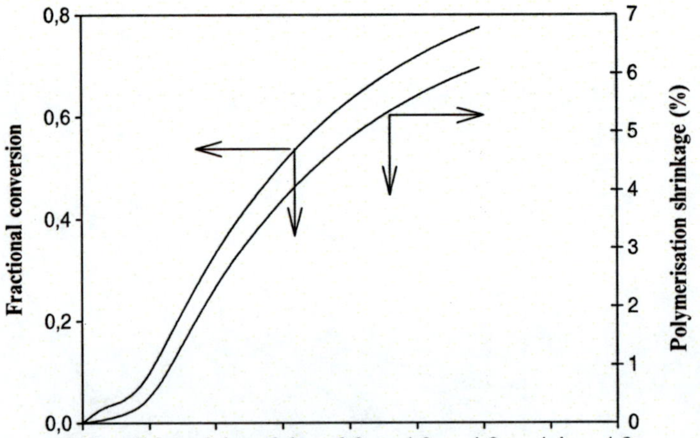

Fig. 9.10 Fractional conversion and polymerization shrinkage vs. irradiation time. Resin sample containing 5 wt% of photoinitiator irradiated with a UV beam of maximum light intensity of 200 mW/cm^2 [56]

polymerization shrinkage increase rapidly, followed by a progressive slowing down. The induction period is more evident at lower light intensities and for resin compositions with low concentration of initiator.

The reduction of both photoinitiator concentration and light intensity strongly decreases the reaction rate and the final amount of fractional conversion as shown in Fig. 9.11. However, the reduction of photoinitiator concentration also increases the light penetration depth. These results were obtained considering a resin sample containing 0.5 wt% of photo-initiator irradiated at room temperature with beam of 28.4 mW/cm^2 produced by a lamp (the variation of light intensity along the resin thickness is illustrated in Fig. 9.12). A resin layer thickness of 0.03 cm and length 3 cm was considered. From Fig. 9.13, that describes the variation of temperature as a function of position after 1.6 min of irradiation, it is possible to observe that the peak temperature no longer occurs at the surface, but at the bottom of the resin layer, due to both the slowing of the reaction at higher conversion, which occurs earlier at the resin surface, and the subsequent effects of the cooling due to conduction and convection.

The heat developed during the exothermic cure reaction and the variation of the rate of gel formation are indicated in Fig. 9.14. The values represented in this figure are average values at each position and they correspond to the curing reaction of a layer thickness of 0.1 and 1.5 mm of length of UP resin containing 5 wt% of

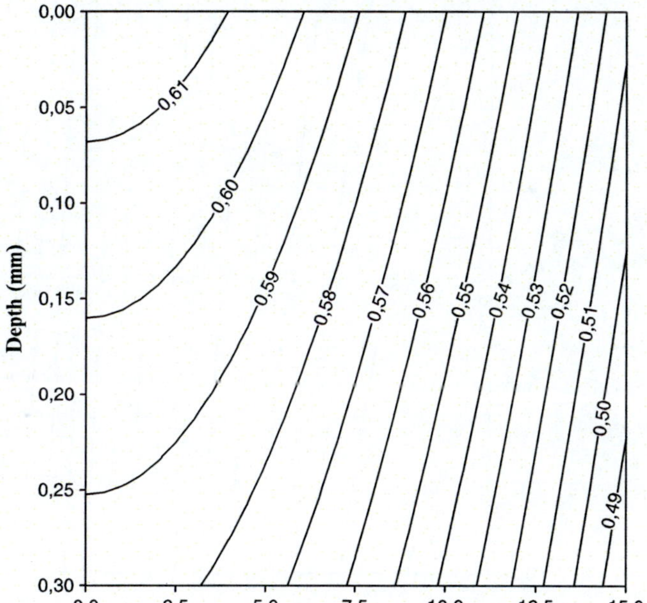

Fig. 9.11 Contour plot showing the variation of fractional conversion as a function of position. Irradiation time: 1.6 min. UP resin containing 0.5 wt% of photo-initiator. Maximum light intensity: 28.4 mW/cm^2 [10, 56]

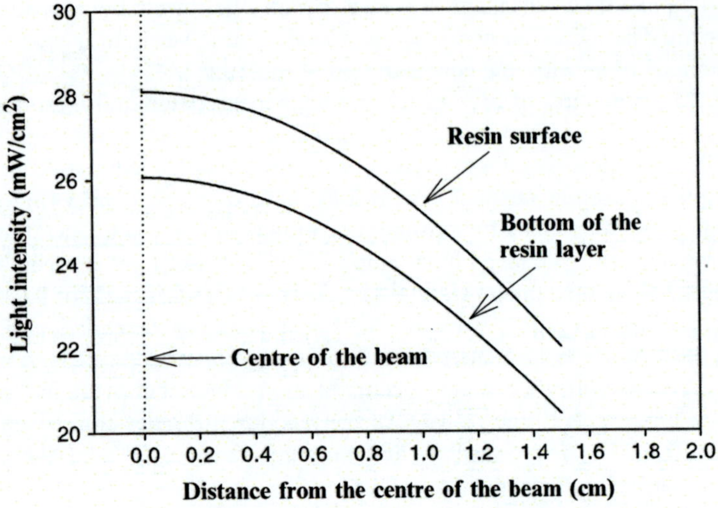

Fig. 9.12 The variation of light intensity along the resin thickness [57]

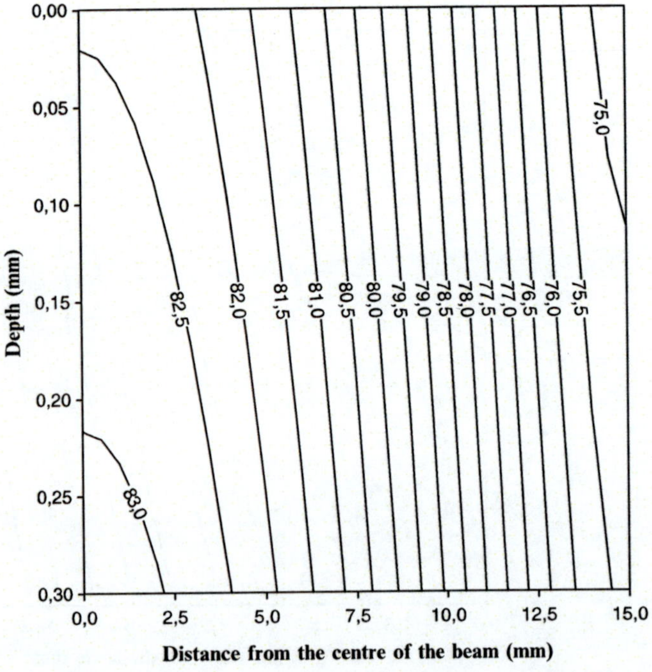

Fig. 9.13 Contour plot showing the variation of temperature as a function of position. Irradiation time: 1.6 min. UP resin containing 0.5 wt% of photoinitiator. Maximum light intensity: 28.4 mW/cm^2 [10, 56]

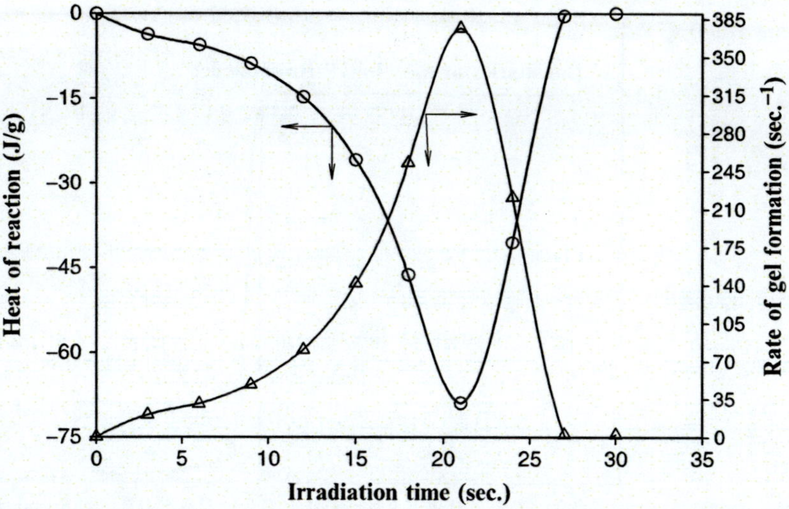

Fig. 9.14 Comparison between the heat developed during the exothermic cure reaction and the variation of the rate of gel formation. UP resin containing 5 wt% of photo-initiator. Maximum light intensity: 28.1 mW/cm^2

photoinitiator. The maximum intensity value of the incident UV radiation was considered to be 28.1 mW/cm^2.

The STLG-FEM code can also be used to study photomask curing processes. Figure 9.15 shows the variation of the fractional conversion at the surface of the resin as a function of position for a square mask (1 cm length) placed 1 cm above the resin surface. The simulation was performed considering a nonuniform mesh. The area directly under the mask was modeled using 10 rectangular elements, where each element has an horizontal dimension of 0.05 cm and a vertical dimension of 0.03 cm. The area immediately after the mask was modeled using 250 elements, considering each element with an horizontal dimension of 0.002 cm and a vertical dimension of 0.03 cm. Finally, to model the rest of the resin layer 40 rectangular elements were considered, defined by an horizontal dimension of 0.05 cm and a vertical dimension of 0.03 cm. This case corresponds also to a very slow cure mechanism as the UP resin contains only 0.5 wt% of photoinitiator and the maximum intensity value of the incident UV radiation was considered to be 28.1 mW/cm^2. Figure 9.15 shows that due to the light diffracted through the mask, deviations from the ideal dimensions, a square model with 1 cm length, are produced. To find the optimum conditions to produce exact models under mask-irradiation processes, a lateral accuracy index (χ) was considered. This index, represented in Fig. 9.15, corresponds to the area defined by the fractional conversion vs. position line and the boundaries of the exact model. High values of χ means less resolution, and exact models are produced when $\chi = 0$ at the end of the curing process.

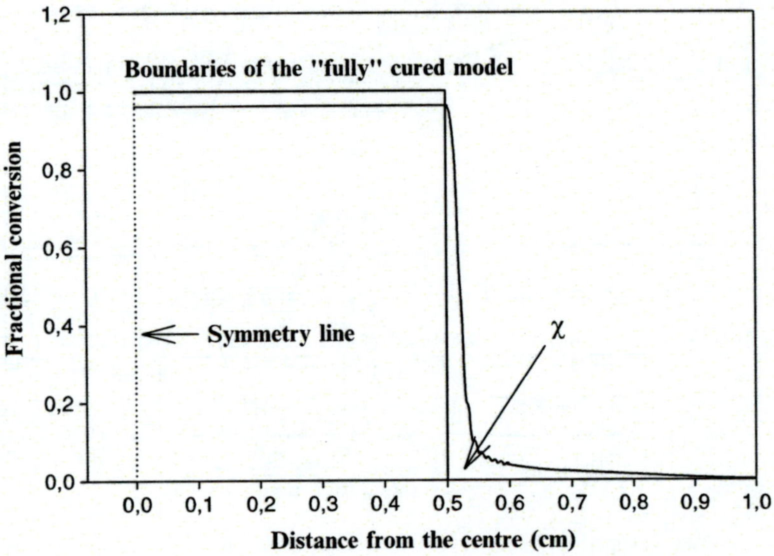

Fig. 9.15 Variation of fractional conversion vs. position for a sample irradiated through a squared mask (1 cm length) placed 1 cm above the resins' surface. Irradiation time: 10 min [10, 58]

Figure 9.16 compares the variation of both fractional conversion and χ as a function of irradiation time. Fractional conversion values represented in this figure are average values at each position (average of the fractional conversion at the surface and bottom of the resin layer for each position). This figure shows that, after vitrification of the resin directly under the mask, continuing exposure to radiation decreases the resolution. Moreover, it can be observed that χ initially increases rapidly due to the conversion of the resin inside and outside the area defined by the mask, but close enough to receive sufficiently high values of diffracted radiation. After the vitrification of this area, χ increases more slowly as the major contribution to this index comes from the resin far from the edges of the mask, subjected to low-intensity diffracted radiation and to slow reaction rates.

The effect of the mask position above the resin surface layer is indicated in Figs. 9.17 and 9.18, which are contour plots showing the variation of χ as a function of mask position and irradiation time or fractional conversion. Fractional conversion values are the average values of the fractional conversion considering all the nodes directly under the mask (which define the exact model). As indicated in these figures, less resolution is obtained by increasing the distance between mask and resin, because more radiation is diffracted and a larger area of resin is covered by this radiation.

An alternative process to improve the accuracy of the obtained models consists in coupling the effects of temperature with photoinitiated curing reactions. Because temperature increases the reaction rate, this effect can be used to "assist" the curing. Figure 9.19 shows the variation of χ with temperature. In this case the values of χ

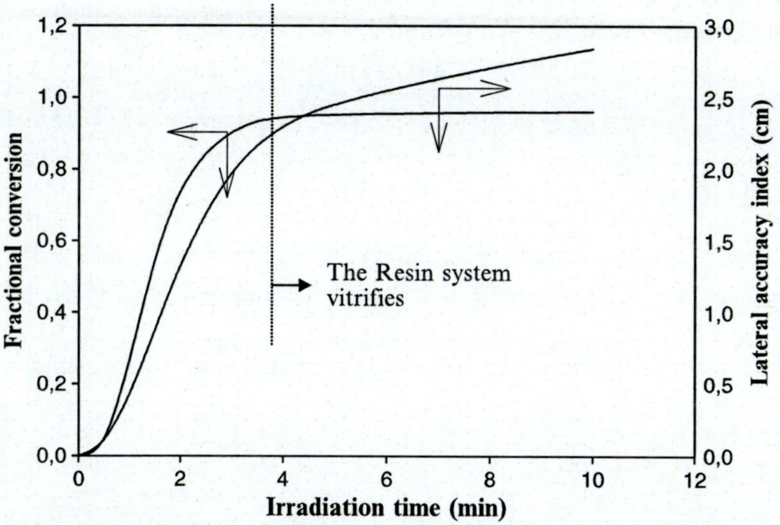

Fig. 9.16 Comparison of the variation of fractional conversion and lateral accuracy index vs. irradiation time. Fractional conversion values are the average values of the fractional conversion at the surface and bottom of the resin layer for each position. Sample irradiated through a squared mask (1 cm length) placed 1 cm above the resin surface [10, 58]

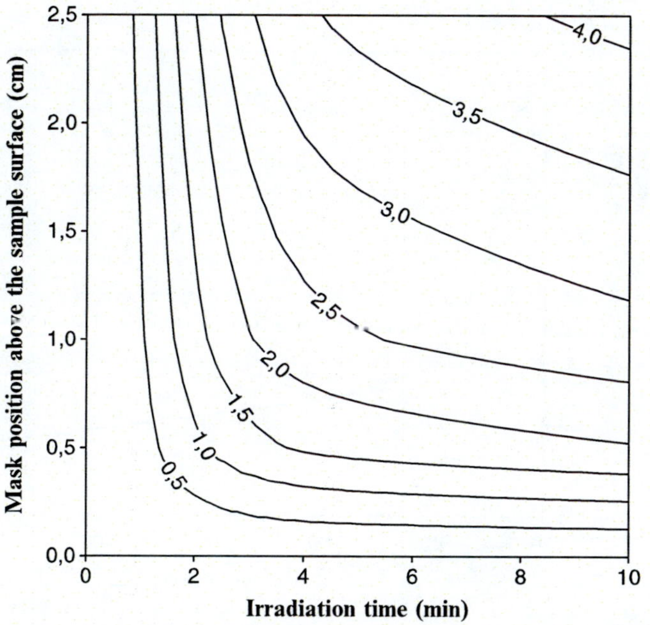

Fig. 9.17 Contour plot showing the variation of χ as a function of mask position and irradiation time [10, 58]

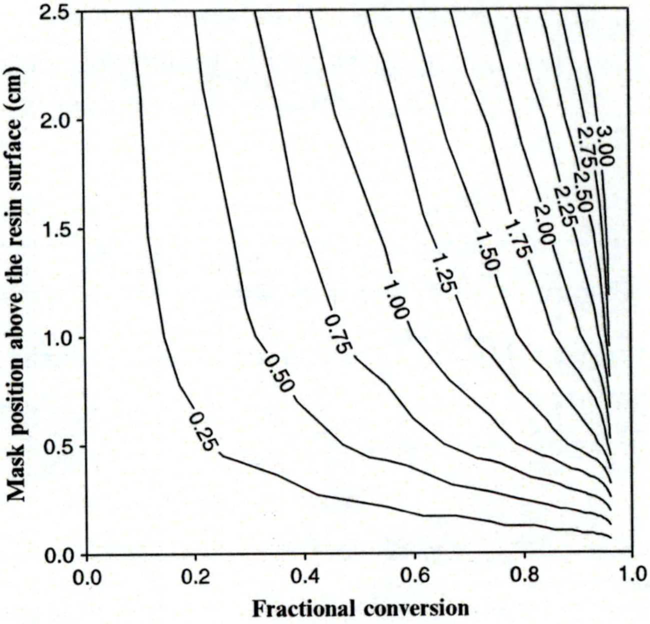

Fig. 9.18 Contour plot showing the variation of χ as a function of mask position and fractional conversion. Fractional conversion values are the average values of the fractional conversion considering all the nodes directly under the mask [10, 58]

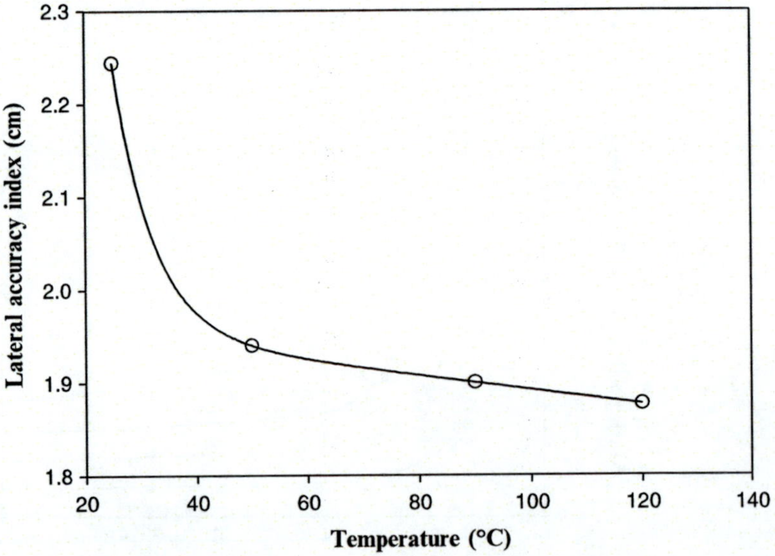

Fig. 9.19 Variation of χ vs. temperature. Samples containing 0.5 wt% of photo-initiator irradiated through a square masked positioned 1 cm above the resin surface. Maximum value of light intensity 28.1 mW/cm^2

were considered for an average value of $\alpha = 0.94$ for the resin directly under the mask under the conditions previously reported. As observed from this figure a small increase in temperature (25–50°C) has a significant effect in terms of accuracy, while for higher temperatures (between 90 and 120°C) the increase in temperature decrease the value of χ in a slower way.

9.6 Conclusions

This Chapter describes a thermal-kinetic model implemented through a computer simulation code named STLG-FEM, which is based on the understanding of the fundamental physical and chemical phenomena governing the behavior of a thermosetting material in streolithographic applications. This model assumes a Gaussian intensity distribution of light at the resin surface, the Beer–Lambert's law to model the decrease in light intensity with depth, and Fresnel's diffraction theory to evaluate the effect of diffraction. The model has been numerically solved using the finite element technique. The finite element model has been developed by using the Galerkin method for spatial discretization and the Crank–Nicolson method for temporal discretization. Linear rectangular elements has been considered and the concept of isoparametric formulation used.

This model enables to get a better understanding of the mechanism of photo-initiated curing reaction and to understand influence of material composition and process parameters. From the simulations, it is possible to observe that:

- The exothermic polymerization reaction and the irradiation process result in a temperature increase of the irradiated region.
- Due to the decrease in light intensity with depth, fractional conversion contours show a conical variation, which can be an important factor to improve the quality of the obtained models. This effect can be minimized by decreasing the concentration of photoinitiator as more light can be absorbed along the thickness of the resin layer.
- Light intensity and temperature increases the rate of gel formation. Consequently, higher values of fractional conversion are obtained and the necessary exposure time is reduced.
- The temperature can be locally used to increase the rate of gel formation of photo-initiated curing reactions. This effect improves the accuracy because the reaction becomes more localized.

References

1. C.M. Cheah and J.Y.H. Fuh, Characteristics of photopolymeric material used in rapid prototypes Part I. Mechanical properties in the green state, Journal of Materials Processing Technology, 67, 41–45, 1997

2. J.Y.H. Fuh, L. Lu, C.C. Tan, Z.X. Shen and S. Chew, Curing characteristics of acrylic photopolymer used in stereolithography process, Rapid Prototyping Journal, 5, 27–34, 1999
3. P.C. Powell and A.J.I. Housz, Engineering with polymers, Stanley Thornes, Cheltenham, 1998
4. I.C. Choy and D.J. Plazek, The physical properties of bisphenol-A-based epoxy resins during and after curing, Journal of Polymer Science Part B: Polymer Physics, 24, 1303–1320, 1986
5. S.L. Simon and J.K. Gillham, Reaction kinetics and TTT cure diagrams for off-stoichiometric ratios of a high-Tg epoxy/amine system, Journal of Applied Polymer Science, 46, 1245–1270, 1992
6. T.G. Fox and P.J. Flory, Second-order transition temperatures and related properties of polystyrene. I. Influence of molecular weight, Journal of Applied Physics, 21 (6), 581–591, 1950
7. X. Ramis and J.M. Salla, Time-temperature transformation (TTT) cure diagram of an unsaturated polyester resin, Journal of Polymer Science Part B: Polymer Physics, 35, 371–388, 1997
8. S. Lunak, J. Vladyka and K. Dušek, Effect of diffusion control in the glass transition region on critical conversion at the gel point during curing of epoxy resins, Polymer, 19, 931–933, 1978
9. K.P. Pang and J.K. Gillham, Competition between cure and thermal degradation in a high Tg epoxy system: effect of time and temperature of isothermal cure on the glass transition temperature, Journal of Applied Polymer Science, 39, 909–933, 1990
10. P.J. Bartolo, Optical approaches to macroscopic and microscopic engineering, PhD Thesis, University of Reading, UK, 2001
11. C.C. Riccardi, H.E. Adabbo and R.J.J. Williams, Curing reaction of epoxy resins with diamines, Journal of Applied Polymer Science, 29, 2481–2492, 1984
12. H.E. Adabbo and R.J.J. Williams, The evolution of thermosetting polymers in a conversion-temperature phase diagram, Journal of Applied Polymer Science, 27, 1327–1334, 1982
13. J.P. Pascault and R.J.J. Williams, Glass transition temperature versus conversion relationships for thermosetting polymers, Journal of Polymer Science Part B: Polymer Physics, 28, 85–95, 1990
14. L.E. Nielsen, Crosslinking effect on physical properties of polymers, Journal of Macromolecular Science Reviews in Macromolecular Chemistry C3, 69–103, 1969
15. J.B. Enns and J.K. Gillham, Time-temperature-transformation (TTT) cure diagram: modelling the cure behaviour of thermosets, Journal of Applied Polymer Science, 28, 2567–2591, 1983
16. S. Montserrat, Effect of crosslinking density on ΔCp(Tg) in an epoxy network, Polymer, 36, 435–436, 1996
17. P.R. Couchman, Thermodynamics and the compositional variation of glass transition temperatures, Macromolecules, 20 (7), 1712–1717, 1987
18. Y.G. Lin, H. Sautereau and J.P. Pascault, Epoxy network structure effect on physical aging behaviour, Journal of Applied Polymer Science, 32, 4595–4605, 1986
19. R.A. Venditti and J.K. Gillham, A relationship between the glass transition temperature (Tg) and fractional conversion for thermosetting systems, Proceedings of the American Chemical Society, Division of Polymeric Materials: Science and Engineering, 69, 434–435, 1993
20. O. Georjon, J. Galy and J.P. Pascault, Isothermal curing of an uncatalyzed dicyanate ester monomer: kinetics and modelling, Journal of Applied Polymer Science, 49 (8), 1441–1452, 1993
21. A. Hale, C.W. Macosko and H.E. Bair, Glass transition temperature as a function of conversion in thermosetting polymers, Macromolecules, 24 (9), 2610–2621, 1991
22. E.A. DiMarzio, On the Second-order transition of a rubber, Journal of Research of the National Bureau of Standards. Section A Physics and Chemistry, 68A(6), 611–617, 1964
23. J.H. Gibbs and E.A. DiMarzio, Nature of the glass transition and the glassy state, Journal of Chemical Physics, 28 (3), 373–383, 1958
24. D.R. Miller and C.W. Macosko, A new derivation of post gel properties of network polymers, Macromolecules, 9 (2), 206–211, 1976

25. K. Dusek, in Development in polymerisation Vol. 3, Edited by R.N. Haward, Applied Science, London, 1982
26. M.A. Zumbrum, G.L. Wilkes and T.C. Ward, in Radiation curing in polymer science and technology, Vol. III: Polymerisation mechanisms, Edited by J.P. Fouassier and J.F. Rabek, Elsevier, London, 1993
27. Y.Y. Chiu and L.J. Lee, Microgel formation in the free radical crosslinking polymerization of ethylene glycol dimethacrylate (EGDMA). I. Experimental, Journal of Polymer Science Part A-1: Polymer Chemistry, 33 (2), 257–267, 1995
28. A. Battisti, A.A. Skordos and I.K. Partridge, Percolation threshold of carbon nanotubes filled unsaturated polyesters, Composites Science and Technology, 70, 633–637, 2010
29. M. Ghiass, A.D. Rey and B. Dabir, Microstructure evolution and simulation of copolymerization reaction using a percolation model, Polymer, 43, 989–995, 2002
30. K. Dusek, in Polymer networks. Principles of their formation, structure and properties. Edited by R.F.T. Stepto, Blackie Academic & Professional, London, 1998
31. P.J. Flory, Molecular size distribution in three dimensional polymers. I. Gelation, Journal of American Chemical Society, 63 (11), 3083–3090, 1941
32. P.J. Flory, Molecular size distribution in three dimensional polymers. II. Trifunctional branching units, Journal of American Chemical Society, 63 (11), 3091–3096, 1941
33. P.J. Flory, Molecular size distribution in three dimensional polymers. III. Tetrafunctional branching units, Journal of American Chemical Society, 63 (11), 3096–3100, 1941
34. W.H. Stockmayer, Theory of molecular size distribution and gel formation in branched-chain polymers, Journal of Chemical Physics, 11 (2), 45–55, 1943
35. W.H. Stockmayer, Theory of molecular size distribution and gel formation in branched polymers II. General cross linking, Journal of Chemical Physics, 12 (4), 125–131, 1944
36. M. Gordon and G.N. Malcolm, Configurational statistics of copolymer systems, Proceedings of the Royal Society London A, 295 (1440), 29, 1966
37. R.S. Whitney and W. Burchard, Molecular size and gel formation is branched poly(methyl methacrylate) copolymers, Makromolecular Chemistry, 181, 869, 1980
38. K. Ito, Y. Murase and Y. Yamashita, Polymerization of diallyl phthalate and gelation of poly (diallyl phthalate) by copolymerization with styrene, Journal of Polymer Science: Polymer Chemistry Edition, 13 (1), 87–96, 1975
39. B. Soper, R.N. Haward and E.F.T. White, Intramolecular cyclization of styrene-p-divinylbenzene copolymers, Journal of Polymer Science Part A-1: Polymer Chemistry, 10 (9), 2545–2564, 1972
40. T. Holt and W. Simpson, Observations on intramolecular reaction in addition polymerization systems, Proceedings of the Royal Society London A, 238 (1213), 154–174, 1956
41. C. Walling, Gel formation in addition polymerization, Journal of American Chemical Society, 67 (3), 441–447, 1945
42. B.H. Zimm, F.P. Price and J.P. Bianchi, Dilute gelling systems. IV. Divinylbenzene–styrene copolymers, Journal of Physical Chemistry, 62 (8), 979, 1958
43. J. Maunsky; J. Klaban and K. Dušek, Vinyl-Divinyl Copolymerization: copolymerization and network formation from styrene and p- and m-Divinylbenzene, Journal of Macromolecular Science, Part A Pure and Applied Chemistry, 5 (6), 1071–1085, 1971
44. D.R. Miller and C.W. Macosko, Substitution effects in property relations for stepwise polyfunctional polymerization, Macromolecules, 13 (5), 1063–1069, 1980
45. G.R. Dobson and M. Gordon, Theory of branching processes and statistics of rubber elasticity, Journal of Chemical Physics, 43 (2), 705–713, 1965
46. M. Gordon, Good's theory of cascade processes applied to the statistics of polymer distributions, Proceedings of the Royal Society London A, 268 (1333), 240–256, 1962
47. J. Bastide and L. Leibler, Large-scale heterogeneities in randomly cross-linked networks, Macromolecules, 21 (8), 2647–2649, 1988
48. H. Boots and N. Dotson, The simulation of free-radical crosslinking polymerization: the effect of diffusion, Polymer Communication, 29, 346, 1988

49. H.M.J. Boots and R.B. Pandey, Qualitative percolation study of free-radical cross-linking polymerization, Polymer Bulletin, 11 (5), 415–420, 1984
50. R. Bansil, H.J. Herrmann and D. Stauffer, Computer simulation of kinetics of gelation by addition polymerization in a solvent, Macromolecules, 17 (5), 998–1004, 1984
51. J.M. Morancho, A. Cadenato, X. Ramis, X. Fernández-Francos and J.M. Salla, Thermal curing and photocuring of an epoxy resin modified with a hyperbranched polymer, Thermochimica Acta, 510, 1–8, 2010
52. P. Cañamero-Martínez, M. Fernández-García and J.L. Fuente, Rheological cure characterization of a polyfunctional epoxy acrylic resin, Reactive and Functional Polymers 70, 761–766, 2010
53. C.E. Corcione, A. Previderio and M. Frigione, Kinetics characterization of a novel photopolymerizable siloxane-modified acrylic resin, Thermochimica Acta, 509, 56–61, 2010
54. B. Jankovíc, The kinetic analysis of isothermal curing reaction of an unsaturated polyester resin: estimation of the density distribution function of the apparent activation energy, Chemical Engineering Journal, 162, 331–340, 2010
55. L. Zhao and X. Hu, Autocatalytic curing kinetics of thermosetting polymers: a new model based on temperature dependent reaction orders, Polymer, 51, 3814–3820, 2010
56. P.J. Bartolo, Computer simulation of stereolithographic curing reactions: phenomenological versus mechanistic approaches, Annals of the CIRP, 55, 221–226, 2006
57. P.J. Bártolo, Photo-curing modelling: direct irradiation, International Journal of Advanced Manufacturing Technologies, 32, 480–491, 2007
58. J.M. Matias, P.J. Bártolo and A.V. Pontes, Modelling and simulation of photofabrication processes using unsaturated polyester resins, Journal of Applied Polymer Science, 114, 3673–3685, 2009
59. C. Popescu and E. Segal, Critical considerations on the methods for evaluating kinetic parameters from nonisothermal experiments, International Journal of Chemical Kinetics, 30 (5), 313–327, 1998
60. A. Yousefi, P.G. Lafleur and R. Gauvin, Kinetic studies of thermoset cure reactions: a review, Polymer Composites, 18 (2), 157–168, 1997
61. M.R. Kamal and M.E. Ryan, in Fundamentals of computer modeling for polymer processing. Edited by C.L. Tucker, Hanser, Munich, 1989
62. G.L. Batch and C.W. Macosko, Kinetic model for crosslinking free radical polymerization including diffusion limitations, Journal of Applied Polymer Science, 44 (10), 1711–1729, 1992
63. J.K. Stevenson, Free radical polymerization models for simulating reactive processing, Polymer Engineering and Science, 26 (11), 746–759, 1986
64. C.D. Han and D.S. Lee, Analysis of the curing behavior of unsaturated polyester resins using the approach of free radical polymerization, Journal of Applied Polymer Science, 33 (8), 2859–2876, 1987
65. Y.J. Huang, J.D. Fan and L.J. Lee, Casting of diffusion-controlled free radical polymerization-experimental and theoretical analysis, Journal of Applied Polymer Science, 33 (4), 1315–1341, 1987
66. S. Swier and B.V. Mele, Mechanistic modelling of the reaction kinetics of phenyl glycidyl ether (PGE) + aniline using heat flow and heat capacity profiles from modulated temperature DSC, Thermochimica Acta, 411, 149–169, 2004
67. C.W. Wise, W.D. Cook and A.A. Goodwin, Chemico-diffusion kinetics of model epoxy-amine resins, Polymer, 38, 3251–3261, 1997
68. F.X. Perrin, T.M.H. Nguyen and J.L. Vernet, Chemico-diffusion kinetic and TTT cure diagrams of DGEBA-DGEBF/amine resins cured with phenol catalysts, Journal of European Polymer, 43, 5107–5120, 2007
69. L. Flach and R.P. Chartoff, A process model for nonisothermal photopolymerization with laser light source I: basic model development, Polymer Engineering Science, 35, 483–492, 1995
70. Y. Tang, C. Henderson, J. Muzzi and D.W. Rosen, Stereolithography cure modelling and simulation, International Journal of Materials and Product Technology, 21, 255–272, 2004

71. M. Abdalla, D. Dean, R. Robinson and E. Nyairo, Cure behavior of epoxy/MWCNT nanocomposites: the effect of nanotube surface modification, Polymer, 49, 3310–3317, 2008
72. L. Zhao and X. Hu, A variable reaction order model for prediction of curing kinetics of thermosetting polymers, Polymer, 48, 6125–6161, 2007
73. Y.T. Chen and C.W. Macosko, Kinetics and rheology characterization during curing of dicyanates, Journal of Applied Polymer Science, 62, 567–576, 1996
74. R.W. Lewis, K. Morgan, H.R. Thomas and K.N. Seetharamu, The finite element method in heat transfer analysis, Wiley, Chichester, 1996
75. J.N. Reddy and D.K. Gartling, The finite element method in heat transfer and fluid dynamics, CRC Press, Boca Raton, FL, 1996
76. G. Comini, S.D. Giudice and C. Nonino, Finite element analysis in heat transfer, Taylor and Francis, London, 1994
77. O.C. Zienkiewicz, The finite element method, McGraw-Hill, London, 1977
78. J.N. Reddy, Introduction to the finite elements method, McGraw-Hill, New York, 1993

Chapter 10
Injection Molding Applications

Russell Harris

10.1 Introduction

The use of plastic tooling in injection molding occurs within the field of Rapid Tooling (RT), which provides processes that are capable of producing injection mold tooling for low volume manufacturing at reduced costs and lead times. Such tooling allows the injection molding of parts in the end-use materials for functional prototype evaluation, short series production, and the validation of designs prior to hard tooling commitment. The term Rapid Tooling is somewhat ambiguous – its name suggests a tooling method that is simply produced quickly. However, the term is generically associated with a tooling method that in some form involves rapid prototyping technologies.

Investigation and application of Stereolithography (SL) to produce mold cavities for plastic injection molding primarily began in the 1990s. Initially the process was promoted as a quick route to soft tooling for injection molding (a tool to produce a relative low number of parts). The advantages of this have been somewhat diluted as other mold production technologies, such as high speed machining, have progressed, but other unique capabilities of the process have also been demonstrated.

Stereolithography has several process capabilities that are particularly advantageous for injection mold tooling, but we should also appreciate that is accompanied by some significant restrictions. This chapter introduces several aspects of the process accompanied by a discussion of its pros and cons, along with examples of work by different parties (Fig. 10.1).

R. Harris (✉)
Loughborough University, Leicestershire, UK
e-mail: R.A.Harris@lboro.ac.uk

P.J. Bártolo (ed.), *Stereolithography: Materials, Processes and Applications*,
DOI 10.1007/978-0-387-92904-0_10, © Springer Science+Business Media, LLC 2011

Fig. 10.1 Injection molding insert generated by stereolithography, shown with part

10.2 Mold Production

In order to discuss the main topic; the direct production of mold cavities, it is first necessary to differentiate this from the indirect route. This is not a significant topic since SL merely provides the master pattern which, irrespective of the process used to produce this, has little influence on the subsequent injection molding.

10.2.1 Indirect Mold Production

The indirect methods involve the use of an initial geometry that has been produced by SL. This geometry is utilized as a pattern in a sequence of process steps that translate into a tool which may be made of a material different to that of the pattern. Cast epoxy tooling represents a common indirect plastic RT method for injection molding. The process begins with a 3D model (i.e. CAD) of the part to be molded. Subsequently this model is produced by SL to provide a master pattern around which the mold will be formed. Traditionally, the part is produced solely without provision for parting lines, gating, etc. Such ancillaries are generated by manual methods (i.e. by fixing additional features to the part). However, the advent of easier CAD manipulation allows the model to be produced including such features. Once the complete master pattern has been produced, the mold halves are created by casting epoxy around the pattern, thus recreating a negative profile of the pattern. The epoxy may include fillers in attempts to improve strength and thermal properties of the mold. Such fillers include metal and ceramic particles in various forms.

The epoxy is commonly cured by exposure to an elevated temperature. Prior to use in injection molding, the epoxy mold halves (inserts) need to be contained within a bolster (an encapsulating metal mold frâme) for both alignment and load bearing. The mold halves may require some further processing to allow fitment and ensure alignment of the mold faces within the bolster.

10.2.2 Direct Mold Production

The direct methods involve a SL system directly generating the tooling cavity inserts in its native material. The accuracy of the SL RP process results in inserts that require few further operations prior to their use in injection molding. Like all RP related techniques the process is dependant on a 3D CAD model of the intended geometry. Unlike indirect techniques, the whole tool insert is generated by SL and so a 3D CAD representation of the whole tool insert is required. This involves creating negatives of the part to form the mold insert bodies, plus the provisions for gating, part ejection, etc. Previously, this extra CAD work would have represented more work required in the preparation. Such input is now minimized as modern CAD manipulation packages (e.g. Materialise's Magics software) allow the automation of such activities. Once generated, the cavity inserts need to be secured in a bolster to withstand clamping forces and to provide alignment to the mold halves.

It should also be mentioned that direct SL tooling for injection molding has also been referred to as Direct AIM. This term was given to the process by 3D Systems (SL system manufacturers) and refers to Direct ACES Injection Molding. (ACES stands for "Accurate Clear Epoxy Solid," which is a SL build style) [1].

10.3 The Requirement of the Process

The introduction of rapid prototyping has allowed engineers and designers to generate physical models of parts very early in the design and development phase. However, the requirements of such prototypes have now progressed beyond the validation of geometry and onto the physical testing and proving of the parts. For such tests to be conducted, the part must be produced in the material and manner (process) that the production intent part will be. For injection molding, this situation highlights the requirement of a rapid mold-making system that can deliver these parts within time and cost boundaries.

Stereolithography provides a possible solution to this by providing the rapid creation of a mold. A negative of the part required plus gating and ejection arrangements are generated in 3D CAD to create a tool that is fabricated by SL. This provides an epoxy mold from which it is possible to produce plastic parts by injection molding.

Both Luck et al. [2] and Roberts and Ilston [3] evaluated SL in comparison with other direct RP mold-generating techniques for producing a typical development quantity of moldings. The SL molding process was found to be a superior alternative for producing design-intent prototypes.

It has also been noted that other alternative techniques involve additional steps to the process, therefore becoming less direct and not really RT [4]. Other advantages of the process have been highlighted beyond the prototype validation phase. Since the tool design has been verified, the lead-time and cost involved in the manufacture of production tooling is also often reduced as the tool design has already been proven [5].

During the early years of SL it was never envisaged that such a RT method would be possible. At first glance the application of SL for injection mold tooling seems unfeasible due to the low thermal conductivity and limited mechanical properties of epoxy, especially at high temperatures. The glass transition temperature of SL materials available was only ~60°C, while the typical temperature of an injected polymer is over 200°C. Despite these supposed limits, successful results were achieved by SL users worldwide, including the Danish Technological Institute, Ciba Geigy, Fraunhofer Institute, the Queensland Manufacturing Institute, and Xerox Corporation [6].

10.4 Mold Design Considerations

In terms of the mold's actual cavity design, relatively little information exists on the specific requirements of SL tooling. The early white paper issued by 3D Systems suggests the incorporation of a generous draft angle, but does not state the amount and recommends the use of a silicone based release agent (every shot) in an attempt to prevent the parts sticking to the inserts [7]. Work has been conducted [8] that quantifies the effects of draft angle on the force exerted on SL tools upon ejection of a molding. It has been shown that an increase in tooling draft angle results in a lower force required to remove a part from the tool. However, the effect of draft angle variation on ejection force is minimal and little compensation for the deviation from intended part geometry caused by the addition or removal of material required to form the draft.

Work has been conducted to establish the cause of core damage during molding. This found that damage was not related to pressure, but to the size of the core features. Smaller core features were broken due to a shearing action caused by polymer melt movement [9].

Experimentation has revealed two modes of wear during the material flow within the cavity. These modes were abrasive at medium flow points (i.e. sharp corners), and ablative at high flow points (i.e. injection points) [10]. Other work [11] has also emphasized the importance of the material flow influenced by mold design, identifying gating, and parting line shut off areas as points of potentially high wear.

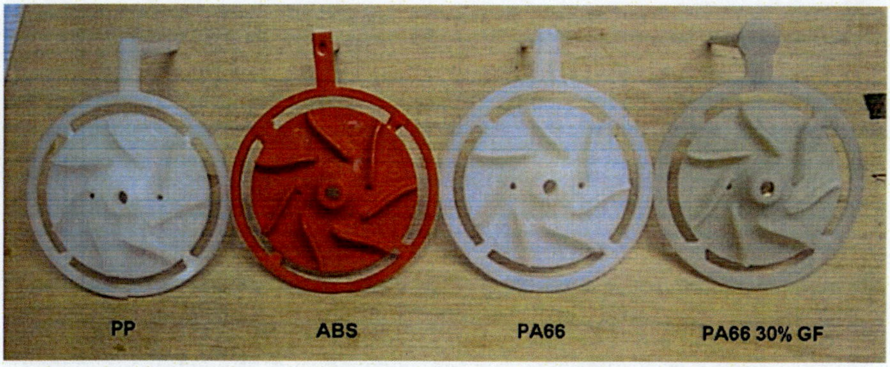

Fig. 10.2 Parts requiring different gating arrangements according to molding material

As with all mold design, thermal expansion of the cavity should also be taken into account as this expansion has a resultant effect on injection molded part accuracy. This should consider the expected expansion rate of the epoxy and/or acrylic composition of the given SL material.

In addition to this, the author would also like to add that the gating arrangement employed can have great influence on localized tool wear. Conventional mold design often dictates the use of minimal gating contacts partly to aid material-flow characteristics and also to enable easier sprue removal. Depending on the temperatures and viscosity of the molding material used, this may lead to localized heat and pressure increases in SL molds, resulting in premature localized failure. Examples are illustrated, where the same part in different molding materials has required different approaches to the gating design including the volume of the land, runner, and gates (Fig. 10.2).

10.5 Process Considerations

The key to successful SL tooling is to understand the demands of the mold design and injection molding parameters, which are very different from those for metal molds. It has also been demonstrated that appropriate choices in mold design and process variables can reduce the risk of failure in SL tooling [12–15].

10.5.1 Stereolithography Mold Production

The physical properties of an SL part and thus its performance in injection molding, is influenced by its state of photopolymer cure, which is related greatly to the degree of energy exposure received during its production [16]. During the SL build process, this amount of energy exposure is specific to the laser type used (which differs according to the SL machine) and the build parameters used. A pulsed mode laser system's degree of curing is dependent upon the pulse frequency and the hatch spacing. Generally a continuous mode laser system allows for greater energy exposure.

With respect to post-curing operations it should be noted that the amount of curing is not greatly affected by UV environment exposure [16]. If thermal post curing is to be used it should also be noted that a large majority of warpage occurs during this stage [17], which may be a concern if thin walled sections are in existence.

The layer thickness of each build slice dictates the SL part's roughness on surfaces parallel to the build direction. When this surface roughness is parallel to a molded part's direction of ejection it has a resultant effect on the force required to remove the part from the mold which in turn applies a force to the insert which could result in damage. This surface roughness and the ejection forces experienced, correspond linearly to the build layer thickness [18]. The solution is to re-orientate the SL build direction or employ a lesser layer thickness.

10.5.2 Injection Molding

During molding, a release agent should be frequently used to lower the force experienced by the tool due to part ejection. In the author's experience, a silicone-like agent is the most successful. Low-injection pressures and speeds should be used whenever feasible. Much lower settings are feasible in comparison to some forms of metal tooling due to SL heat transfer characteristics as discussed within this section.

Early recommendations for SL injection molding stated that since damage occurs during part ejection it was appropriate to allow as much cooling prior to mold opening as possible. This reduced the tendency of the parts to stick to the inserts [6]. The author has trialled this approach, which often leads to greater success, but the part-to-part cycle times are extremely long.

More recent work [19] has demonstrated that it is advantageous to eject the part as soon as possible (when part strength allows) before the bulk mass of cavity features have exceeded their glass transition point, when their physical strength is greatly reduced. This greatly reduces the heat transmitted into the tool and the cycle time for each part.

Subsequently, it is also critical to monitor the mold temperature throughout the molding cycle to avoid exceeding the glass transition temperature (T_g) of epoxy, where tool strength is reduced. This entails each molding cycle beginning with the epoxy insert at ambient temperature and the part being ejected prior to T_g of the majority of the inserts volume being reached. This has been achieved in practice by inserting thermocouples from the rear of the cavity insert into the most vulnerable mold features such that the probe lies shortly beneath the cavity surface.

Allowing the polymer to remain for sufficient time within the mold, while also avoiding critical T_g, is possible due to the very low thermal conductivity of SL materials.

In addition, the low thermal conductivity of SL materials has been demonstrated to be advantageous in this application for injection mold tooling. It has been shown that the low thermal conductivity of SL tooling allows the use of low injection speeds and temperatures which are required due to the limited mechanical properties of SL materials [20]. Traditional metal tooling needs these high pressures and speeds to prevent the injected polymer freezing prior to the mold completely filling. This is due to the rapid cooling of the injection melt when it comes into contact with the high thermal conductivity mold surface.

Also, the SL tooling process has shown itself to be capable of producing parts that would not be possible under the same conditions using a metal mold. The thermal characteristics of SL tooling have made it possible to completely mold crystalline polyether ether ketone (PEEK), which has an injection temperature of 400°C (752°F). An equivalent steel mold would require a premolding temperature of about 200°C (392°F). An impeller geometry was successfully molded with vastly lower injection speeds and pressures were utilized, as shown in the Table 10.1 and Fig. 10.3.

Table 10.1 Polyether ether ketone molding variables in SL mold vs. steel mold [21]

	Polymer melt temperature (°C)	Mold preheat temperature (°C)	Injection pressure (bar)	Injection speed (mm/s)
SL mold	400	23 (room temp)	30	10
Requirements in an equivalent steel mold (approx)	400	200	650	70

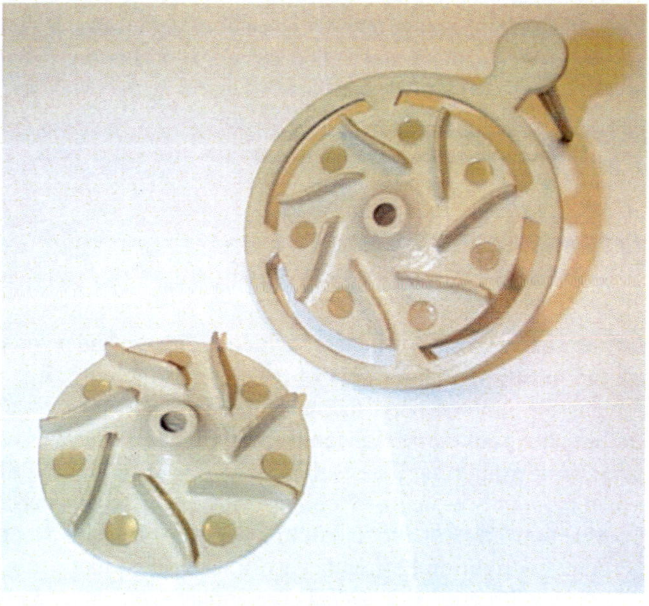

Fig. 10.3 PEEK impeller molded by stereolithography tools [21]

A particularly illustrative account of the cooling conditions is shown in the above image. It can be seen that the polymer is primarily gray in color where it contacts SL surfaces indicating crystalline formation. Whereas where it comes into contact with the steel ejector pins it is brown, indicating localized amorphous areas. This is due to the difference in heat transfer of the two materials and hence the cooling rate experienced by the contacting polymer.

10.6 Process Considerations

Various polymers have been successfully molded by SL injection molding. These include polyester, polypropylene (PP), polystyrene (PS), polyamide (PA), polycarbonate, PEEK, acrylonitrile styrene acrylate, and acrylonitrile butadiene styrene [2, 17, 22].

The greatest material limitation encountered has been the use of glass filled materials. All evidence indicates that the SL molding technique does not cope well with glass filled materials due to severe problems of abrasion to the SL cavity surface [2, 17, 22, 23]. This leads to poor quality, inaccurate parts, and undercuts in the cavity, which eventually result in the destruction of the SL insert. This abrasive nature has been quantified with a comparative SL molding study of PA 66 and PA 66 with 30% glass fiber content. The PA 66 enabled 19 shots prior to damage, while the glass filled variant allowed only 6 shots before the same level of damage was incurred [17]. These findings are supported by work conducted by the author, with PA 66 with a 30% glass fiber content inducing high mold wear. However, it has been demonstrated that appropriate choices in mold design and process variables reduced the rate of wear. The use of appropriate settings has allowed the successful molding of a low number of parts as large as 165 × 400 × 48 mm (6.5 × 16 × 2 in.) with high geometrical complexity in PA66 with 30% glass content. The tool and parts are shown in Figs. 10.4 and 10.5.

10.7 Molded Part Properties

During the course of my work with SL tooling, I have endeavored to investigate and pursue the most important aspect of tooling and molding; it is a means to an end. The end is the molded parts themselves. These are the products and if they are unsuitable, then tool performance is entirely irrelevant.

Early work examining the resultant parts produced by the SL injection process described them only as being of a poor quality, effected by warping, and requiring a longer time to solidify due to the mold's poor heat transfer producing a nonuniform temperature distribution [24]. Other work also noted that using differing materials in a mold's construction (i.e. a steel core and a SL cavity) led to warping of the part due to the different thermal conductivities of the mold materials [10].

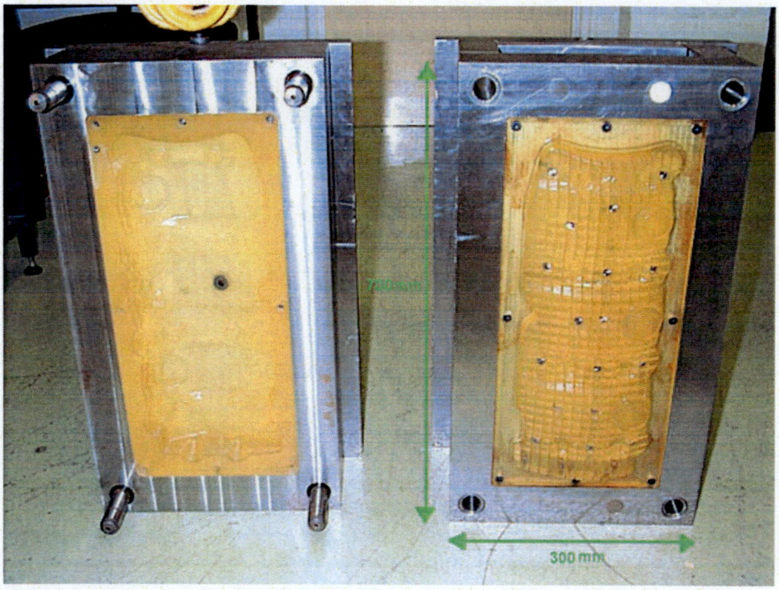

Fig. 10.4 Large stereolithography molding tool

Fig. 10.5 Subsequent parts produced in polyamide 66 (30% glass fiber)

The low thermal conductivity, and hence the low cooling rate, of the mold has a significant influence on the material properties of the molded parts. It was shown that parts from an epoxy mold exhibit a higher strength, but a lower elongation; around 20% in both cases [22].

The differing mechanical properties of parts produced from SL molds as compared to those from metal tools is also demonstrated in other work [10]. This showed that the parts manufactured by SL molding had a lesser value of Young's Modulus compared to those produced in a steel mold but possessed a greater maximum tensile strength and percentage elongation at break. These different part properties were attributed to a slow rate of heat transfer of the tool. This slow rate of heat transfer produces longer part cooling times giving a greater strength but less toughness [25].

Research performed at Georgia Institute of Technology further investigated the mechanical properties of parts produced by the SL molding process. This work showed that noncrystalline and crystalline thermoplastic parts produced by the SL molding technique displayed differing mechanical characteristics than parts from traditional molds. Noncrystalline material parts possessed similar all-round mechanical properties compared to those produced in identical steel molds. However, crystalline thermoplastic parts demonstrated higher tensile strength, higher flexural strength, and lower impact properties compared to those manufactured in identical steel molds [26].

More so with crystalline polymers than with amorphous materials, the mechanical properties of the plastic parts are influenced by the cooling conditions. These differing effects on mechanical properties have been demonstrated with PS (amorphous) and PP (crystalline). When the respective part's mechanical properties were compared when produced by steel and by SL molds, the PS parts showed very little change while the PP parts demonstrated a great difference [17].

In addition to differences in mechanical properties it has also been identified that some polymers exhibit different shrinkage according to the cooling conditions of the part during molding [27–33]. These works indicate that crystalline polymers are susceptible to greater shrinkage when subjected to a slow cooling time.

These differences in part properties have been attributed to the degree of crystallinity developed in the molded parts. This has been demonstrated by microscopic comparisons of parts produced by SL and metal alloy tooling [34]. This revealed the spherulites (a crystal structure consisting of a round mass of radiating crystals) to be considerably larger from the SL tooling parts due to the higher temperatures and slower cooling involved during molding.

In the wider field of general injection molding and plastics research, work has been conducted to identify and assess the variables that influence parts properties [35–37]. These papers report a common theme, they identify the thermal history of the part to be a critical variable responsible for the parts resulting attributes.

Recent work has shown that the slower molded part cooling imposed by SL tooling provides an opportunity to make some variations in the molding parameters for crystalline polymers which allow the control of critical morphological factors (level of crystallinity) [38]. The subsequent level of crystallinity dictates many of the resultant part properties. The process modifications in this work were realized without changes to the machine, tool, or molded material (i.e. external cooling control, different polymer etc). This demonstrates a possible "tailoring" of molded part properties that would allow certain desirable part properties to be altered. These revelations demonstrate an advantage of SL tooling that was shown to not be possible in metal tooling.

In summary, we must consider that the thermal characteristics of SL molds have an influence on the morphological structure of some parts. This may lead to a difference in the morphology of parts from SL tools as compared to those from metal tools. Such morphological differences can affect the shrinkage and mechanical properties of the molded part. When using SL tooling, one must decide if these differences are critical to the functionality of the part.

10.8 Conclusion

In conclusion, SL molding is a viable process for some, but by no means all, injection molding tooling applications. Most important, is that the user should be informed of the alternate design and processing requirements compared to conventional tooling, and be aware of the difference in resultant part characteristics, thus enabling realistic expectations and a more assured project outcome.

References

1. Harris, R.A., 2001, Direct AIM Tooling Section, Wohlers Report 2001, Rapid Prototyping & Tooling State of the Industry Annual Worldwide Progress Report, Wohlers Associates Inc., USA, Part 3: Tooling, Page 61.
2. Luck, T., Baumann, F., and Baraldi, U., 1995, Comparison of Downstream Techniques for Functional and Technical Prototypes – Fast Tooling with RP, Proceedings of 4th European RP Conference, 13th–15th June, Belgriate, Italy, pp 247–260.
3. Roberts, S.D. and Ilston, T.J., 1998, Direct Rapid Prototyping Injection Moulding Tools, Proceedings of 7th European Conference on Rapid Prototyping & Manufacturing, 7th–9th July, Aachen, Germany, Ed. by Campbell, R.I., pp 457–470.
4. Janyanthi, S., 1997, Stereolithography Injection Moulds for Direct Tooling, Proceedings of the 9th European Stereolithography Users Group Meeting, 2nd–3rd November, Florence, Italy.
5. Heath, J., 1996, Proceedings of the Society of Manufacturing Engineers Rapid Prototyping and Manufacturing Conference, 23rd–25th April, Dearborn, USA.
6. Jacobs, P., 1997, Recent Advances in Rapid Tooling from Stereolithography, 7th International Conference on RP, 31st March–3rd April, San Francisco, USA, pp 341–354.
7. Jacobs, P.F., 1996, Recent Advances in Rapid Tooling from Stereolithography, Proceedings of the 2nd National Conference on Rapid Prototyping and Tooling Research, 18th–19th November, Buckinghamshire College, UK, pp 135–152.
8. Harris, R.A. and Dickens, P.M., 2001, Mould Variables in Stereolithography Injection Moulds, Proceedings of the Society of Plastics Engineers (SPE) Annual Technical Conference (ANTEC), CRC Press, Dallas, Texas, USA.
9. Rahmati, S. and Dickens, P.M., 1997, Stereolithography Injection Mould Tool Failure Analysis, Proceedings of Solid Freeform Fabrication symposium, 11th–13th August, Austin, USA, pp 295–306.

10. Jayanthi, S., Hokuf, B., McConnell, R., Speer, R.J., and Fussell, P.S., 1997, Stereolithography Injection Moulds for Direct Tooling, Solid Freeform Fabrication Symposium, 11th–13th August, Austin, USA, pp 275–286.
11. Greaves, T. and Troy, M.I., 1997, Using Stereolithography to Directly Develop Rapid Injection Mold Tooling, TCT Conference, 29th–30th September, Gaydon, UK, pp 127–133.
12. Harris, R.A., 2002, Direct AIM Tooling. Rapid Prototyping & Tooling State of the Industry Annual Worldwide Progress Report, Wohlers Report 2002, Wohlers Associates Inc., USA, Part 3: Tooling, Page 70.
13. Palmer, A. and Colton, J., 1999, Design Rules for Stereolithography Injection Moulding Inserts, Proceedings of the Society of Plastics Engineers (SPE) Annual Technical Conference (ANTEC), New York, USA, pp 4002–4006.
14. McDonald, J.A., Ryall, C.J., and Wimpenny, D.I., 2001, Rapid Prototyping Casebook, Professional Engineering Publishing, London.
15. Hilton, P.D. and Jacobs, P.F., 2000, Rapid Tooling: Technologies and Industrial Applications, Dekker, New York.
16. Jayanthi, S., 1998, Influence of Laser Systems on the Physical Properties of Parts Imaged Using Stereolithography, North American Stereolithography Users Group meeting, 1st–5th March, San Antonio, USA.
17. Schulthess, A., Steinmann, B., and Hofmann, M., 1996, Cibatool™ SL Epoxy Resins and Some New Applications, Proceedings of the 1996 North American Stereolithography Users Group Meeting, 10–14 March, San Diego, USA.
18. Harris, R.A., Newlyn, H.A., and Dickens, P.M., 2002, The Selection of Mould Design Variables in Direct Stereolithography Injection Mould Tooling, Journal of Engineering Manufacture, IMechE Proceedings Part B, vol. 216, no. B4, ISSN 0954-4054, pp 499–505.
19. Hopkinson, N., 1999, Investigation into Part Ejection and Heat Transfer in the Direct AIM™ Process, PhD thesis, De Montfort University, Leicester, UK.
20. Harris, R.A., 2002, The Injection Moulding of PEEK Using Stereolithography Moulds, RAPTIA Newsletter No.7, April 2002, Grafisk Profil, Denmark, pp 6–7.
21. Harris, R.A., Hague, R.J.M., and Dickens, P.M., 2004, Thermal conditions in stereolithography injection mould tooling and their use for polyether-ether-ketone moulding, International Journal of Production Research, vol. 42, no. 1, pp 119–129.
22. Eschl, J., 1997, Experiences with Photopolymer Inserts for Injection Moulding, European Stereolithography Users Group meeting, 2nd–5th November, Florence, Italy.
23. Stierlen, P., Dusel, K.-H., and Eyerer, P., 1997, Materials for Rapid Tooling Technologies, 8th European Conference on RP&M, Nottingham, UK.
24. Li, Y., Keefe, M., and Gargiulo, E., 1997, Studies in Direct Tooling by Stereolithography, Proceedings of the 6th European Conference on Rapid Prototyping in Manufacturing, 1–3 July, Nottingham, UK.
25. Dusel, K.-H., 1997, Materials for Rapid Tooling Technologies, Society of Manufacturing Engineers Rapid Prototyping and Manufacturing conference, 22nd–24th April, Dearborn, USA.
26. Dawson, K., 1998, The Effect of Rapid Tooling on Final Product Properties, Proceedings of North American Stereolithography Users Group Meeting, 1st–5th March, San Antonio, USA.
27. Damle, M., Mehta, S., Malloy, R., and McCarthy, S., 1998, Effect of Fibre Orientation on the Mechanical Properties of an Injection Molded part and a Stereolithography-Insert Molded Part, Proceedings of the Society of Plastics Engineers (SPE) Annual Technical Conference (ANTEC), Atlanta, USA, pp 584–588.
28. Velarde, D.A. and Yeagley, M.J., 2000, Linear Shrinkage Differences in Injection Moulded Parts. Plastics Engineering, The Society of Plastics Engineers, December 2000, pp 60–64.
29. Gipson, P., Grelle, P., and Salamon, B., 1999, The effects of process conditions, nominal wall thickness, and flow length on the shrinkage characteristics of injection molded polypropylene, The Journal of Injection Molding Technology, vol. 3, no. 3. pp 117–125.

30. Patel, P., 1997, Effect of Processing Conditions on the Shrinkage and Crystallinity of Injection Moulded Parts, Proceedings of the Society of Plastics Engineers (SPE) Annual Technical Conference (ANTEC), Toronto, Canada, pp 3632–3635.
31. Tursi, D. and Bistany, S.P., 2000, Process and tooling factors affecting sink marks for amorphous and crystalline resins, Journal of Injection Molding Technology, vol. 4, no. 3, pp 114–119.
32. Segal, J.I. and Campbell, R.I., 2001, A review of research into the effects of rapid tooling on part properties, Rapid Prototyping Journal, vol. 7, no. 2, pp 90–98.
33. Pierick, D. and Noller, R., 1991, The Effect of Processing Conditions on Shrinkage, Proceedings of the Society of Plastics Engineers (SPE) Annual Technical Conference (ANTEC), Montreal, Canada, pp 252–253.
34. Langen, M. and Michaeli, W., 1997, Temperature Distribution in Rapid Tooling Cavities – How Close are Injection Moulded Prototypes to Series Parts? Proceedings of the Society of Plastics Engineers (SPE) Annual Technical Conference (ANTEC), Toronto, Canada, pp 939–945.
35. Yang, T. and Nunn, R., 1992, Injection Molding Processing Effects on the Crystallinity of Polypropylene, Proceedings of the Society of Plastics Engineers (SPE) Annual Technical Conference (ANTEC), Detroit, USA, pp 48–53.
36. McLeod, M.A. and Baird, D.G., 1999, The Influence of Processing Variables on the Mechanical Properties of Injection Moulded Pregenerated Composites, Composites: Part B, vol. 30, pp 297–308.
37. Kantz, M.R., 1974, The Effects of Melt Processing Variables on the Morphology and Properties of Injection Moulded Polypropylene, International Journal of Polymeric Materials, vol. 3, pp 245–258.
38. Harris, R.A., Hague, R.J.M., and Dickens, P.M., 2003, Crystallinity Control in Parts Produced from Stereolithography Injection Mould Tooling, Proceedings of the Institution of Mechanical Engineers, Part L, Journal of Materials: Design and Applications, vol. 217, no. L4, pp 269–276.

Chapter 11
Photonic and Biomedical Applications of the Two-Photon Polymerization Technique

Aleksandr Ovsianikov, Maria Farsari, and Boris N. Chichkov

11.1 Introduction

Since first experimental demonstration of microstructuring using two-photon polymerization (2PP) [1], the technology has experienced rapid development. The unique capability of this technique to create complex 3D structures with resolution, reproducibility, and speed superior to other approaches paved its way to applications in many areas. Figure 11.1a shows some SEM images of structures fabricated by 2PP for demonstrational purposes. Microvenus statues fabricated from negative photoresist SU8 [2] material are presented in comparison to the human hair. Each statue is about 50 μm tall and 20 μm wide, the overall fabrication time is just few minutes. Figure 11.1b shows an array of microspiders fabricated on a glass slide. Each structure is about 50 μm wide and the spider's body is supported by eight 2 μm thick legs. Finally, a fragment of a windmill array (Fig. 11.1c), produced by 2PP using Ormocore [3] is shown. Fabricated in a single step, the structure consists of two physically separate parts – windmill body and propeller, which are interlocked in such way that the propeller can be rotated around the shaft. Therefore, using 2PP microfabrication it is possible to produce functional micromechanical components in a single step, without the necessity of tedious assembly procedure. Looking at these images, one can see the strength of 2PP technology and envision many potential applications.

Despite the fact that the 2PP is a relatively new technology, it has been successfully established as a microprocessing tool and its application area has been rapidly expanding in the recent years. The most prominent example is the fabrication of 3D photonic crystals. It was first demonstrated in 1999 [4, 5] and nowadays it is employed by different groups in the world. The submicrometer structural resolution of 2PP microfabrication has allowed the realization of 3D

B.N. Chichkov (✉)
Laser Zentrum Hannover, Hannover, Germany
and
Institute of Electronic Structures & Laser (IESL), Heraklion, Greece
e-mail: b.chichkov@lzh.de

P.J. Bártolo (ed.), *Stereolithography: Materials, Processes and Applications*, 257
DOI 10.1007/978-0-387-92904-0_11, © Springer Science+Business Media, LLC 2011

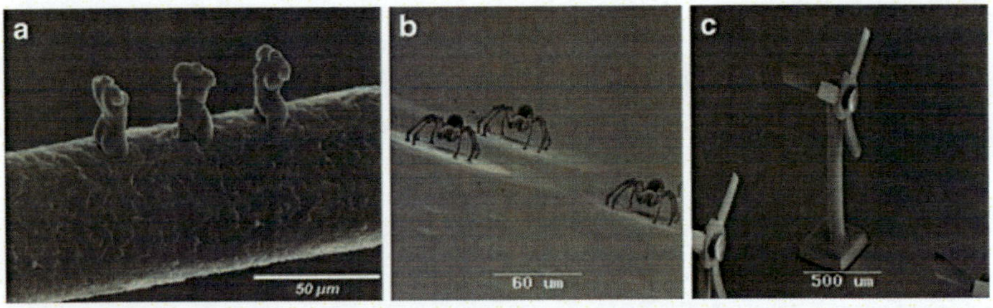

Fig. 11.1 SEM images of demo structures fabricated by 2PP technique: (**a**) microvenus statue shown in comparison to a human hair; (**b**) micro-spider array; (**c**) functional micro-windmill

photonic crystals operating in the near-IR spectral range [6–8]. Most importantly, in contrast to most rival methods, 2PP has the capability to create arbitrary structures and therefore to introduce defects at any desired location of a 3D photonic crystal.

2PP microfabrication has also been used for the fabrication of less demanding micro-optical components such as waveguides [9], diffractive optics, [10] and refractive micro-optical elements [11]. A distributed feedback microlaser has been demonstrated by 2PP of an optical gain medium by Yokoyama et al. [12]. Apart from that, 2PP is also used for the fabrication of micromechanical systems [13], plasmonic components [14], biomedical devices [15], and scaffolds for tissue engineering [16]. One of the most exciting application examples for the field of biotechnology is direct structuring of biomolecules. The group of Campagnola experimentally demonstrated cross-linking of proteins by use of fs lasers [17, 18]. Functionalization of surfaces with biomolecules via photoinduced process using fs lasers has also been considered and recently demonstrated [19]. Optically actuated micromechanical devices fabricated by 2PP have been extensively investigated for applications in microfluidic systems and cell manipulation [20, 21]. The photosensitive materials currently used for 2PP microfabrication are intrinsically dielectric. Recent reports demonstrated various approaches for metallization of 2PP produced structures [22–25]. Microelectronical components such as microinductor coils have been created this way [26].

In the following paragraphs we will describe in detail some applications of 2PP in photonics and in biomedicine. More specifically, we will discuss the fabrication of micro-optical elements, woodpile, and spiral photonic crystals as well as microneedles for transdermal drug delivery, microprosthesis, and scaffolds for tissue engineering.

11.2 Fabrication of Micro-Optical Elements

Conventional methods used for fabrication of refractive micro-optical elements, such as lithography or reflow, often lack flexibility when it comes to redefining the element design or implementing new solutions [11]. In the initial micro-optical element array prototype fabrication and testing stage, application of 2PP provides considerable advantages. Many materials that can be structured by 2PP are actually designed for optical applications, and therefore possess superior optical properties. Produced structures exhibit very low surface roughness. During fabrication design of single elements can be freely adjusted directly from CAD input and series of identical patterns can be produced. Patterns fabricated by 2PP technique can be accurately replicated using established techniques as soft lithography, injection molding, or electroplating. Therefore the original structure can be microfabricated by 2PP, while replication provides cost-effective route to mass production.

Here, we use microprisms as an example for the fabrication of refractive micro-optical elements by 2PP. Since 2PP microfabricated structures consist of single voxels, "surface discretization" is one of the practical aspects that has to be taken into account. Most often 2PP structures are produced in a layer-by-layer manner, with layers parallel to the substrate. This approach is inherited from stereolithography and by no means is the only one possible by 2PP. Figure 11.2 presents different slicing strategies that can be used to fabricate a prism; single polymerized voxels are represented by ellipsoids. Conventional slicing approach illustrated in Fig. 11.2b would result in unacceptable step-like surface. In order to improve surface quality one can let the layers overlap, as it is shown in Fig. 11.2c. On one hand enhancing the voxel overlap in vertical direction improves surface quality; on the other hand the overall fabrication time is increased accordingly. The best solution for the prism slicing is presented in the Fig. 11.2a. Here the sliced layers are oriented parallel to the prism surface. The overlap of the voxels and the distance between the slices is the same in both cases (Fig. 11.2a, b); resulting fabrication time will also be similar. Nevertheless the surface will be smoother for the structure sliced parallel to the outer prism facet.

To illustrate the approach presented in Fig. 11.2c, complex microprisms with facets inclined at few different angles were fabricated. Here a 100× immersion oil objective (NA = 1.4) was used for the fabrication. The slicing is performed parallel to the substrate plane. The parameters used for fabrication are as follows: in-layer scan distance of 370 nm, interlayer distance of 300 nm, and the average laser power of 70 mW at the scan velocity of 320 μm/s. The disadvantage of this approach is readily observed by comparing the quality of the visible facets having four different inclination angles. While most surfaces are of high quality and exhibit low roughness, the step-like features emerge as the angle between the facet and the substrate

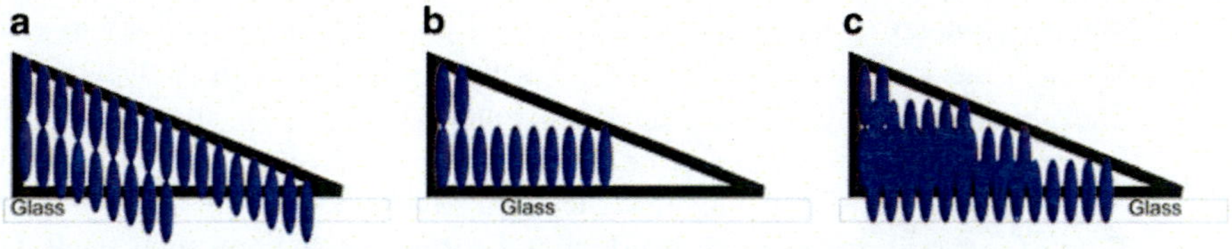

Fig. 11.2 Schematic illustrations of different slicing strategies on the example of the microprism (side view)

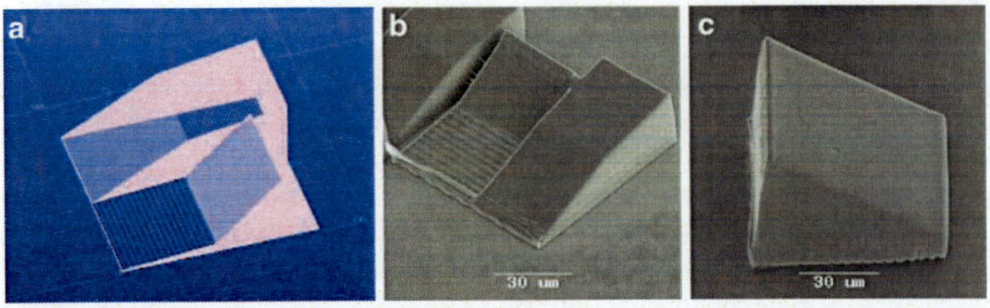

Fig. 11.3 Refractive element consisting of microprisms with few different facet inclination parameters (**a**) original design, (**b**) SEM image of according microstructre fabricated by 2PP technique; (**c**) SEM image of single simple microprism fabricated by 2PP technique

becomes very small (left-most part of the prism in Fig. 11.3b). This feature is not surprising, since it already appears in the slices of the original design. One way to improve the situation would be to further reduce the distance between the slices, which will result in a proportional increase in fabrication time. A better solution is to use the approach illustrated in the Fig. 11.2a. Using this approach, a series of microprisms were fabricated. An example of such 2PP produced microprism is shown in Fig. 11.3c. The fabrication of such structures does not require very high resolution in the vertical direction, which allows the use of 50× microscope objective (0.8 NA) lens instead of high NA optics, to focus the laser beam. The material used in this case is a standard Ormocore formulation. Key performance parameters of the micro-optical elements are the deviations of the surface profile in terms of peak to valley deviation and surface roughness. From the SEM image some conclusions about the surface roughness can be drawn; it is however not easy to analyze the overall conformity of the produced prism with the original design.

11.3 Photonic Crystals: Engineering the Propagation of Light

Photonic crystals are artificial dielectric microstructures exhibiting periodic modulation of refractive index [27, 28]. Due to this periodicity they can diffract electromagnetic waves in a way that prevents their propagation. In analogy to solid-state physics, the range of wavelengths, in which no propagating states exist, can be interpreted as a gap in the dispersion function; it is also known as a photonic bandgap. The most prominent example of one-dimensional photonic crystals are dielectric mirrors or multi-layered dielectric stacks, exhibiting high reflectivity for light of a certain wavelength incident at a certain angle upon such stack. Here the periodicity along one direction is achieved by the variation of dielectric indices of stacked dielectric layers.

The main distinction of a 3D photonic crystal is that one can design structures where photonic bandgaps for different light propagation directions overlap, resulting in a complete or omnidirectional photonic bandgap – a frequency range for which the propagation of light is forbidden in any direction. The central frequency of the bandgap or the photonic crystal operation wavelength is roughly on the order of the half of the dimension of periodicity. Therefore, by tailoring the structures dimensions and dielectric index contrast, the position of the central frequency of the photonic bandgap can be precisely adjusted. For 3D photonic crystals the periodicity has to be achieved for any given propagation direction inside the structure. It is commonly done by sequential succession of air and dielectric regions inside the structure.

For the past decades, semiconductor technology has played a role in almost every aspect of our daily lives. The drive towards miniaturization and high-speed performance of integrated electronic circuits has stimulated considerable research around the world. Unfortunately, miniaturization result in circuits with increased resistance and higher levels of power dissipation. In an effort to further the progress of high density integration and system performance, scientists are now turning to light instead of electrons as the information carrier. Devices based on photons instead of electrons as information carriers have potential to revolutionize the information technology [29, 30].

Since the introduction of the principle of the photonic bandgap, the design and construction of 3D photonic devices has been the subject of much research [31–35]. Devices based on photonic crystals allow tailoring the propagation of light in a desired manner. Many fascinating physical phenomena occur in such structures: control of spontaneous emission [36], sharp bending of light [37], lossless guiding [38], zero-threshold lasing [39], trirefrigence [40]. Futuristic prospects include not only applications in telecommunications as all-optical signal processing, but also as "transistors" for light and optical computers.

However, the fabrication of devices, operating in the near-IR and visible spectral range is a major technological issue, since the requirement is for structures of millimeter scale to be fabricated with nanometre resolution. Already back in 1990 some believed that photonic bandgaps in this spectral range are impossible to achieve [41]. First successful attempts to produce 3D photonic crystals were based on intrinsically planar semiconductor technology approaches, therefore a layer-by-layer structure, known as woodpile (see Sect. 3.2.2 for more detail), was first to be realized. Lin et al. used repetitive deposition and selective etching of multiple dielectric films to fabricate a five layer silicon woodpile, exhibiting a bandgap centered at around 11 μm [42]. Chutinan et al. also used deposition of multiple semiconductor films [43, 44]. In order to obtain the rods, the upper layer, consisting of GaAs, was structured by electron beam lithography or reactive ion etching. The stripe patterns were than assembled on top of each other and joined using wafer bonding technique. The carrier substrate material was then removed by wet etching, revealing a GaAs woodpile containing four layers. Later, using this method, the first woodpile operating at the telecommunication range has been realized [37]. An even more tedious approach was undertaken by

Garcia-Santamaria et al. [45], they have reproduced a 3D diamond lattice by assembling latex and silica spheres using a nanorobot and an SEM for live observation. In a subsequent step, latex spheres were removed to reveal a 3D lattice consisting of silica spheres only, no bandgap effect was demonstrated for this structure.

Currently, the most promising approaches for photonic crystal fabrication are colloidal self-assembly, holographic photofabrication, and direct laser writing by 2PP. In colloidal self-assembly, spherical beads from silica, polystyrene, or polymethylmethacrylate, are used to produce 3D structures also known as opals. This method is based on natural tendency of such particles to organize into closely packed 3D structures characterized by a face-centered cubic (fcc) symmetry. Theoretically, large scale photonic crystals can be produced in a fast and potentially inexpensive way. The main problems originate from unavoidable presence of disorder, which can occur due to polydispersity of the microspheres or inherent stacking defects, developing during the opal growth process. Many different self-assembly methods have been developed in order to improve the situation. Currently, the most popular approach is vertical sedimentation, where a carrier substrate is submerged into a colloidal suspension. While solvent is allowed to evaporate, the meniscus, which is formed at the liquid–air interface on the substrate, helps to deposit the spheres in a crystalline arrangement [46]. Such opal photonic crystals allowed to study some properties of 3D photonic crystals experimentally [47]. Despite the fact that by definition opals are photonic crystals, due to low porosity, direct configuration cannot have a complete photonic bandgap. These structures are mostly used as a template for obtaining an inverted opal photonic crystals, which, provided that the refractive index contrast exceeds 2.8, can exhibit a complete photonic bandgap [48, 49]. For infiltration, methods that allow filling the air gaps of an opal with high refractive index materials were developed [50]. After such infiltration procedure, the opal itself is etched way, revealing a porous 3D mesh consisting of high refractive index material. Inverse opal photonic crystals consisting of silicon with an omnidirectional bandgap at 1.5 μm were first successfully demonstrated by Blanco et al. [51]. Many photonic crystal applications, such as low-loss waveguides or zero-threshold optical cavities, rely on the presence of line or point defects, introduced into the 3D structures in a controlled manner. These defects cannot be formed by self-assembly. Lee et al. have used 2PP to write waveguides inside an opal whose voids were filled with a liquid photosensitive material [52]. In order to produce opals shaped as prisms [53] or microfibers [54], patterned surfaces can be used as substrates.

Holographic recording is a method that allows the fast fabrication of up to few millimeter large 3D crystals with long range periodicity. The 3D interference patterns, produced by a multiple-laser beams, is a well studied subject, as it is used for the trapping of atoms. The idea is to record spatial intensity distribution, created by two or more beams, into a photosensitive material. If the beam parameters are chosen properly, it is possible to polymerize photosensitive resin at the locations with higher intensity, while leaving the rest unpolymerized. That is, one has to choose intensities at the maxima to be larger than the threshold intensity for polymerization, while intensities at the minima have to be lower than the polymerization threshold. Tuning the intensity inside this working window changes the filling factor and therefore has an effect on stability of the structure. The nonpolymerized material is then washed out to reveal a solid 3D mesh in air. Depending on the number of beams employed and their arrangement, different lattices can be obtained. The period of the produced photonic crystals depends on the wave vectors of the used laser beams. Campbell et al. have reported results on using interference of four UV laser beams to create periodic 3D pattern in 30 μm thick commercial lithographic SU8 photo resist. Structures with sub-micrometer features were produced by a single 6 ps pulse of a UV laser [55]. Shortly after that Shoji and Kawata have reported fabrication of 3D hexagonal photonic crystal lattice by a two step process: in a first step, three-beam interference was used to obtain 2D hexagonal arrangement of rods while in a second, immediate step, two more interfering beams formed additional layers which are perpendicular to the rod array [56]. In 2003 Shoji et al. also demonstrated the fabrication of a woodpile structure by two-step four-beam interference [57]. Several groups used single-step exposure four-beam interference to fabricate fcc lattices with different filling factors and have optically characterized them [58, 59]. In these works umbrella-like beam configuration was used, where a central beam symmetrically surrounded by three other beams.

The fabrication of 3D periodic lattices by holographic illumination of photosensitive materials has a great potential as a technology for mass production. Since the required illumination time is quite short, the requirements for the mechanical stability of the system are not as high as in the case of 2PP microfabrication. Similarly to self-assembled opal structures, the main drawback of holographic photofabrication is that it is not possible to introduce defects in a first fabrication step. As in the case of opals an additional step, where waveguides or point defects are introduced by 2PP direct writing is possible. Since the sample is already a photosensitive material, no additional effort is required in order to match refractive indices. Also final sample developing is a single common step. Common to most available photosensitive materials are values of refractive index that do not allow realization of complete photonic bandgap in such holographically recorded 3D crystal. In order to overcome this limitation an additional infiltration procedure, similar to that of opal structures, is necessary.

Fabrication of photonic crystals by 2PP was first suggested by Cumpston et al. back in 1999 [5]. The structural resolution of this technology is sufficiently high for the fabrication of photonic crystals operating at the telecommunication wavelength. Compared to intrinsically serial methods, such as self-assembly or holographic microfabrication, 2PP is relatively slow, since here each structure is produced by direct laser writing. The main advantage is that 2PP allows realization of any structure, be it periodic or not. Therefore, the introduction of defects at desired locations or structure shaping do not present any difficulties [60]. This approach is more flexible in a sense that structural parameter variations can be introduced without effort. For example, in order to change the period of fabricated 3D photonic crystal, it is enough to change the design provided as input for the microfabrication setup. In case of opal self-assembly different bead size has to be used, which will eventually require adjusting the overall fabrication process parameters and conditions. In holographic microfabrication it will require changing the illumination wavelength, and therefore again readjusting the fabrication conditions. In the same manner, by providing appropriate input, different 3D structures such as woodpile [4], diamond [61], spiral-based [8] or even quasiperiodic [62] photonic crystals, can be produced by 2PP microfabrication. The refractive index of most photosensitive materials is comparably low. Just as in the case of holographic photofabrication, structures directly produced by 2PP do not possess omnidirectional photonic bandgap. Nevertheless, they can be used as a template for later infiltration with high refractive index material, in order to produce 3D structure with a complete photonic bandgap. This approach has been successfully demonstrated on woodpile structures produced by 2PP [63].

11.4 Fabrication of 3D Photonic Crystals Using 2PP

The woodpile configuration is still on of the most frequently used 3D photonic crystals geometries. Fabrication of a woodpile structure by 2PP has been demonstrated by several groups [4, 64, 65]. One important advantage of this particular geometry is the possibility of straight-forward visual analysis of possible defects and fabrication process failures. In contrast to previously proposed fabrication methods, in the case of 2PP the cross-section of the fabricated rods is elliptical. This is an intrinsic property of 2PP microfabrication, related to the fact that the two-photon point-spread function is elongated in the axial direction, owing to the finite numerical aperture of the focusing lens. Since every voxel reflects the intensity distribution in the laser focus, it will resemble an ellipsoid stretched along a beam propagation direction. In addition, the woodpile rods of neighboring layers overlap to a certain degree in order to provide structural stability. This contributes to a change in the dielectric filling fraction. Generally, for the 2PP microfabrication of woodpiles, the distances between the rod geometrical axes are defined in the positioning algorithm. So, while keeping the in-layer rod distance and interlayer distance fixed, it is possible to independently tune the rods cross-section parameters by adjusting the illumination conditions – laser pulse energy and scanning speed. It is clear that increase in the rods cross-section area will lead to increase in the dielectric filling fraction. The rod obtained by a single scan has typical cross-section height to width (h/w) aspect ratios exceeding 2.5. At the expense of lateral resolution, it is possible to achieve lower aspect ratios by placing two such single-scan rods side by side. The circular spirals based fcc lattice photonic crystals, proposed by Chutinan and Noda [66] in 1998, is an example of a structure with optical properties superior to those of a woodpile. Three different configurations consisting of spiral-shaped rods arranged in simple-cubic, fcc, or body-centered cubic lattices were proposed. The fcc-based structure was predicted to exhibit maximal relative complete gap of the 28% when fabricated in silicon ($n = 3.5$) [66], which is the second largest bandgap predicted to-date. The idea of spiral configuration has emerged from connecting the lattice points of structures known to possess a

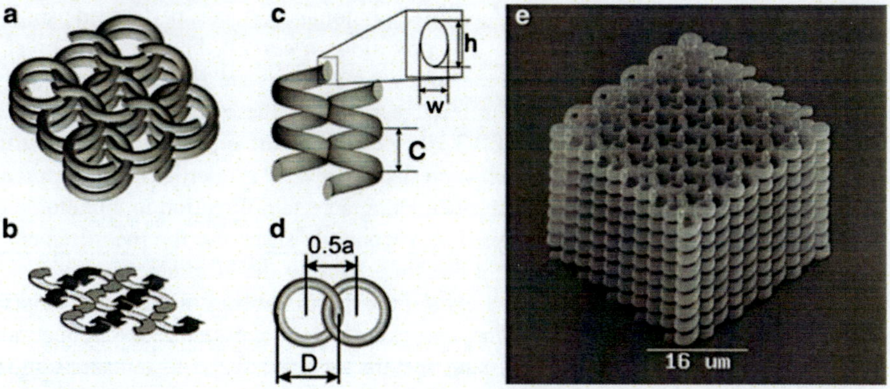

Fig. 11.4 Spiral photonic crystal: (**a**) Schematic representation; (**b**) spiral termination pattern in the upper/lower structure facet illustrates the phase shift between the neighboring spirals composing the structure; (**c**) and (**d**) show relevant structural parameters: spiral pitch *C*, diameter *D*, distance between the spirals *a*, spirals cross-sections width *w* and height *h*; (**e**) SEM image of a spiral structure fabricated by 2PP microfabrication

photonic band gap, such as diamond or woodpile. In the same manner spiral, lines can be found in all of the known three dimensional photonic crystals structures [67]. Figure 11.4a shows a schematic structural representation of such a structure. Each spiral is characterized by its pitch *C*, diameter *D*, and lateral lattice period *a* corresponding to the double distance between two neighboring spirals. Parameters *C* and *a* correspond accordingly to the height and the width of a unit cell. Two further parameters, width *w* and height *h*, describe the cross-section of spiral rods by a plane coinciding with the spiral axis. The fcc configuration is achieved by shifting adjacent spirals in by half a period as they wind in the vertical direction. Figure 11.4b shows an example of spiral photonic crystal fabricated by 2PP.

It was reported that the structural parameters that would yield the largest gap is obtained when the rods have a diameter of 0.22a (here "*a*" is the lateral lattice constant), and the spirals have a diameter of 0.32a and a pitch of *a* [66]. The fabrication of this structure, is quite complex, mainly due to the half a period shift between the adjacent objects. As a result, despite its attractiveness, the structure received minor attention since its proposal.

11.5 Photonic Crystal Characterization

The fundamental attribute of photonic crystal concept is the possibility to "design" a structure with desired optical properties. Following from the equation of Maxwell scaling, the position of the PBG can be adjusted by proportionally varying the structure dimensions. The flexibility of the 2PP technique provides the opportunity to precisely adjust and define the optical properties of fabricated photonic crystals.

Optical characterization of fabricated photonic crystals is performed using Fourier transform infrared (FTIR) spectroscopy along a certain direction in a crystal. The photonic crystals presented in the following sections are fabricated from polymeric photosensitive materials. The refractive index of these materials does not exceed $n = 2$, therefore no complete PBG is expected for the directly fabricated structures. Nevertheless, it is possible to characterize the optical properties of such structures by measuring the bandgaps, which open along a certain direction in a photonic crystal, referred further to as bandstops. For a woodpile structure, the measurement is performed along the layer stacking direction. Due to the "forbidden frequency range" effect of PBG, the spectra shall exhibit the dip in the transmission and an according peak in the reflection spectra. The reflection spectra are normalized to reflection from a silver mirror, and the transmission spectra are normalized to transmission of glass cover slip on which the structures are fabricated.

11.6 Characterization of Woodpile Photonic Crystals Fabricated by 2PP

For a woodpile photonic crystal with fcc symmetry, the ratio between the in-layer rod period and a unit-cell height is fixed to $c/d = \sqrt{2}$. The distance between the adjacent layers is $dz = (1/4) \times c = (1/4) \times d \times \sqrt{2}$. Here, this ratio is varied by introducing a scaling factor s, i.e. $dz = s \times \text{fcc}$ denotes $dz = s \times (1/4) \times d \times \sqrt{2}$.

Figure 11.5 presents the results of the FTIR measurements on a woodpile structure fabricated using a zirconium-containing hybrid material [68]. The in-layer photonic crystal period d was varied between 1.2 and 1.8 μm. The interlayer distance was chosen to be 0.6 × fcc, therefore here the unit cell "width to height" ratio is preserved for all structures and essentially the dimension of the unit cell is adjusted. The theoretically predicted bandstop position for such structure is at 0.66 (normalized frequency). At scanning speed of 200 μm/s, the applied laser power was adjusted in the range of 8–12 mW, such that the dielectric filling fraction for structures with different in-layer rod distance is approximately constant.

FTIR spectra, obtained for the fabricated structures, indicate two clear bandstop positions. In accordance to Maxwell equation scaling – the central frequency of a bandstop is shifting to shorter wavelengths as the unit cell size is reduced. In addition, the spectra show appearance of the higher order bandgaps for all the structures (e.g. at 1.5 μm for $d = 1.6$ μm), indicating the high quality of the fabricated samples.

The transmission suppression of up to 60% is achieved in fabricated structures. Since the reflection peak measurement is much more sensitive to scattering inside the photonic crystal and on the woodpile-glass/woodpile-air interfaces, the reflection peak amplitudes do not exceed 20%. The absorption bands at around 3 and 3.4 μm originate from the absorption of the material (C–H stretch vibrations in the polymer network), as has been confirmed by measurements on the flat, unstructured layers.

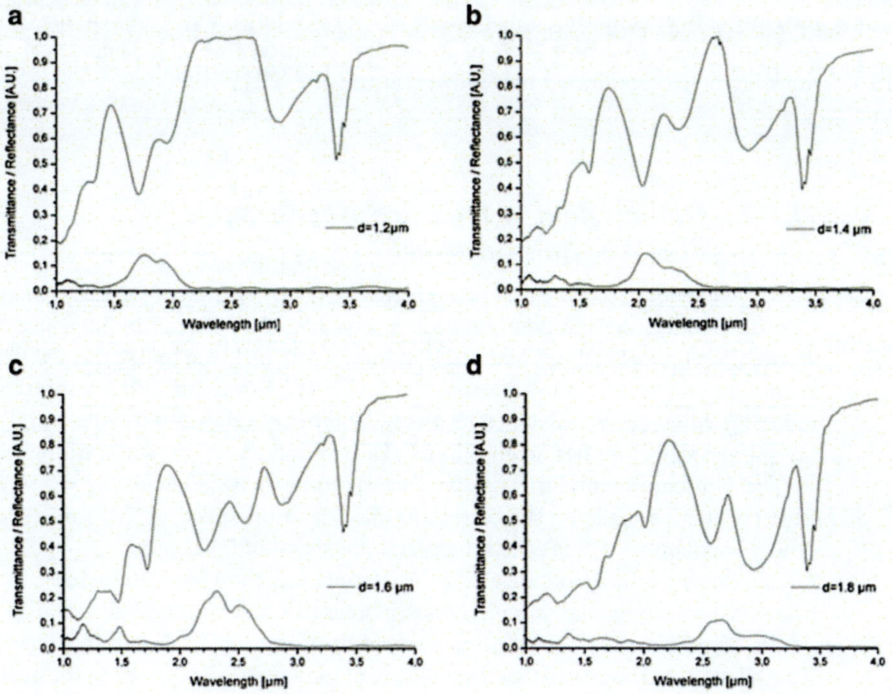

Fig. 11.5 Measured reflection/transmission FTIR spectra of the fabricated woodpile structures. The central stopband position is adjusted between 1.7 and 2.7 µm by variation of the lattice constant "*d*"

The non-structured material is fully transparent in the range of 550–2,700 nm (see Figs. 11.3.3 and 11.3.25). The relative drop in the transmission values at higher frequencies (to the shorter wavelength side from the bandstop dips) is associated with the known effect of light coupling to higher modes in the photonic crystal structure [69].

The values for central stopband frequency obtained from the theoretical simulations are accordingly 1.8, 2.1, 2.4 and 2.7 µm for the spectra shown in Fig. 11.5a–d. Compared to these values, the observed central position of the measured reflection peak maxima and transmission dip minima is blue shifted by around 100 nm. The observed blue shift, and splitting of the absorption and transmission peaks (e.g. for the period of 1.6 µm this splitting occurs around 2.4 µm) can be explained by taking a closer look at the experimental setup used for the FTIR transmission measurements. In order to focus the beam to the size of the fabricated photonic crystal in a wide range of measured wavelengths, a Cassegrain reflector optical assembly is used. In contrast to the ideal case, where collimated measuring beam is perpendicular to the surface of the structure, this assembly provides illumination of the structure with a hollow light cone having an acceptance angle between 15 and 30°. Previous studies on 3D photonic crystals have shown that scattering of the measuring beam entering the photonic crystal at a large angle leads to the reflection peak splitting and a blue shift of its central position [69–71]. Theoretical simulations have also confirmed these observations [69, 72].

11.6.1 Optimisation of the Interlayer Distance of a Woodpile Structure

Conventionally, the fcc configuration is considered to be the optimal one for the woodpile photonic crystal. Variation of interlayers distance in the vertical direction is another way to adjust the relative width of PBG. In case of 2PP microfabrication, reducing the interlayer distance results in a higher overlap between the neighboring layers, and therefore has an additional effect of increased dielectric filling fraction.

The fabrication conditions as well as the in-layer rod distance ($d = 0.9$ μm) are kept constant for all the structures. At the scanning speed of 200 μm/s, the laser power was kept at 4.5 mW. Each crystal consisted of 24 layers.

"Stretching" the woodpile along the vertical direction leads to increase in reflection peak amplitude and simultaneous increase of absorption peak (see Fig. 11.6). At the same time the central frequency of the bandstop is shifted to longer wavelengths. It is also observed that the splitting of the spectra disappears as the quantity of s is increased. Similar behavior, at bandstop positions around 2 μm, has been reported previously (also see [64]). The maximum reflection peak amplitude is obtained for the s value of 1.05, while the most expressed transmission dip is found for $s = 0.9$.

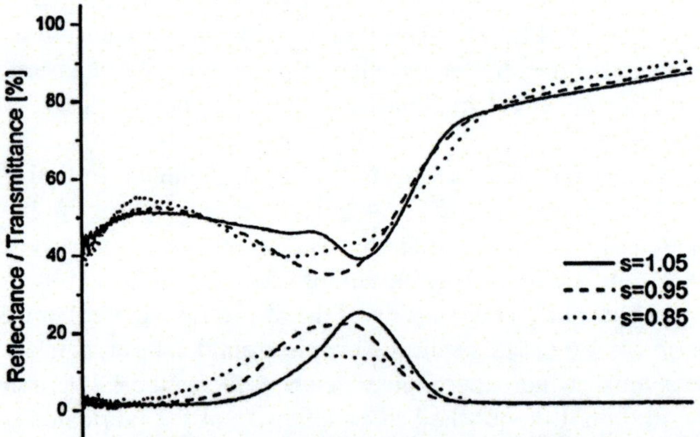

Fig. 11.6 Measured reflectance/transmittance spectra of the fabricated woodpile structures with varying distance between the layers

11.6.2 *Effect of the Dielectric Filling Fraction Value on the Quality of the Bandstop*

As it has been described in the theoretical section, the dielectric filling fraction has a great effect on the size and the position of the PBG. It is therefore important to investigate the effect of dielectric filling fraction (Φ) variation during 2PP microfabrication of realistic microstructures. In 2PP microfabrication, Φ is directly adjusted by the illumination parameters (scanning speed, pulse energy), since they are in direct relation to polymerized voxel size, i.e. the woodpile rods cross-sections dimensions. Figure 11.7 shows the reflection spectra for woodpile photonic crystal structures having different Φ values. The in-layer photonic crystal period is fixed at 1.0 μm and the distance between the layers in *z*-direction corresponds to an fcc configuration. Each woodpile consists of 32 layers. The Φ is changed by varying the laser power in the range 5–7 mW in steps of 0.5 mW. The scanning speed is kept at constant value of 200 μm/s. The spectra indicate the clear bandstop positions, with central frequency shifting to shorter wavelengths as the Φ is decreased to 46%. Simultaneously, the reflectance is increased eightfold when compared to that measured with a structure having an Φ value of 76%.

Photonic structures with Φ values below 46% were found to be too weak to survive. This is connected to the fact that further reduction of feature size for this woodpile dimensions results in structures that do not possess sufficient mechanical strength and collapse during the developing step. Therefore, here we approach the lower limit of this technique when using current materials at the described structural parameters.

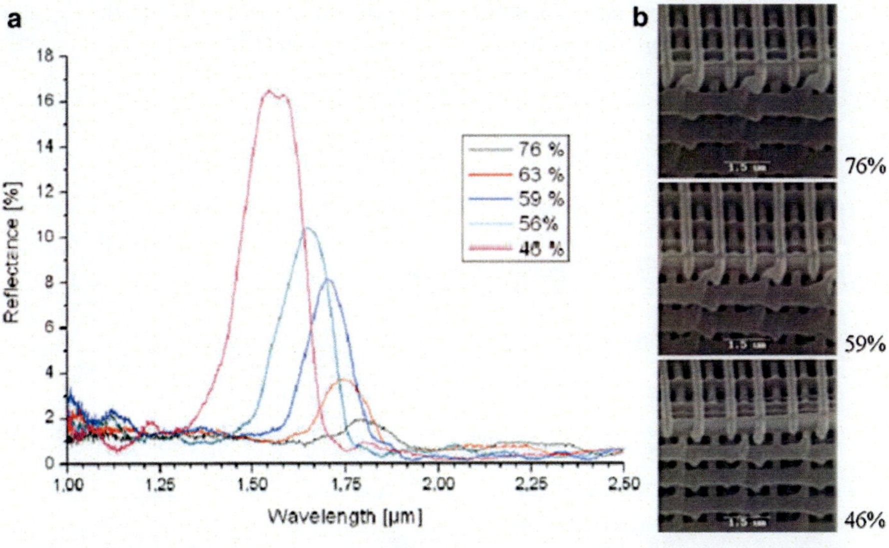

Fig. 11.7 (**a**) Reflection spectra for woodpile structures with different fill factors; (**b**) SEM side/top view images of according woodpile structures with Φ values 76, 59, and 46%

11.6.3 *Effect of the Number of Woodpile Layers on the Quality of the Bandstop*

Although for the case of 2PP microfabrication, increasing number of layers does not present any significant challenge. For considerations of practicality, with prospective on other materials and infiltration methods, it is interesting to know what the minimum number of layers a crystal should contain in order to still exhibit the considerable photonic bandstop effect. Figure 11.8 illustrates the measured reflectance spectra for woodpile structures containing different number of layers in vertical direction. The number of layers is changed by the multiples of 4, i.e. a new unit cell is added every time in the vertical direction. The rest of the relevant structure parameters are kept constant: in-layer-period is 1 μm, the distance between the layers in the vertical direction corresponds to 1.05 × fcc, and the $\Phi = 46\%$. The fabrication parameters are set to 5 mW for the average laser power and 200 μm/s for the scanning speed. To avoid possible effects from structure termination and to make sure that none of the layers is truncated by the glass substrates on which the crystal is attached, the structures were fabricated above the glass substrates supported by a set of poles. This feature also helps to avoid possible anisotropic structure deformation due to shrinkage, and, which is more critical in this case, partial photonic crystal truncation by the carrying substrate. An array of such "free-standing" structures containing different number of layers were analyzed by FTIR.

It is observed already that a structure containing only eight layers shows almost 12% reflectance peak at the position of the bandstop. In this case a second peak appears at the shorter wavelength, this feature manifests presence of a higher order bandstop. Evolution of reflectance peak amplitude vs. number of layers is presented in the Fig. 11.8b. Starting with 16 layers and up the reflectance

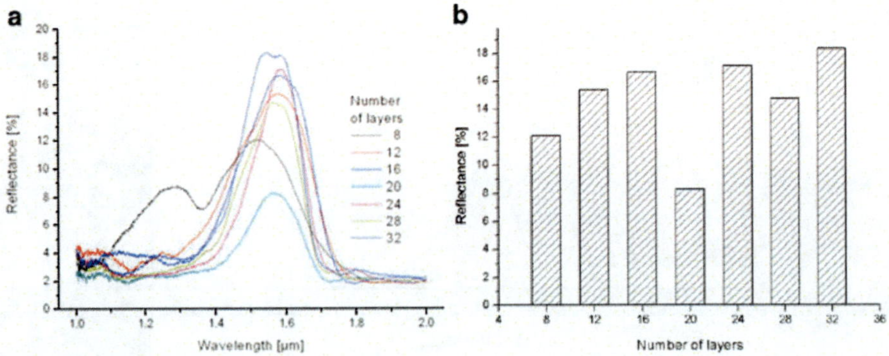

Fig. 11.8 (**a**) Reflection spectra for woodpile structures with different number of layers; (**b**) amplitude of reflectance peak vs. number of unit cells

increases a mere 3%. The sudden drop in reflectance for the structures with 20 and 28 layers could be explained after detailed SEM analysis of the sample. It revealed that due to some deformation of support structure, the upper facet of these structures were not exactly parallel to the sample surface. As a result, the incidence angle of testing beam deviates from the optimal value and results in reduced transmission signal.

11.7 Characterization of Spiral Photonic Crystals Fabricated by 2PP Technique

11.7.1 Optimisation of the Dielectric Filling Fraction Value

A few factors have direct influence on the size of PBG, one of them is the dielectric filling fraction (Φ) determining the ratio of air to dielectric material in a photonic crystal. Spiral structures are fabricated by "pinpoint" illumination, where every single voxel is produced separately.

During 2PP microfabrication of photonic crystals with fixed structural parameters, the value of Φ is adjusted by setting the illumination parameters, since they are in direct relation to polymerized voxel size, i.e. the spirals cross-section dimensions w and h. The relevant fabrication parameters in this case are the average laser power and the illumination duration. In order to adjust Φ, the average laser power was varied in the range 5.5–3 mW, while the illumination duration is fixed to 1 ms. Each spirals parameters are fixed to have a radius $r = 0.65\ \mu m$ and distance between the neighboring spirals of $d = 1.2\ \mu m$, i.e. the neighboring spirals overlap to a certain extent. Each structure consists of 18×18 single spirals completing five revolutions in the vertical direction. The pitch size value is set to 1,200 nm. Supporting structures are provided to each spiral in order to avoid anisotropic deformation caused by possible shrinkage of the whole photonic crystal. Figure 11.9 presents transmission spectra for spiral photonic crystal structures characterized by different Φ values. The according side/top view SEM images of analyzed structures are shown in the inset at the top of the figure. Compared to the case of woodpile photonic crystals, structures with much lower Φ values can be experimentally realized for spiral configuration.

Naturally, the variation of Φ also influences the central position of the measured bandstop – the position of the reflection peak is shifted between 1.3 and 1.8 μm for fabricated structures. Decrease in Φ value results in a blue shift of central bandstop position. The variation of Φ directly influences the amplitude of the reflection peak, i.e. the quality of the bandstop. It is observed that a $\Phi = 16\%$ represents an optimum value for direct configuration spiral photonic crystal, characterized by a maximum reflectance of $R = 17\%$.

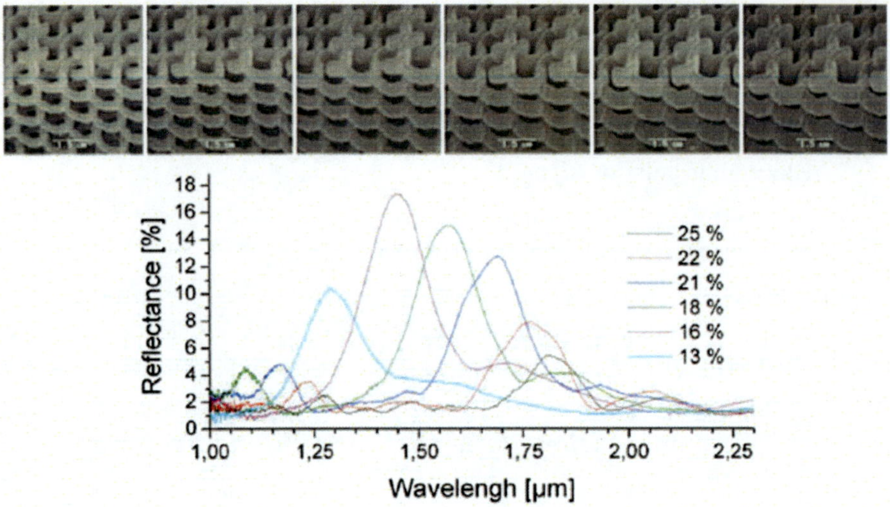

Fig. 11.9 Measured reflection spectra of spiral photonic crystal structures with different dielectric filling fractions (Φ). *Inset* at the *top* shows according SEM images of the structures, from left to right: 13, 16, 18, 21, 22, 25%

11.7.2 Tuning the Bandstop Position by Adjustment of Spirals Pitch Size Value

The bandstops measured along the vertical direction are highly sensitive to the variation in spirals pitch size. Figure 11.10 presents the reflection spectra for the spiral photonic crystals with different pitch sizes. The structures were fabricated using constant illumination parameters, i.e. the spiral cross-sections width and height are equal in all structures. The average laser power is set to 4 mW, the illumination time to 1 ms. Also in this case each spiral has a radius $r = 0.65\,\mu\text{m}$ and distance between the neighboring spirals of $d = 1.2\,\mu\text{m}$. Each structure consists of 18×18 single spirals completing five revolutions in the vertical direction. Again supporting structures are provided in order to avoid possible deformations due to shrinkage and to make sure that no part of the photonic crystal is truncated by the carrier substrate.

For pitch values between 1,060 and 1,935 nm, the central position of the measured reflection peaks is shifted by about 850 nm. By increasing the value of the pitch size, while the rest of the parameters are fixed, one is simultaneously reducing the relative value of the dielectric filling fraction Φ. For pitch size values of 1,060, 1,220, 1,430, 1,620, and 1,935 nm, the calculated values of Φ are accordingly 23, 20, 16.5, 13, and 11%. Comparison of the results presented in Figs. 11.9 and 11.10 show that the directions of the shifts resulting from increasing the pitch size and decreasing the Φ only are opposite. Therefore, in comparison to a variation of the value of Φ only, larger relative shift of the central bandstop position can be obtained by tuning the spirals pitch.

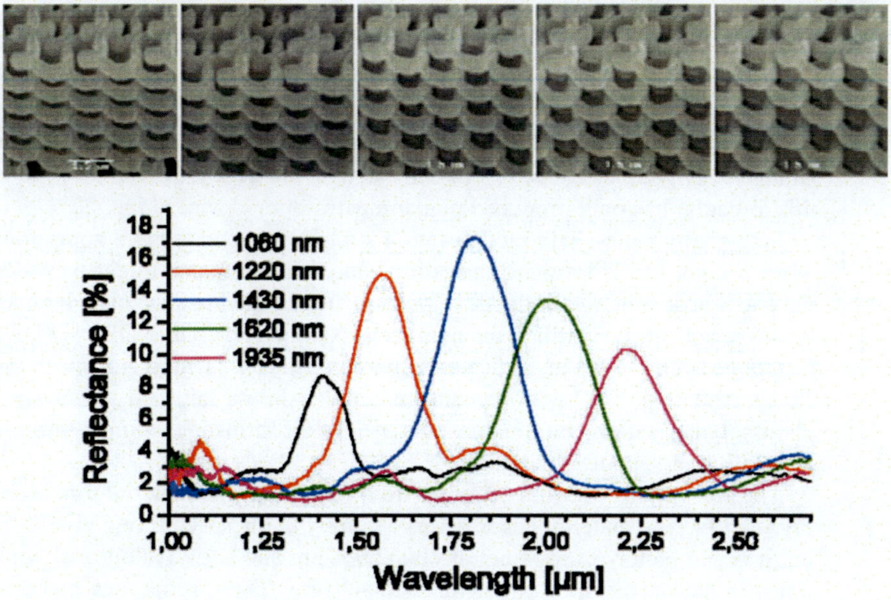

Fig. 11.10 Measured reflection spectra for spiral photonic crystal with different pitch sizes; *Inset* at the *top* shows according SEM images of the structures, from left to right: 1,060, 1,220, 1,430, 1,620, 1,935 nm

The maximum reflection signal is obtained for the spirals with pitch size value of 1,430 nm. This photonic crystal has a filling fraction of 16.5%, which is close to the optimum value of 16% obtained from experiment on Φ optimisation. From here it is concluded that the quality of the bandgap is mainly influenced by the parameter Φ, while the central position of the bandstop is best defined by the pitch size.

11.8 Fabrication of 3D Photonic Crystals Containing an Active Nonlinear Optical Chromophore

Photonic crystals are considered to be the optical equivalent of semiconductors, as they modify the properties of light in the same way a semiconductor does for electrons. However, in contrast with electrons, photons cannot be easily dynamically tuned; for this, it is necessary fabricate photonic crystals made of nonlinear materials, whose optical response depends on the intensity of light that propagates through them. The unique properties of nonlinear photonic crystals would allow the creation of fast and compact all-optical devices. Despite efforts over the last decade, investigations into nonlinear photonic crystals have been limited to theoretical simulations and the fabrication of one or two dimensional semiconductor systems.

Nonlinear optical (NLO) silica sol–gel materials have attracted much interest because of advantages such as high nonlinearity, low cost, high poling efficiency and stable alignment of chromophore molecules, leading to high temporal and thermal stability of their nonlinear performance [73, 74]. Incorporating NLO chromophores into sol–gels in either a guest-host or a side-chain–main-chain strategy has produced electro-optically active sol–gels [75]. Research to date in this direction has only produced planar electro-optic devices [76, 77].

Side-chain–main-chain incorporation strategy presents a more appropriate way when it comes to 2PP microfabrication, since NLO dopants might be washed out the guest-host structure in the development step involving solvents. Here, a photosensitive sol–gel is modified to contain the NLO chromophore DR1 [78]. The 2PP technique is employed to fabricate 3D photonic crystals with bandgaps in the near-IR spectral region. This composite material exhibits minimal shrinkage during photopolymerization, eliminating the need for shrinkage compensation or the fabrication of support structures.

The absorption spectrum acquired from this composite material film is shown in Fig. 11.11a. The material absorbs very strongly in the spectral region 400–550 nm, but it is completely transparent at 600–2,000 nm and has an additional window of transparency in the spectral region 300–400 nm. These properties make SGDR1 ideal for 2PP structuring with a Ti:Sapphire laser, emitting at 780 nm. The IR transparency allows the focusing of the laser within the volume of the material, while the relatively high UV transparency means that there will be two-photon absorption mostly by the photoinitiator, and not by the NLO chromophore. In addition, as the material has no natural absorption in the near-IR spectral range, it can be used for the fabrication of photonic crystals exhibiting PBG centered at the telecommunication wavelengths, and consequently any nonlinearity will be off resonance.

Figure 11.11b shows an optical microscope image of an array of photonic crystals; they have the bright red color of Disperse Red 1. SEM images of a DR1-containing photonic crystal fabricated by the 2PP are presented in the Fig. 11.12.

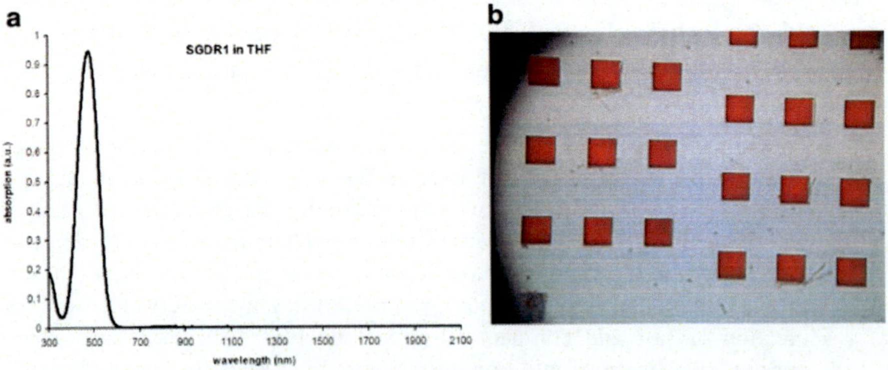

Fig. 11.11 (**a**) Absorption spectra of the material containing DR1; (**b**) optical microscope image of an array of fabricated woodpile photonic crystals (single structure size is 30 μm)

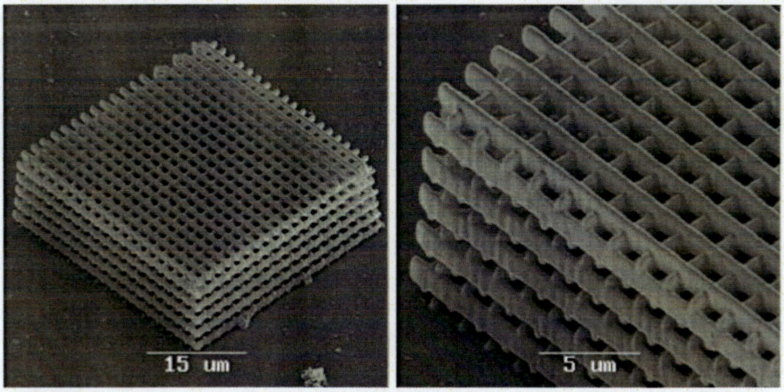

Fig. 11.12 SEM images of woodpile photonic crystal fabricated by 2PP of NLO material

As it can be seen, SGDR1 can be structured very accurately and without defects. The highest resolution achieved in this material is 250 nm.

Previous studies have shown that Disperse Red 1 sol–gel copolymers exhibit second order nonlinearity, with d33 = 30–55 pm/V at 1,300 nm, depending on the DR1 content [79, 80]. To manifest this nonlinearity, the symmetry in the material needs to be removed. This can be done by corona poling the samples at elevated temperatures. Once corona poled, the samples have been reported to exhibit excellent optical as well as mechanical stability.

11.9 Biomedical Applications of 2PP Microfabrication

11.9.1 Microneedles for Transdermal Drug Delivery

Advances in genetic engineering and proteomics have provided protein- and nucleic acid-based treatments for cancer and other chronic diseases [81, 82]. Unfortunately, many novel drugs and vaccines cannot be administered in oral or transdermal form, because they may be metabolized by the kidneys or liver before reaching systemic circulation [83]. These agents may be administered intravenously using hypodermic needles; this route provides complete and instantaneous absorption [84]. Use of conventional hypodermic needles involves trauma at the injection site, pain to the patient, medical skill to administer the injection, and difficulty in providing sustained delivery of a pharmacologic agent over extended period of time.

Transdermal drug delivery is an alternative method that has experienced a rapid development in the past two decades, and has often shown improved efficiency over the other delivery routes [85, 86]. It avoids many issues associated with intravenous drug administration, including pain to the patient, trauma at the injection site, and difficulty in providing sustained release of pharmacologic agents. In addition, precise dosing, safety, and convenience are also addressed by transdermal drug delivery. However, only a small number of pharmacological substances are delivered in this manner today. The most commonly known example is nicotine patches. The main reason for transdermal drug delivery is the significant barrier to diffusion of substances with higher molecular weight provided by the upper layers of the skin.

The top layer, called stratum corneum, is composed of dead cells surrounded by lipid. This layer provides the most significant barrier to diffusion to approximately 90% of transdermal drug applications [87, 88]. A few techniques, enhancing the substance delivery through the skin have been proposed. Two of the better-known active technologies are iontophoresis and sonophoresis. The rate of product development, involving these technologies has been relatively slow [89]. This is partly conditioned by the relative complexity of the resulting systems, compared to the passive transdermal systems. One of the passive technologies is based on microneedle enhanced drug delivery. These systems use arrays of hollow or solid microneedles to open pores in the upper layer of the skin and assist the drug transportation. Figure 11.13 shows examples of such microneedle arrays fabricated by 2PP technique. The length of the needles is chosen such that they do not penetrate into the dermis, which is pervaded with nerve endings, and thus do not cause pain. In order to penetrate the stratum corneum, microneedles for drug delivery have to be longer than 100 μm, and are generally 300–400 μm long, since the skin exhibits thickness values that vary with age, location, and skin condition. Application of microneedles has been reported to greatly enhance (up to 100,000-fold) the permeation of macromolecules through the skin [90]. Hollow microneedles allow diffusion- or pressure-driven transport of pharmacologic agents through the needle bore to be adjusted over an extended period of time. The microneedles for withdrawal of blood must exceed lengths of 700–900 μm in

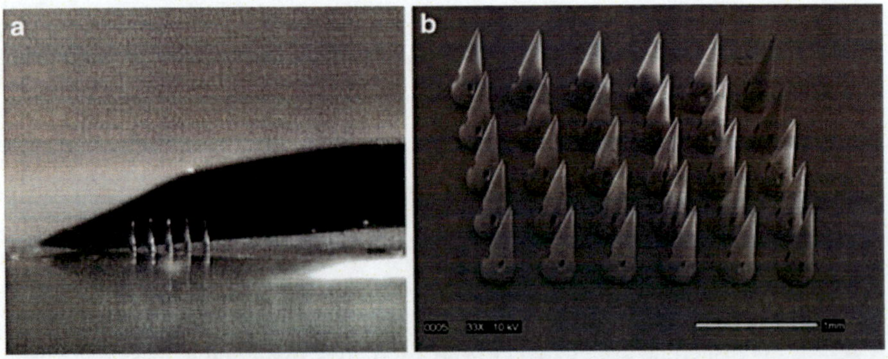

Fig. 11.13 Microneedles fabricated by 2PP technique: (**a**) an optical image of five Ormocore microneedles placed next to a conventional hypodermic needle; (**b**) SEM image of an array of microneedles with an offcenter channel

order to penetrate the dermis, which contains blood vessels. Most importantly, microneedle devices must not fracture during penetration, use, or removal.

Finally, arrays of microneedles may be used to provide pharmacologic agent injection or biological fluid extraction at higher rates and over a wider area than individual microneedles. Microneedle arrays provide redundancy if individual needles are fractured or obstructed during the microneedle insertion process. In addition, arrays of microneedles provide greater possibilities for directly reaching vasculature in the dermis for extraction of blood. Microneedle arrays are also less prone to fracture if exposed to shear forces, because these forces are distributed over a wider area [91].

The flexibility and high resolution of the 2PP technique allows rapid fabrication of microneedle arrays with various geometries, and study its effect on the tissue penetration properties. In order to study the effect of microneedles geometry on its penetration properties, square arrays of 25 identical 500 μm spaced needles were fabricated by 2PP of Ormocore material. Since these structures do not require very high resolution in the vertical direction, a conventional 10× microscope objective (0.2 NA) lens was used to focus the laser beam into a volume containing photosensitive resin. The average length of a fabricated needle is 800 μm and the base diameters are in a range of 150–300 μm. The length of the needle enables its application for both drug delivery and blood drawing. Tip sharpness was changed by adjusting the position of the channel relative to the central symmetry axis in the original design (see Fig. 11.14). The diameter of the channel is fixed for all the samples. The four openings visible at the base of each needle were fabricated in order to guarantee the complete removal of the non-cross-linked polymer from the channel during the developing fabrication step, since one end of the needle would be blocked by the glass coverslip otherwise. These openings are omitted in the

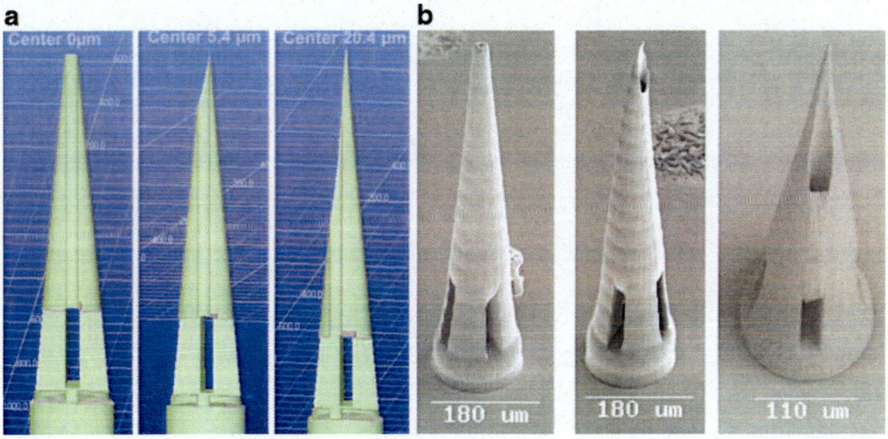

Fig. 11.14 Hollow microneedles for transdermal drug delivery: (**a**) cross-section of original CAD design of microneedles with various positions of the channel center. The channel displacement allows to control the tip sharpness; (**b**) SEM images of according microneedles fabricated by 2PP technique

needles actually used for flow studies and drug delivery, since in that case both channel ends are open.

Naturally it is desirable to fabricate needles that require least load on order to penetrate the skin. Microneedles undergo several forces during insertion into the skin, including compressive forces, bending forces, shear forces, buckling forces, and skin resistance. Fracture testing and penetration testing of microneedle arrays was determined using compression load testing (ELF 3200, Bose EnduraTEC Systems Group, Minnetonka, MN, USA). In these studies, 50 and 500 g load cells were driven at 0.008 mm/s displacement rate against polytetrafluoroethylene (duPont, Wilmington, DE, USA) and cadaveric porcine adipose tissue (Nahunta Pork, Pikeville, NC, USA). Cadaveric porcine material was chosen for this study because porcine epidermal and dermal layers closely resemble their human counterparts. Although cadaveric porcine adipose tissue is softer than whole porcine skin, it exhibits a homogeneous, uniform surface that is appropriate for mechanical testing. A video capture device was used to examine the microneedle penetration behavior and determine the mode of failure.

Figure 11.15a contains load-displacement curves for in-plane and out-of-plane hollow microneedle arrays with several flow channel-microneedle tip displacement values (indicated in micrometer on the graph). Similar load vs. displacement curves were observed for needles with different base diameters. As indicated by the asymptotes, the slopes on the curves can be divided into three regions (from left to right), exhibiting different trends: initial needle contact with the porcine adipose tissue, prior its surface puncture; tissue puncture by a single, few or complete microneedles array; the final displacement value attributed to the contact between the tissue and the cover slip substrates carrying the microneedle array. The local maximum values indicate the points at which the load threshold was reached. At these values, the porcine adipose tissue was punctured by one or few microneedles

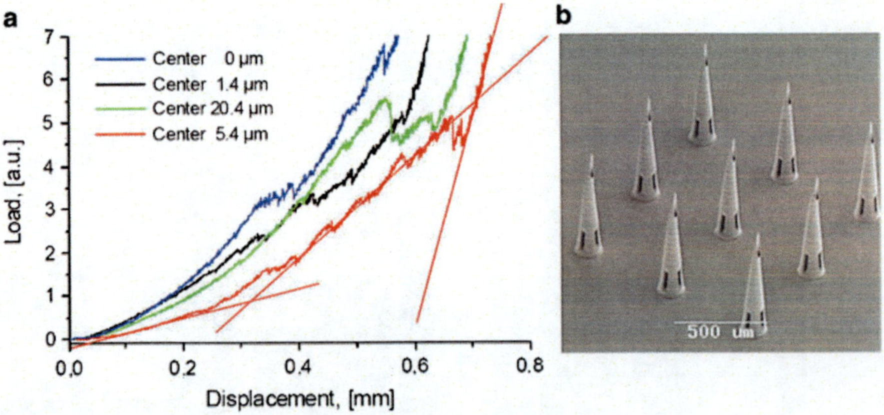

Fig. 11.15 (**a**) Load-displacement curves obtained for the needles with different flow channel-microneedle tip displacement (indicated in micrometer for each deign); (**b**) SEM image of a 3 × 3 off-center microneedles array in the optimum configuration (channel is displaced by 5.4 µm relative to the needle axis)

in the microneedle array. Local minimum values were observed immediately adjacent to these local maximum values. The "off-center microneedles" with channel positioned at 20.4 μm relative to the needles axis exhibited the sharpest needle tips values among the out-of-plane microneedles (see rightmost images in Fig. 11.15a, b). However, series of experiments consistently confirmed the lowest penetration threshold for the microneedles array with channel to tip displacement value of 5.4 μm, and not 20.4 μm which has the sharpest tip. Optical microscopy analysis of the microneedles tips helped to clarify the situation. Apparently very long and sharp but less rigid tips of the microneedles with the channel shifted by 20.4 μm relative to the needles axis, are bent during tissue penetration. Needle-tip bending is associated with the presence of methacrylate groups in the cross-linked Ormocore, as a consequence the structures stay somewhat elastic. During the test the tip bends even further, resulting in increased penetration resistance and therefore the larger values of load threshold. Overall results indicate minimal dependence between geometry and skin penetration properties for polymeric microneedles fabricated using two photon polymerization.

Similar load vs. displacement curves acquired for microneedles with different aspect ratios indicated a minor difference in the load values for different needles. The tests against polytetrafluoroethylene surfaces, however, suggested dependence in failure. In general, the larger the base of the microneedle, the higher the load it can endure before breaking or bending. Majority of the microneedles bend, and eventually tip off the glass when tested against polytetrafluoroethylene. Therefore by improving the needle attachment to the substrate one can further increase failure load threshold. This can be done by applying adhesion promoter to the substrate or by producing substrate from the same material as the microneedles.

The 2PP microfabrication allows the introduction of precise structural details into the needle design. Figure 11.16 shows a microneedle whose design was inspired by female mosquito fascicle. Mosquito needle is a well known "nature-created" tool for

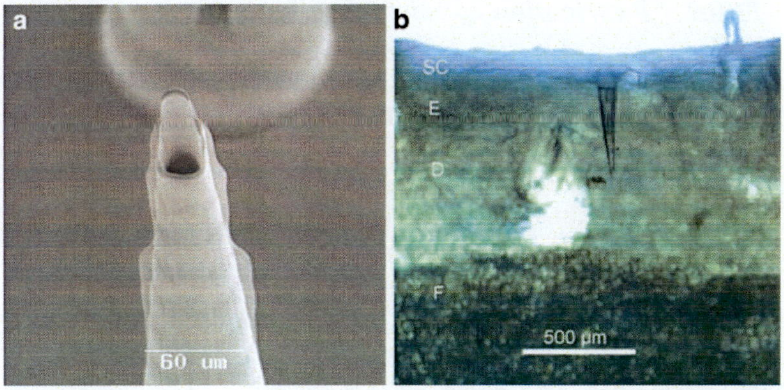

Fig. 11.16 (**a**) SEM image of bioinspired microneedle with mosquito-like tip; (**b**) optical micrograph of 2PP fabricated microneedle on full thickness porcine skin. Abbreviations on the *left* side of the image indicate different skin layers: the stratum corneum (SC), epidermis (E), dermis (D), and fat (F)

effective and painless "blood sampling." Mosquito fascicle has a somewhat flat tip and additional microstructures on its sides. It has been suggested that by applying small amplitude vibrations the mosquito cuts through the skin, rather than just statically puncturing it, this way reducing the force needed for penetration. Fracture and penetration study of 2PP fabricated microneedles with such saw-toothed walls showed no significant differences in load response. A separate trial was performed with such biomimetic microneedle arrays mounted on 3D piezoelectric actuator (in x, y, and z) in order to mimic mosquito fascicle movement. In this case vibrations, induced by piezo actuator, resulted in high noise levels in load response, which did not allow to draw any evidential conclusions, when compared to no piezoelectric actuation. Therefore a more sensitive apparatus has to be used in order to verify the advantages of such tip geometry over the other proposed designs.

Figure 11.16b shows an optical microscope image of a single microneedle on full thickness porcine skin. It is observed that the inserted 2PP fabricated needle penetrates all the way through epidermis, and will allow substance injection directly into dermis. Controlled administration of fluorescein-conjugated biotin (a water-soluble B-complex vitamin) solution into the dermal and hypodermal regions of skin was studied using off-centered microneedles. Figure 11.17 shows DIC-fluorescence micrographs of skin sections injected with biotin and frozen at various times. The images containing the skin cross-sections with the microneedle-induced perforations show injected substance penetration through diffusion at different times after injection. After 5 min, the biotin is localized in the perforation created by the microneedles. After 20 min, the diffusion of biotin into the dermal/epidermal junction is evident. At 60 min, the entire dosage of biotin diffused as deep as the dermal/hypodermal region of the skin. At the same time the control sample (biotin applied on the surface of the skin) showed no diffusion beyond the stratum corneum even after 60 min.

11.9.2 Fabrication of Microprosthesis

The malleus, incus, and stapes bones compose a so-called ossicular chain that serves to transmit sounds from the tympanic membrane to the inner ear (see Fig. 11.18a, b). Ear diseases may cause discontinuity or fixation of the ossicles, which results in conductive hearing loss [92, 93]. Ossiculoplasty is a technique for reconstruction of the bones of ossicular chain within the middle ear [94–96]. The traditional method for restoring sound conduction in the middle ear involves the use of the remaining ossicles [97]. The most commonly used material is the patient's own incus bone. However, autograft materials are not always available in patients suffering from chronic diseases. In addition, shaping of these materials requires additional operative time. There are also concerns regarding loss of rigidity, resorption, or fixation of the implant to the wall of the middle ear. Homograft materials, which are obtained from cadavers, are now rarely used due to the risk of infection [98, 99]. Alternatively, artificial prostheses are used in ossiculoplasty.

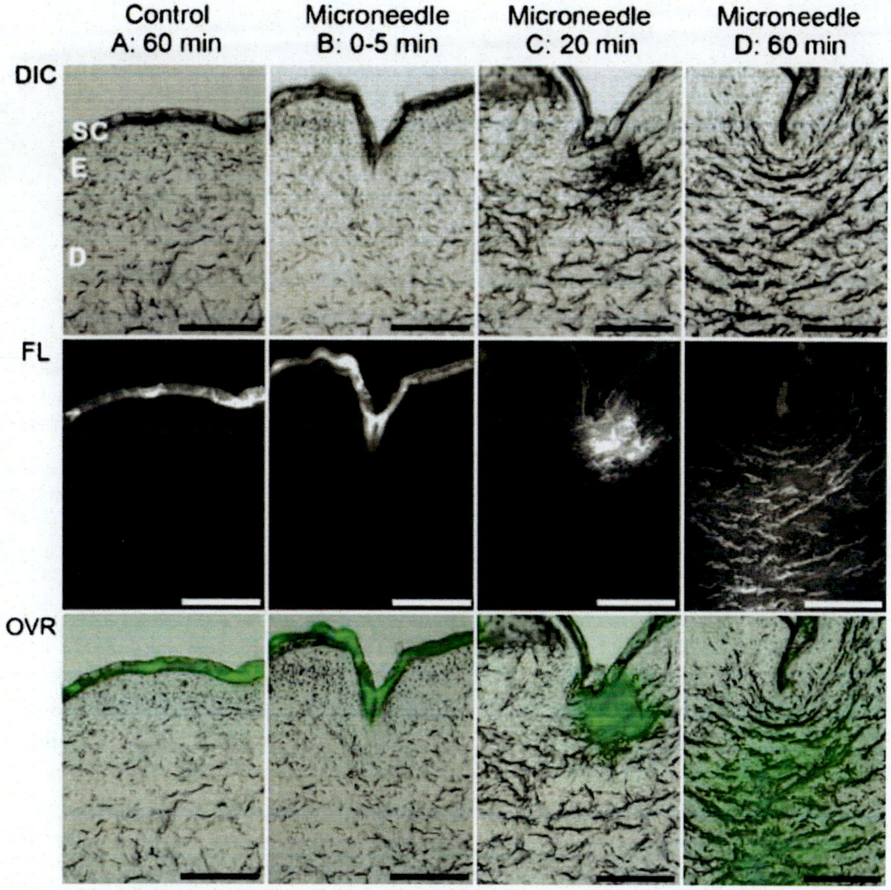

Fig. 11.17 DIC-fluorescence microscopy of skin after administration of fluorescein-conjugated biotin using microneedles fabricated by 2PP technique. Treatment duration with microneedle (5, 20, and 60 min) is indicated on *top* of each column and may be compared with control at 60 min (without microneedle). *Top row*: DIC single-channel provides clear view of the various skin layers. The stratum corneum (*SC*), epidermis (*E*), and dermis (*D*) layers are indicated. *Middle row*: fluorescence images (*FL*) of fluorescein-biotin emission. *Bottom row*: DIC-fluorescence overlay (*OVR*) with the fluorescence channel (*Green*) shows distribution of biotin within the skin layers. Scale bar equals 100 μm in all images

Figure 11.3.33c shows schematics of such reconstruction where a part of the remaining stapes bone is used along with the partial ossicular replacement prosthesis. Several alloplastic materials have been considered for use in ossicular reconstruction. Polyethylene, high-density polyethylene sponge, polytetrafluoroethylene (PTFE), and Proplast (PTFE-carbon composite) pistons were used in middle-ear reconstruction [Bra86]. These materials demonstrated migration, extrusion, penetration into the inner ear, and significant middle-ear reactivity, resulting in a necessity of a repeated surgery. As a result, these solid polymers are not currently

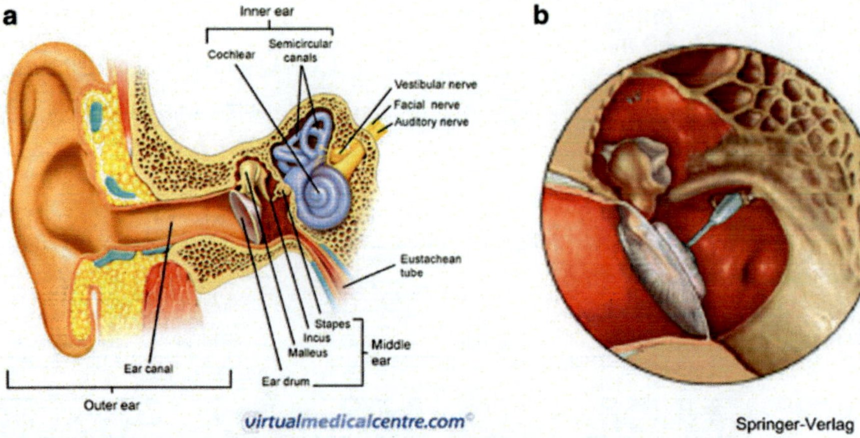

Fig. 11.18 Schematic representation of middle ear with: (**a**) intact ossicular chain (courtesy Virtual Medical Center); (**b**) damaged ossicular chain repaired with Total Ossicular Replacement Prosthesis (TORP) [112]

used for restoration of sound conduction. Titanium, alumina, stainless steel, and gold are biocompatible materials that have been considered for use in ossicular reconstruction [100–105]. For example, titanium prostheses have demonstrated biocompatibility, rigidity, biostability, low ferromagneticity, low density, and good sound conduction. However, recent studies have indicated that a thin cartilage graft is needed to cover the head plate of the prosthesis and prevent prosthesis extrusion. Bioactive materials (Bioglass, Ceravital, hydroxyapatite) promote direct chemical bonding with body tissues, and bioactive prostheses were developed with the hope that these materials would have a lower rate of extrusion [106–110]. These materials can be placed directly under the tympanic membrane, and do not demonstrate significant extrusion. However, bioactive materials have not gained wide acceptance due to difficulty in shaping the prostheses and mechanical instability, especially at infected middle ear sites.

Novel materials and prostheses that provide improved sound transmission for longer periods of time are demanded by patients and surgeons. These materials must demonstrate a high degree of cell compatibility, ease of use, and minimal extrusion [111–113]. Mass-produced implants are produced in several shapes and sizes; however, these designs do not take individual patient anatomy into account. One way of increasing the success of ossicular prostheses is the development of patient-specific implants [114, 115]. Such implants possess the appropriate design features, including head size, weight, and footplate attachment, for a particular patient.

Ormocore surfaces fabricated using 2PP approach demonstrated acceptable cell viability and cell growth profiles. Here we demonstrate the use of 2PP technique for rapid prototyping of Ormocore middleear bone replacement prostheses. The design used in this study was similar to that of a commercially available TORP prosthesis

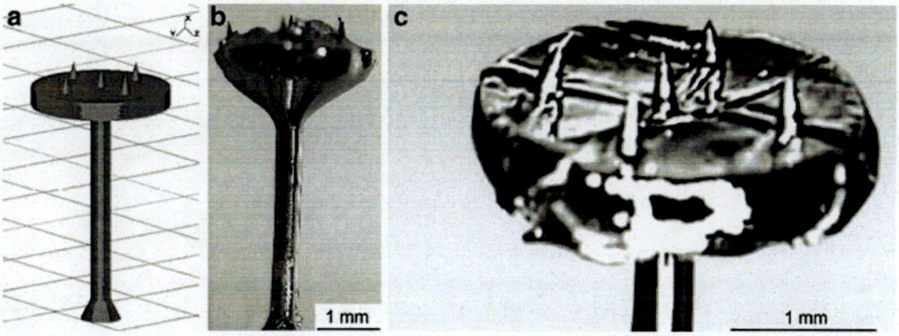

Fig. 11.19 Total Ossicular Replacement Prosthesis (*TORP*): (**a**) original design; (**b**) optical image of a structure fabricated by 2PP of Ormocore; (**c**) close-up on the fabricated TORPs head plate

(total ossicular replacement prosthesis, Kurz Medical Inc., Dusslingen, Germany). It contains a CAD and optical image of the TORP-shaped Ormocer ossicular prostheses. Such structure was produced by 2PP (see Fig. 11.19b, c) directly from CAD input in a layer-by-layer approach, with a distance of 25 μm between the layers. The prostheses stand 4.3 mm tall, 300–400 μm wide at shaft, 2.6 mm wide, and 400–500 μm thick at the diskshaped headplate. The disk-shaped headplate of the prosthesis was designed to be placed under the tympanic membrane. The shaft of the prosthesis is designed to connect the footplate of the stapes with the under-surface of the tympanic membrane in order to enable sound conduction. The head-plate also contained five microneedle-like structures protruding at 400 μm with 160 μm base width. These structures are introduced to improve cellular adhesion, avoid device migration, or provide means for possible cartilage attachment before the operation.

An important criterion for such prosthesis is the handling in a real operational room environment. In order to verify it, in vitro implantation test was performed into the right ear of commercially obtained human head. First the chorda tympani were partly removed in order to provide access to the middle ear and a possibility of optical imaging. The natural incus and the head of the stapes, initially present in the middle ear, were removed using surgical tools. The 2PP fabricated Ormocore prosthesis was then carefully aligned against the posterior canal wall (see Fig. 11.20a). Finally, the implant was positioned in situ on the footplate of the stapes below the tympanic membrane to provide sound conduction, as found in natural middle ear (see Fig. 11.20b). The prosthesis was handled without fracture during the entire procedure. The detailed description of this in vitro implantation study can be found elsewhere [116].

For the mechanical tests the headplate was separated from the fabricated TORP. Nanoindentation tests performed on the surface of the headplate of the ossicular-prosthesis were compared with thin-film Ormocore material (Fig. 11.21). As the load increased, the measured values of Hardness (H) and Modulus (E) of the Ormocore thin-film were found to be significantly higher than the headplate surface. Both values were consistently higher by a factor of four in the thin-film material.

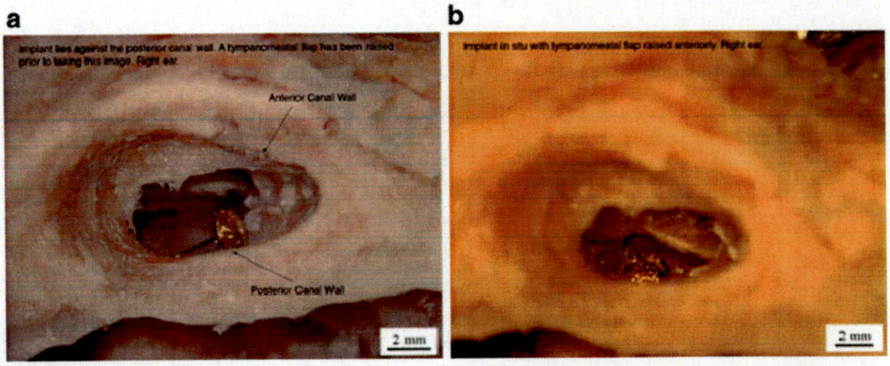

Fig. 11.20 The implantation process of the 2PP fabricated TORP through the right external auditory canal: (**a**) prosthesis aligned against the posterior canal wall; (**b**) the implant positioned in situ on the footplate of the stapes below the tympanic membrane to provide sound conduction, as found in natural middle ear

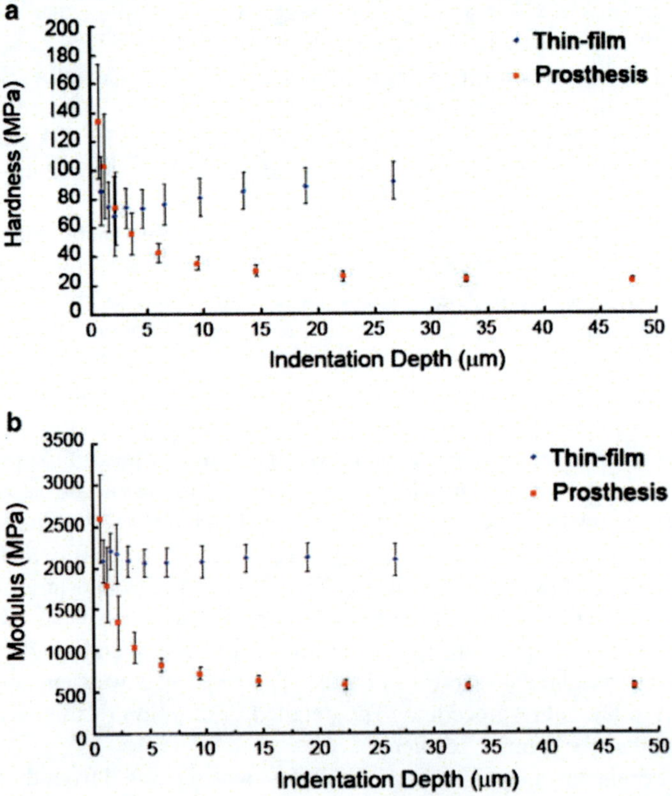

Fig. 11.21 Nanoindentation test results of UV photopolymerized Ormocore thin film and Ormocore ear prosthesis fabricated by 2PP technique: (**a**) hardness (MPa) vs. indentation depth; (**b**) modulus (MPa) vs. indentation depth. Error bars indicate standard deviation of mean

11.9.3 Scaffolds for Tissue Engineering

Tissue engineering is an interdisciplinary field that applies the principles of engineering and life sciences toward the development of biological substitutes that restore, maintain, or improve tissue function or a whole organ. Tissue engineering has also been defined as "understanding the principles of tissue growth, and applying this to produce functional replacement tissue for clinical use" [117]. Cells in the body are exposed to a complex milieu regulated by their interactions with other cells, the surrounding cell matrix and soluble factors. A key element of this microenvironment is the 3D architecture of the extracellular matrix (ECM). Furthermore, upon removing cells from this microenvironment, many cell types, such as liver cells, quickly lose their function. One of the most popular approaches for tissue engineering relies on the application of scaffolds. Scaffolds help to guide cell growth and create certain milieu required for cells to form a functional 3D tissue. Generally, scaffolds have to fulfill several basic requirements. They should promote cell attachment and migration, enable diffusion of vital cell nutrients and expressed products, and have appropriate mechanical properties [118]. A number of studies have demonstrated that cells in artificial 3D matrices have an improved function relative to their culture on 2D substrates. Ideally the 3D matrices should mimic the structure and biological function of native ECM as much as possible. In an artificial scaffold, pore shape, size, their distribution, and connectivity play an important role. These factors influence the regeneration process [119], and simultaneously define the mechanical properties of the structure [120].

Most conventional scaffold fabrication approaches such as freeze-drying, phase separation, solvent casting, etc. have limitations when it comes to precise control of the scaffold internal architecture. Using these techniques it is now feasible to control pore connectivity and pore size. Porosity levels of around 90% can be realized, this way providing sufficient pore connectivity for cell attachment and nutrition factors diffusion. However, no active control over the size and the position of each individual pore is provided. As a consequence, it is virtually impossible to produce structures in accordance to a predefined blueprint or series identical scaffolds.

With emergency of solid free-form fabrication and rapid prototyping technologies, it became possible to party solve these problems. Such techniques as stereolithography, 3D printing, selective laser sintering, or fused deposition modeling allow realization of structures using a predefined CAD model. Many materials, relevant in biomedical and specifically tissue engineering fields have been successfully tested. Some of these techniques even allow deposition of living cells and proteins directly during the fabrication process. Of equal importance is the fact that also the external shape of such scaffold can be defined in the initial CAD design. Using the 3D data input, provided for example by computer tomography, patient-specific scaffolds conforming to individual anatomic structure at the implantation sight can be realized.

A main difference between described passive and active technological approaches is the size scale. Passive technologies can produce scaffolds with pore sizes in the range from a few microns to hundreds of micrometers. In solid freeform fabrication and rapid prototyping, due to limited structural resolution, large scaffolds with pore sizes on the order of hundreds of microns are produced. In order to overcome this barrier a combination of both, three-dimensional printing and salt-leaching has been applied [121]. In this way the large features inside the scaffold are defined in accordance to initial CAD design, while smaller pore size and distribution cannot be controlled precisely.

On the other hand, the structural resolution of the 2PP microfabrication technique can be scaled up by using a different focussing optics. Therefore, 2PP technique allows to close the existing gap in structural parameters of scaffolds produced by different techniques. Structures with feature sizes ranging from submicrometer to hundreds of microns can be now produced using a single technique. The particular pores have diameter of around advantage of 2PP for the fabrication of scaffolds for tissue engineering is the possibility to precisely control both external shape and internal porosity of the structure while delivering high structural resolution. Scaffolds with a sophisticated pore size distribution can be realized by designing the pores directly. For the first time it is possible to create a "designer scaffolds" with feature sizes on the order of that of the natural ECM. The position of pores with sizes ranging from hundreds of microns down to submicrometers can now be exactly defined inside the scaffold. Figure 11.22a shows a CAD design of such scaffold having porosity of more than 80%, an inset in the image presents three side projections of the structure. The through pores, which can be accessed from each scaffold surface, are large enough to accommodate cells. A second set of pores, having diameter much smaller than the cell dimension, is introduced in order to enhance fluid transport inside the scaffold. In addition, the mechanical properties

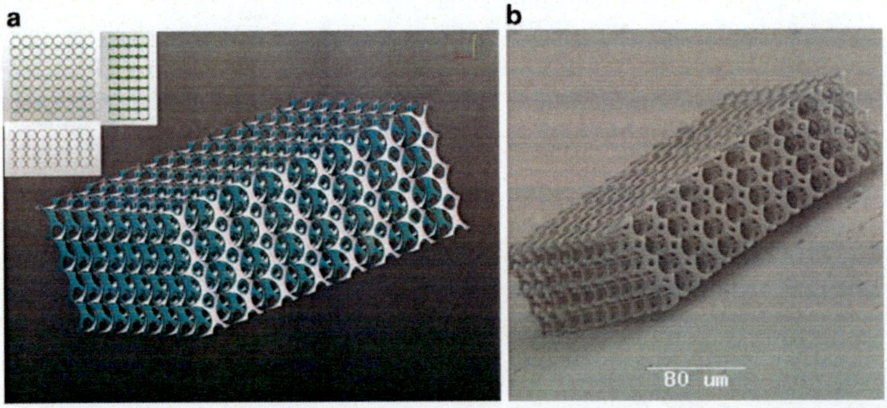

Fig. 11.22 Highly porous 3D scaffold with two distinct pore dimensions: **(a)** CAD concept (*inset* shows three side projections of the structure); **(b)** SEM image of according structure fabricated by 2PP technique

of such scaffold can be controlled by adjusting the applied material properties, varying the porosity, and introducing changes in the internal geometry of the structure. An SEM image of according scaffold fabricated by 2PP of a Zr-based hybrid material is shown in Fig. 11.22b. Despite high porosity the structure is mechanically stable. Large 20 μm Interconnected micropores, of around 5 μm, while not accessible by cells, provide means for liquid transport inside this 3D scaffold. The high reproducibility of 2PP microfabrication, for the first time allows the fabrication of large series of identical samples, and therefore, enables systematic studies of scaffold-cell and cell–cell interactions in a 3D system.

One of the issues associated with the repair of large volume tissue defects is vascularization. Since 2PP allows introducing pores of desired size at any location in the scaffold, it has a great potential for defining the future vascular network already in the initial scaffold fabrication step. This way it would be possible to integrate the newly formed vascular network, with the one already present around the particular defect in the body. Figure 11.23 shows an original CAD design of a fragment of a branching microcapillary network model and an according structure fabricated by 2PP of SU8 material. The overall structure diameter is about 100 μm, the smallest feature size is around 2 μm. Similarly, to most mammalian branching vascular and respiratory networks, the Murray's law governing the relation between the branching vessel diameters has been taken into account here.

The properties of material applied for scaffold fabrication can play a crucial role in tissue formation mechanisms. For adherent cells, the proliferation is only possible if the cells attach to the cultivation surface. Different properties of the culture material affect the adherence of the cells to the surface. It is known that a favorable distribution of charges influences the interaction between cellular adhesion molecules, such as integrins, and material [122].

In order to test the biocompatibility of materials applied for 2PP microfabrication, various cells were seeded onto flat cover slips coated with these materials.

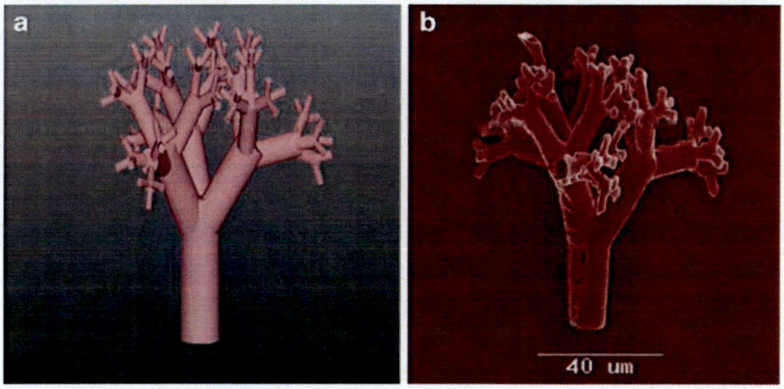

Fig. 11.23 An original CAD design (**a**) and a scanning electron microscope (SEM) image of a fabricated structure (**b**), which resembles a fragment of pulmonary alveoli – microcapillaries responsible for gas exchange in the mammalian lungs

By comparing the proliferation rates of cells grown on flat material surfaces and under control conditions, it was demonstrated that hybrid organic inorganic and SU8 materials are not cytotoxic. Thus, these polymers do not affect the cell proliferation in a negative manner, which demonstrates that they fulfill the minimum requirement for a scaffold material. However, there still can be effects associated with the modifications on a genetic level. If the adherent cells, used in this study, cannot attach, they react by reduction of the proliferation rate and even by inducing cell death, which correlates with an increase of the DNA damage effect. Additional tests comparing the DNA strand breaking of the cells grown on the Ormocore and in control conditions have shown no statistically significant influence by the presence of Ormocore. The presence of Ormocore does not alter the proliferation parameter of the cells and does not increase the tail moment. Therefore, one can conclude that Ormocore sustains the adhesion of cells.

When the intercellular junctions of a new tissue are formed, the gap junctions are the last junctions which are established. Their formation is conditioned by cell-to-cell adhesion structures, composed from other molecules, such as cadherins and cadherin-associated molecules [123]. Gap junction are cell-to-cell adhesion structures composed of direct cell-to-cell channels which allow a direct exchange of ions and small metabolites such as second messengers between neighboring cells. An evidence of gap junction formation is therefore a good merit for evaluation of tissue formation. The compatibility of Ormocore with the formation of cellular junctions appropriate for the living tissue, such as gap junction coupling, was studied using the double whole patch-clamp technique [124]. The gap-junction conductance measurements reveal that Ormocore does not alter formation of cell-to-cell junctions, critical for functional tissue growth [125]. Based on these results one can conclude that the cultivation of cells on Ormocore is compatible with the formation of tissue. The full description of the results of this study can be found in a following publication [16].

11.10 Summary and Outlook

Section 11.2 is dedicated to fabrication of micro-optical elements by 2PP technique. Regarding the variety of possible surface profiles 2PP technique is superior to conventional approaches for fabrication of refractive optics. Furthermore, many polymeric materials, designed for optical applications, can be structures by 2PP technique. In the perspective to mass production of such micro-optical elements, it is meaningful to use structures fabricated directly by means of 2PP microfabrication in order to prepare elastomeric molds for further replication with UV-micromolding.

Fabrication of 3D photonic crystals with various structural parameters, geometries, and even shapes has been demonstrated by 2PP microfabrication. By setting the structural parameters, measured bandstops for these experimentally realized photonic crystals, can be precisely positioned in the near-IR spectral range at the telecommunication wavelengths. Higher order features observed in the measured FTIR spectra indicate high optical quality of the fabricated photonic crystal structures. In the example of woodpile structures it was shown that quality of the fabricated photonic crystals is sufficient to achieve a considerable PBG effect already for four unit cells along the testing direction. Dielectric filling fraction (Φ) is a very important parameter, influencing the size and, to some extent, the exact position of the PBG. In 2PP microfabrication the Φ value is controlled independently of structural parameters of a photonic crystal, by adjusting the illumination dose. For woodpile architecture a minimal Φ value of 46% is limited by the stability of the produced structures. Further reduction of Φ value, i.e. woodpiles rods cross-section, results in mechanically weak structures, which collapse during the developing step. It was observed that much lower values of Φ can be experimentally realized for spiral photonic crystals. This can be explained by the fact that compared to woodpile, spiral structure resembles the diamond architecture much closer, while requiring less material. In addition, substantial spatial overlap between individual spirals contributes to better mechanical stability and lower Φ values for the spiral architecture. For some future applications it is important to provide photonic crystals with NLO response. DR1 chromophore was chemically modified and added in a side-chain–main-chain manner to a conventionally used ORMOSIL based sol–gel. Such material containing NLO chromophore was successfully structured by 2PP technique. It is seen that the NLO chromophore stays inside the material throughout the processing. This method provides important guidelines for alternative material modifications with other nonlinear, electrooptic or photoactive materials. In conclusion, 2PP provides important rapid prototyping tool for the microscale all-optical circuit realization.

Applications of 2PP microfabrication in biomedicine are presented in Sect. 11.9. Microneedles produced by 2PP of hybrid photosensitive materials were tested for use in transdermal drug delivery. The optimisation of needle geometry for the lowest skin resistance during needle insertion has been performed. Flow studies using florescent dye solution provide evidence of controlled subcutaneous injection possibility through the microneedle. Results of this study indicate that microneedles created using the 2PP technique are suitable for in vivo use, and for the integration with the next generation MEMS- and NEMS-based drug delivery devices.

Another related application area is fabrication of microprosthesis. While conventional mass-produced implants do not sufficiently take individual patient anatomy into account, CAD/CAM rapid prototyping by 2PP may permit development of patient-specific geometries. Middle ear ossicular replacement prosthesis, fabricated by 2PP, has shown good mechanical properties and ease of handling. The applied Ormocore material can be photostructured with high precision using 2PP technique, and has shown good biocompatibility. This study indicates high potential of 2PP technique for prototyping and fabrication of such patient-specific microimplants.

Finally, flexibility of 2PP technique allows fabrication of complex 3D structures that can be used for guided cell growth. The fabrication of scaffolds with a desired structure in 3D is a very important task in tissue engineering. It has been demonstrated that CAD designed complex 3D structures with varying pores sizes can be generated by 2PP technique. For a given material the mechanical properties of such scaffold can be adjusted by changing the size and the positions of individual pores. These results demonstrate the great potential of 2PP technology for the fabrication of 3D scaffolds and test structures for tissue engineering. Structures with feature size similar to natural ECM can be produced. With respect to other techniques conventionally used for scaffold fabrication, 2PP technique allows to close the existing gap in structural parameters of resulting structures. The high reproducibility of 2PP microfabrication, for the first time allows the fabrication of large series of identical samples, and therefore, enables systematic studies of scaffold-cell and cell–cell interactions in a 3D system.

In perspective to further development of the 2PP technology, it is clear that further downscaling of the structural resolution is necessary in order to expand the number of applications of 2PP in the sector of nanotechnology. For example, photonic crystals with the photonic bandgaps in the visible spectral range can be realized under the condition that structural parameters of these 3D structures can be scaled down accordingly. Indeed, recently, a lot of efforts by scientific groups worldwide have been dedicated to this subject. Currently, 3D structures with features size as small as 65 nm have been reported by 2PP using femtosecond laser at 520 nm [126]. Naturally, use of even shorter wavelengths should allow better focusing and result in smaller feature sizes. For this, materials and photoinitiators sensitive in the according spectral range have to be considered as well. Some possible solutions, based on elaboration of optical system, which were developed in order to improve the spatial resolution of multiphoton microscopy, can be readily applied for 2PP technique. Use of simple shaded ring filters helps to make the intensity distribution in the focus less asymmetric, and therefore improve axial resolution of fluorescence microscopy [127]. Similarly, 4-Pi illumination is a more sophisticated method, where two high NA microscope objectives are juxtaposed in order to allow constructive interference of their illumination wavefronts inside the sample. As a result, a close-to-spherical sharp excitation spot providing high spatial resolution can be realized [128]. Stimulated emission depletion (STED) is another possibility allowing to reduce the size of the excitation region in the fluorescence microscopy [129]. Here pairs of synchronized laser pulses at different wavelengths are applied. First pulse produces regular diffraction-limited excitation spot, while second donut shaped STED-pulse quenches excited molecules back to the ground state by stimulated emission. In the center of the doughnut, fluorescence remains unaffected. Theoretically, using this approach, the excitation spot can be narrowed down to the size of a single molecule. Practically, STED-microscopy resolution as high as few tens of nanometres has been reported [130]. In 2PP, STED can be applied to confine the photopolymerized region by quenching the photoinitiator molecules. Combination of both STED and 4-Pi microscopy is another, rather expensive, but possible route.

Potentially less expensive and more universal methods for improving the structural resolution rely on the modification of the materials applied for 2PP nanofabrication. Application of photoinitiators, specially designed to provide effective two-photon absorption has lead to improved structural resolution [131]. Repolymerization patterns (see Sect. 3.1.2) indicate that diffusion of radical molecules, away from the region where they were produced by the laser pulse, lead to unwanted increase of the polymerised volume. By adding radical quenchers chemical equivalent of STED effect, where excited photoinitiator molecules are spatially confined by local gradients of quencher concentrations, was realized [132, 133]. Here the diffusion of quencher molecules play a critical role. Alternative solution is to reduce the diffusion and so provide better confinement of the polymerized volume. For this, solid materials, similar to ORMOSILs presented in this thesis can be used. Due to mineral network, formed during prebaking step, the mobility of the material molecules is already quite limited. Additional immobilization of photoinitiator molecules, e.g. by chemical bonding onto such material network should provide considerable improvement of the structural resolution.

Eventually the minimal resolution of 2PP technique will be limited by molecular size of applied materials. Further progress will rely on application of novel, possibly purely inorganic, photosensitive material systems.

With the development of turnkey femtosecond laser systems, 2PP technique enters the realm of industrial applications. First all-in-one laboratory scale systems for 2PP microfabrication are already commercially available [134]. Due to comparatively low throughput, it is very likely that such systems will be first used for rapid prototyping or fabrication of small series of microstructures. Few methods, allowing to increase throughput of 2PP by parallel processing with multiple laser spots, have already been demonstrated experimentally [135, 136]. In cases where large series of identical samples are required, structures produced by 2PP can be replicated by means of UV-micromolding or nanoimprinting [137, 138]. While UV-micromolding presents a unique route for replication of 3D structures, well-established industrial scale techniques such as hot embossing or injection molding are most likely to be used for micro-optical element fabrication.

The number of applications of 2PP technique will continue to grow. Micromechanical components produced by 2PP combined with on-chip microfluidic systems have already been demonstrated and are likely to get wider acceptance. A considerable advantage of 2PP for this application is a possibility to fabricate complex interlocking micromechanical components in a single step, this way eliminating the assembly step [20]. Metallization of structures produces by 2PP opens new possibilities through additional functionality, for applications where e.g. conductive 3D structures are required [26, 139].

Realization of 3D photonic crystal is a topic where 2PP still presents a unique technology capable of the widest number of possible solutions. The progress in this field will depend on the real demand of such structure realization from the side of industry.

Without doubt, biomedical field will become one of the most important application areas of microstructures produced by 2PP. Scalability of this technique can provide solutions for a wide range of applications. Further reduction of the fabrication times will enable fast production of patient-specific prosthesis directly in the operational room environment. A set of data obtained by micro-coherence tomography of implantation site can be used as an input for 2PP microfabrication system. Next generation intelligent transdermal drug delivery and blood sampling μTAS-based (micro total chemical analysis system) solutions will rely on the use of microcapillaries or microneedles integrated into microfluidic chip. Here, 2PP presents a unique tool for rapid prototyping or manufacturing of small series of samples. Increasing throughput, e.g. by parallel processing or substantially accelerating the speed of fabrication, will bring 2PP approach closer industrial to scale manufacturing.

Fabrication of scaffolds for tissue engineering is without doubt an application where 2PP offers enormous number of opportunities. When it comes to variation of possible 3D scaffold design and fabrication of series of identical samples, no other technique possesses such flexibility as 2PP. Using 2PP microfabrication, for the first time, enables systematic studies of scaffold-cell and cell–cell interactions in a 3D system. Eventually, scaffolds fabricated from synthetic biodegradable polymers will be required. Such photosensitive materials are already available; their investigations are currently in progress at the Laser Zentrum Hannover. In perspective, 2PP can also provide means for adjustable surface-cell interaction mediated by nanostructures or biomolecules deposited by photografting [19]. Since the 2PP processing takes place at room temperature, the denaturation of proteins due to high temperatures can be avoided.

11.11 Conclusions

Recent research on the photonics and biomedical applications of 3D structuring by two-photon polymerization of photosensitive materials has been reviewed.

The first subject to be examined was the fabrication of micro-optical elements by 2PP. Regarding the variety of possible surface profiles, fabrication by 2PP is superior to conventional approaches for fabrication of refractive optics. Furthermore, many polymeric materials, designed for optical applications, can be structures by 2PP.

In addition, the fabrication of 3D photonic crystals with various structural parameters and geometries by 2PP microfabrication was reviewed. By setting the structural parameters, the bandstops for the fabricated photonic crystals can be precisely positioned in the near-IR spectral range, at the telecommunication wavelengths. When it comes to woodpile structures, it was shown that the quality of photonic crystals fabricated by 2PP is sufficient to achieve a considerable PBG effect already for four unit cells along the testing direction. The dielectric filling fraction (Φ) is very important parameter, influencing the size and, to some extent, the exact position of the PBG. In 2PP microfabrication the Φ value is controlled independently of the structural parameters of a photonic crystal, by adjusting the illumination dose. For the woodpile architecture a minimal Φ value of 46% is limited by the stability of the produced structures. Further reduction of Φ value, i.e. woodpiles rods cross-section, results in mechanically weak structures, which collapse during the developing step. It was shown that much lower values of Φ can be experimentally realized for spiral photonic crystals. This can be explained by the fact that compared to woodpile; spiral structure resembles the diamond architecture much closer, while requiring less material. In addition, substantial spatial overlap between individual spirals contributes to better mechanical stability and lower Φ values for the spiral architecture.

For some future applications it is important to provide photonic crystals with NLO response. It was shown that materials containing NLO molecules can be successfully structured by 2PP. This method provides important guidelines for alternative material modifications with other nonlinear, electrooptic or photoactive materials. In conclusion, 2PP provides important rapid prototyping tool for the microscale all-optical circuit realization.

Applications of 2PP microfabrication in biomedicine were further reviewed. Microneedles produced by 2PP of hybrid photosensitive materials were tested for use in transdermal drug delivery, and the optimization of their geometry for the lowest skin resistance during needle insertion was discussed. Flow studies using florescent dye solution provided evidence of controlled subcutaneous injection possibility through the microneedle. The microneedles created using the 2PP technique are suitable for in vivo use, and for the integration with the next generation MEMS- and NEMS-based drug delivery devices.

Another related application area discussed is the fabrication of microprosthesis. While conventional mass-produced implants do not sufficiently take individual patient anatomy into account, CAD/CAM rapid prototyping by 2PP may permit development of patient-specific geometries. Middle ear ossicular replacement prosthesis, fabricated by 2PP, has shown good mechanical properties and ease of handling. This study indicates high potential of 2PP technique for prototyping and fabrication of such patient-specific microimplants.

Finally, the flexibility of 2PP microstructuring allows the fabrication of complex 3D structures that can be used for guided cell growth. The fabrication of scaffolds with a desired structure in 3D is a very important task in tissue engineering. It has been demonstrated that CAD designed complex 3D structures with varying pores sizes can be generated by 2PP technique. For a given material, the mechanical properties of such scaffold can be adjusted by changing the size and the positions of individual pores. Structures with feature size similar to natural ECM can be produced. With respect to other techniques conventionally used for scaffold fabrication, 2PP allows to close the existing gap in structural parameters of resulting structures. The high reproducibility of 2PP microfabrication, for the first time, allows the fabrication of large series of identical samples, and therefore, enables systematic studies of scaffold-cell and cell–cell interactions in a 3D system.

The future of the 2PP technique looks bright. The fact that there is still a lot of room for improvement in both fundamental as well as technical aspects of this approach, indicates that it will continue to provide new solutions in different application areas.

Acknowledgements The authors gratefully acknowledge very important contribution from their colleagues, who have been involved in different part of this work: R. Kiyan, S. Schlie, A. Ngezahayo, M. Vamvakaki, and C.Fotakis. Biomedical applications have been studied in cooperation with A. Doraiswamy, T. Patz, R. Narayan, R. Modi, R. Auyeung, and O. Adunka. This work has been supported by the Excellence Cluster ReBirth, and DFG Transregio project TR37.

References

1. S. Maruo, O. Nakamura, S. Kawata, Opt. Lett., 22, 132 (1997).
2. http://www.michrochem.com
3. http://www.microresist.de
4. H. B. Sun, S. Matsuo, and H. Misawa, Appl. Phys. Lett., 74, 786 (1999).
5. B. Cumpston, S. Ananthavel, S. Barlow, D. Dyer, J. Ehrlich, L. Erskine, A. Heikal, S. Kuebler, I. Lee, D. McCord-Maughon, J. Qin, H. Rockel, M. Rumi, X. Wu, S. Marder, J. Perry, Nature, 398, 51–54 (1999).
6. M. Deubel, G. von Freymann, M. Wegener, S. Pereira, K. Busch, C. M. Soukoulis, Nat. Mater., 3, 444 (2004).
7. M. Straub, L. H. Nguyen, A. Fazlic, M. Gu, Opt. Mater., 27, 359 (2004).
8. K. K. Seet, V. Mizeikis, S. Matsuo, S. Juodkazis, H. Misawa, Adv. Mater., 17, 541 (2005).
9. S. Klein, A. Barsella, H. Leblond, H. Bulou, A. Fort, C. Andraud, G. Lemercier, J. C. Mulatier, K. Dorkenoo, Appl. Phys. Lett., 86, 211118 (2005).
10. R. Guo, S. Xiao, X. Zhai, J. Li, A. Xia, W. Huang, Opt. Express, 14, 810 (2006).
11. F. C. Wippermann, D. Radtke, U. Zeitner, J. W. Duparre, A. Tunnermann, M. Amberg, S. Sinzinger, C. Reinhardt, A. Ovsianikov, and B. N. Chichkov, Proc. SPIE Vol. 6288, "Current developments in lens design and optical engineering VII", San Diego, 14–15. 8 (2006).
12. T. Yokoyama et al., Appl. Phys. Lett., 82, 3221 (2003).
13. H. Sun, T. Kawakami, Y. Xu, J. Ye, S. Matuso, H. Misawa, M. Miwa, and R. Kaneko, Opt. Lett., 25, 1110 (2000).
14. C. Reinhardt, S. Passinger, B. Chichkov, C. Marquart, I. Radko, and S. Bozhevolnyi, Opt. Lett., 31, 1307 (2006).
15. A. Doraiswamy, C. Jin, R. J. Narayan, P. Mageswaran, P. Mente, R. Modi, R. Auyeung, D. B. Chrisey, A. Ovsianikov, B. Chichkov, Acta Biomater., 2, 267–275 (2006).
16. S. Schlie, A. Ngezahayo, A. Ovsianikov, T. Fabian, H.-A. Kolb, H. Haferkamp, B. N. Chichkov, J. Biomater. Appl., 22, 275–287 (2007).
17. J. D. Pitts, P. J. Campagnola, G. A. Epling, S. L. Goodman, Macromolecules, 33, 151 (2000).
18. S. Basu, P. J. Campagnola, Biomacromolecules, 5, 572 (2004).
19. V. Dinca, E. Kasotakis, J. Catherine, A. Mourka, A. Ranella, A. Ovsianikov, B. N. Chichkov, M. Farsari, A. Mitraki, and C. Fotakis, Nano Lett., 8, 538–543 (2008).
20. S. Maruo, K. Ikuta, H. Korogi, Appl. Phys. Lett., 82, 133 (2003).
21. S. Maruo, Polym. Prepr. Am. Chem. Soc., Div. Polym. Chem., 94, 101 (2006).
22. F. Formanek, N. Takeyasu, T. Tanaka, K. Chiyoda, A. Ishikawa, S. Kawata, Opt. Express, 14, 800 (2006).
23. F. Formanek, N. Takeyasu, T. Tanaka, K. Chiyoda, A. Ishikawa, S. Kawata, Appl. Phys. Lett., 88, 083110 (2006).
24. Y.-S. Chen, A. Tal, D. B. Torrance, S. M. Kuebler, Adv. Funct. Mater., 16, 1739 (2006).
25. V. Mizeikis, S. Juodkazis, R. Tarozaite, J. Juodkazyte, K. Juodkazis, and H. Misawa, Opt. Express, 15, 8454–8464 (2007).
26. R. A. Farrer, C. N. LaFratta, L. Li, J. Praino, M. J. Naughton, B. E. A. Saleh, M. C. Teich, J. T. Fourkas, J. Am. Chem. Soc., 128, 1796 (2006).
27. E. Yablonovitch, Phys. Rev. Lett., 58, 2059–2062 (1987).
28. S. John, Phys. Rev. Lett., 58, 2486–2489 (1987).
29. K. Busch, S. Lölkes, R. B. Wehrspohn, and H. Föll, Photonic Crystals, Wiley, Berlin (2004).
30. J.-M. Lourtioz, H. Benisty, V. Berger, J.-M. Gerard, D. Maystre, A. Tchelnokov, Photonic Crystals: Towards Nanoscale Photonic Devices, Springer, Berlin and Heidelberg (2005).
31. S. G. Johnson, J. D. Joannopoulos, Photonic Crystals: The Road from Theory to Practice, Springer, Berlin (2001).
32. J. D. Joannopoulos, R. D. Meade, and J. N. Winn, Photonic Crystals: Molding the Flow of Light, 2nd Edition, Princeton University Press, Princeton (2008).
33. K. Sakoda, Optical Properties of Photonic Crystals, Springer, Berlin (2005).

34. K. Inoue and K. Ohtaka (eds.), Photonic Crystals, Physics, Fabrication and Applications, Springer, Berlin (2004).
35. S. Noda and T. Baba (eds.), Roadmap on Photonic Crystals, Kluwer Academic Publishers, Boston (2003).
36. J. D. Joannopoulos, P. R. Villeneuve, and S. Fan, Nature, 386, 143–149 (1997).
37. S. Noda, K. Tomoda, N. Yamamoto, and A. Chutinan, Science, 289, 604–606 (2000).
38. A. Chutinan, S. John, and O. Toader, Phys. Rev. Lett., 90, 123901 (2003).
39. P. Markowicz, Ch. Friend, Y. Shen, J. Swiatkiewicz, P. N. Prasad, O. Toader, S. John, and R. W. Boyd., Opt. Lett., 27, 351 (2002).
40. M. C. Netti, A. Harris, J. J. Baumberg, D. M. Whittaker, M. B. D. Charlton, M. E. Zoorob, and G. J. Parker, Phys. Rev. Lett., 86, 1526 (2001).
41. J. Maddox, "Photonic band-gaps bite the dust," Nature, 348, 481 (1990).
42. S. Y. Lin, J. G. Fleming, D. L. Hetherington, B. K. Smith, R. Biswas, K. M. Ho, M. M. Sigalas, W. Zubrzycki, S. R. Kurtz, and J. Bur, Nature, 394, 251–253 (1998).
43. N. Yamamoto, S. Noda, and A. Sasaki, Jpn. J. Appl. Phys., 36, 1907 (1997).
44. S. Noda, N. Yamamoto, M. Imada, H. Kobayashi, M. Okano, J. Lightwave Technol., 17, 1948 (1999).
45. F. García-Santamaría, H. T. Miyazaki, A. Urquía, M. Ibisate, M. Belmonte, N. Shinya, F. Meseguer, C. López, Adv. Mat., 14, 1144 (2002).
46. P. Jiang, J. F. Bertone, K. S. Hwang, and V. L. Colvin, Chem. Mater., 11, 2132 (1999).
47. J. F. Bertone, P. Jiang, K. S. Hwang, D. M. Mittleman, and V. L. Colvin, Phys. Rev. Lett., 83, 300 (1999).
48. R. Biswas, M. M. Sigalas, G. Subramania, and K. M. Ho, Phys. Rev. B, 57, 3701 (1998).
49. K. Busch and S. John, Phys. Rev. E, 58, 3896 (1998).
50. D. J. Norris, Y. A. Vlasov, Adv. Mat., 13, 371–376 (2001).
51. A. Blanco, E. Chomski, S. Grabtchak, M. Ibisate, S. John, S. W. Leonard, C. Lopez, F. Meseguer, H. Miguez, J. P. Mondla, G. A. Ozin, O. Toader, and H. M. van Driel, Nature, 45, 437 (2000).
52. W. M. Lee, S. A. Pruzinsky, and P. V. Braun, Adv. Mater., 14, 271 (2002).
53. Y. Yin and Y. Xia, Adv. Mater., 14, 605 (2002).
54. H. Míguez, S. M. Yang, N. Tétreault, and G. A. Ozin, Adv. Mater., 14, 1805 (2002).
55. M. Campbell, D. N. Sharp, M. T. Harrison, R. G. Denning, and A. J. Turberfield, Nature (London), 404, 53 (2000).
56. S. Shoji and S. Kawata, Appl. Phys. Lett., 76, 2668 (2000).
57. S. Shoji, H.-B. Sun, and S. Kawata, Appl. Phys. Lett., 83, 608 (2003).
58. Y. Miklyaev, D. Meisel, A. Blanco, G. von Freymann, K. Busch, W. Koch, C. Enkrich, M. Deubel, and M. Wegener, Appl. Phys. Lett., 82, 1284 (2003).
59. X. Wang, J. F. Xu, H. M. Su, Z. H. Zeng, Y. L. Chen, H. Z. Wang, Y. K. Pang, and W. Y. Tam, Appl. Phys. Lett., 82, 2212–2214 (2003).
60. H. B. Sun, V Mizeikis, Y. Xu, S. Juodkazis, J.-Y. Ye, S. Matsuo, H. Misawa, Appl. Phys. Lett., 79, 1 (2001).
61. K. Kaneko, H. B. Sun, X. M. Duan, and S. Kawata, Appl. Phys. Lett., 83, 2091 (2003).
62. A. Ledermann, L. Cademartiri, M. Hermatschweiler, C. Toninelli, G. A. Ozin, D. S. Wiersma, M. Wegener, G. von Freymann, Nat. Mater., 5, 942 (2006).
63. N. Tétreault, G. von Freymann, M. Deubel, M. Hermatschweiler, F. Pérez-Willard, S. John, M. Wegener, G. A. Ozin, Adv. Mat., 18, 457 (2006).
64. M. Straub and M. Gu, Opt. Lett., 27, 1824–1826 (2002).
65. J. Serbin, A. Ovsianikov, and B. Chichkov, Opt. Express, 12, 5221–5228 (2004).
66. A. Chutinan, S. Noda, Phys. Rev B, 57, R2006 (1998).
67. M. Maldovan and E. L. Thomas, Nat. Mater., 3 593–600 (2004).
68. A. Ovsianikov, J. Viertl, M. Oubaha, B. MacCraith, I. Sakellari, A. Giakoumaki, D. Gray, M. Vamvakaki, M. Farsari, C. Fotakis, and B. Chichkov, ACS Nano, 2, 2257–2262 (2008).

69. M. Deubel, M. Wegener, S. Linden, and G. von Freymann, Appl. Phys. Lett., 87, 221104 (2005).
70. S. Romanov et al., Phys. Rev. E, 63 056603–056605 (2001).
71. S. Romanov et al., Appl. Phys. Lett., 90 133101–133103 (2007).
72. A. Lavrinenko, S. Romanov, PECS-VII Monterey, USA, April 8–11 (2007).
73. H. X. Zhang, D. Lu, N. Peyghambarian, M. Fallahi, J. D. Luo, B. Q. Chen, and A. K.-Y. Jen, Opt. Lett., 30, 117–119 (2004).
74. H. Goudket, M. Canva, Y. Levy, F. Chaput, and J. P. Boilot, J. Appl. Phys., 90, 6044–6047 (2001).
75. Y. Enami, C. T. Derose, D. Mathine, C. Loychik, C. Greenlee, R. A. Norwood, T. D. Kim, J. Luo, Y. Tian, A. K. Y. Jen, and N. Peyghambarian, Nat. Photonics, 1, 180–185 (2007).
76. H. X. Zhang, D. Lu, T. Liu, M. Mansuripur, and M. Fallahi, Appl. Phys. Lett., 85, 4275–4277 (2004).
77. H. X. Zhang, D. Lu, M. Fallahi, Opt. Mater., 28, 992–999 (2006).
78. M. Farsari, A. Ovsianikov, M. Vamvakaki, I. Sakellari, D. Gray, B. N. Chichkov, and C. Fotakis, Appl. Phys. A, 93, 11–15 (2008).
79. D. H. Choi, J. H. Park, T. H. Rhee, N. Kim, and S.-D. Lee, Chem. Mater., 10, 705–709 (1998).
80. D. Riehl, F. Chaput, Y. Levy, J. P. Boilot, F. Kajzar, and P. A. Chollet, Chem. Phys. Lett., 245, 36–40 (1995).
81. V. G. Zarnitsyn, M. R. Prausnitz, Y. A. Chizmadzhev, Biol. Membr., 21, 355–373 (2004).
82. G. J. Opiteck, J. E. Scheffler, Expert Rev. Proteomics, 1, 57–66 (2004).
83. R. O. P. Potts, R. A. Lobo, Obstet. Gynecol., 105, 953–961 (2005).
84. M. R. Prausnitz, Adv. Drug Deliv. Rev., 56, 581–587 (2004).
85. M. R. Prausnitz, S. Mitragotri, R. Langer, Nat. Rev. Drug Discov., 3, 115–124 (2004).
86. S. Chong, H. L. Fung, "Transdermal Drug Delivery Systems: Pharmacokinetics, Clinical Efficacy, and Tolerance Development". In: J. Hadgraft, R. H. Guy (eds.), Transdermal Drug Delivery: Developmental Issues and Research Initiatives, Dekker, New York, 135 (1989).
87. G. L. Flynn, "Cutaneous and Transdermal Delivery: Processes and Systems of Delivery", In: G. S. Banker, C. T. Rhodes (eds.), Modern Pharmaceutics. Dekker, New York, 239–299 (1996).
88. R. K. Sivamani, B. Stoeber, G. C. Wu, H. Zhai, D. Liepmann, H. Maibach, Skin Res. Technol., 11, 152–156 (2005).
89. S. Mitragotri, J Control. Release, 71, 23–29 (2001).
90. B. W. Barry, Eur. J. Pharm. Sci., 14, 101–114 (2001).
91. P. Griss, G. Stemme, J. Microelectromech. Syst., 12, 296–301 (2003).
92. R. L. Daniels et al., Laryngoscope, 108, 1674–1681 (1998).
93. A. De la Cruz et al., Otolaryngol. Clin. North Am., 27, 799–811 (1994).
94. S. Albu et al., Am. J. Otolaryngol., 19, 136–140 (1998).
95. A. Y. Bayazit, Laryngoscope, 110, 176–177 (2000).
96. R. L. Goode et al., Otolaryngol. Clin. North Am., 27, 663–675 (1994).
97. V. Colletti et al., Otolaryngol. Head Neck Surg., 120, 437–444 (1999).
98. O. Cura et al., Rev. Laryngol. Otol. Rhinol., 121, 87–90 (2000).
99. M. Glasscock et al., Arch. Otolaryngol. Head Neck Surg., 114, 1252–1255 (1988).
100. K. Jahnke et al., Biomaterials, 4, 137 (1983).
101. K. Schwager, Eur. Arch. Otorhinolaryngol., 255, 396–401 (1998).
102. C. Stupp et al., Laryngorhinootologie, 78, 299–303 (1999).
103. S. Schmerber et al., Eur. Arch. Otorhinolaryngol., 263, 347–354 (2006).
104. X. Wang et al., Otolaryngol. Head Neck Surg., 121, 606–609 (1999).
105. T. Zahnert et al., Am. J. Otol., 21, 322–328 (2000).
106. J. Grote et al., Ann. Otol. Rhinol. Laryngol., 123, 1–5 (1986).
107. J. Grote, Am. J. Otol., 6, 269–271 (1985).
108. C. van Blitterswijk et al., J. Biomed. Mater. Res., 20, 1197–1217 (1986).
109. C. Mangham et al., Ann. Otol. Rhinol. Laryngol., 99, 112–116 (1990).

110. R. Goldenberg et al., Otolaryngol. Head Neck Surg., 122, 635–642 (2000).
111. S. Merchant et al., Am. J. Otol., 18, 139–154 (1999).
112. W. Moretz, Laryngoscope, 108, 1–12 (1998).
113. E. Murugasu et al., Otol. neurotol., 26, 572–582 (2005).
114. D. Mahoney, Comput. Graph. World, 18, 42–48 (1995).
115. C. Lim et al., Int. J. Adv. Manuf. Technol., 20, 44–49 (2002).
116. A. Ovsianikov, B. N. Chichkov, O. Adunka, H. Pillsbury, A. Doraiswamy, R. J. Narayan, Appl. Surf. Sci., 253, 6603 (2007).
117. http://en.wikipedia.org/wiki/Tissue_engineering Wikipedia® is a registered trademark of the Wikimedia Foundation.
118. D. W. Hutmacher, J. Biomater. Sci. Polym. Ed., 12, 107–124 (2001).
119. B. Harley, H.-D. Kim, M. Zaman, I. Yannas, D. Lauffenburger, L. Gibson, Biophys. Biophys. J., online first (2008).
120. D.-M. Liu, J. of Mater. Sci. Lett., 15, 419–421 (1996).
121. T. Roy, J. Simon, J. Ricci, E. Rekow, V. Thompson, J. Parsons, J. Biomed. Mater. Res., 66A, 283–291 (2003).
122. S. Seyfert, A. Vogt, D. Kabbeck-Kupijai, Biomaterials, 16, 201–207 (1995).
123. K. Fujimoto, A. Nagafuch, S. Tsukita, A. Kuraoka, A. Ohokuma, Y. Shibata, J. Cell Sci., 110, 311–322 (1997).
124. A. Ngezahayo, B. Altmann, M. Steffens, H. Kolb, J. Membr. Biol., 204 137–144 (2005).
125. A. Ovsianikov, S. Schlie, A. Ngezahayo, A. Haverich, and B. N. Chichkov, J. Tissue Eng. Regen. Med., 1, 443–449 (2007).
126. W. Haske, V. Chen, J. Hales, W. Dong, S. Barlow, S. Marder, and J. Perry, Opt. Express, 15, 3426–3436 (2007).
127. M. Martinez-Corral, C. Ibáñez-López, G. Saavedra, and M. Caballero, Opt. Express, 11, 1740–1745 (2003).
128. S. W. Hell, S. Lindek, C. Cremer, and E. H. K. Stelzer, Appl. Phys. Lett., 64, 1335–1338 (1994).
129. T. A. Klar, E. Engel, and S. W. Hell, Phys. Rev. E, 64, 066611–066619 (2001).
130. T. A. Klar, S. Jakobs, M. Dyba, A. Egner, S. W. Hell, Proc. Natl. Acad. Sci. U S A, 97, 8206 (2000).
131. J.-F. Xing, X.-Z. Dong, W.-Q. Chen, X.-M. Duan, N. Takeyasu, T. Tanaka, and S. Kawata, Appl. Phys. Lett., 90, 131106 (2007).
132. C. A. Leatherdale, R. J. DeVoe, Proc. SPIE Int. Soc. Opt. Eng., 5211, 112 (2003).
133. K. Takada, H.-B. Sun, and S. Kawata, Appl. Phys. Lett., 86, 071122 (2005).
134. A. Ostendorf and B. N. Chichkov, Photonics Spectra, 40, 72 (2006).
135. S. Matsuo, S. Juodkazis, and H. Misawa, Appl. Phys. A, 80, 683–685 (2005).
136. J. Kato, N. Takeyasu, Y. Adachi, H.-B. Sun, and S. Kawata, Appl. Phys. Lett., 86, 044102 (2005).
137. Y. Xia, E. Kim, X. M. Zhao, J. A. Rogers, M. Prentiss, and G. M. Whitesides, Science, 273, 347–349 (1996).
138. C. LaFratta, L. Li, J. Fourkas, Proc. Natl. Acad. Sci. U S A, 103, 8589 (2006).
139. R. T. Hill, J. L. Lyon, R. Allen, K. J. Stevenson, J. B. Shear, J. Am. Chem. Soc., 127, 10707 (2005).

Chapter 12
Practical Use of Hydrogels in Stereolithography for Tissue Engineering Applications

Karina Arcaute, Brenda K. Mann, and Ryan B. Wicker

12.1 Introduction

In recent years, additive manufacturing (AM) or rapid prototyping (RP) technologies, initially developed to create prototypes prior to production for the automotive, aerospace, and other industries, have found applications in tissue engineering (TE) and their use is growing rapidly. RP technologies are increasingly demonstrating the potential for fabricating biocompatible 3D structures with precise control of the micro- and macro-scale characteristics. Several comprehensive reviews on the use of RP technologies, also known as solid freeform fabrication, Additive Manufacturing, direct digital manufacturing, and other names, have been published recently [1–4].

The 3D dispensing of liquids and pastes, known as 3D plotting, was one of the first RP technologies used in TE. Agarose, chitosan, hydroxyapatite, and poly (lactic-co-glycolic acid) are some of the materials used in the creation of scaffolds with 3D plotters [5–7]. Laser sintering is another RP technology explored in TE, using materials in powder form, such as poly(caprolactone), polyetherketone, and hydroxyapatite [8–10]. Fused deposition modeling (FDM) has been used to fabricate scaffolds with controlled porosity for bone regeneration using poly(caprolactone). Pellets of poly(caprolactone) or blends of poly(caprolactone) and hydroxyapatite are processed into filaments in order to be extruded through the head of an FDM machine [11–13].

Alternatively, researchers have developed 3D photopatterning techniques for the formation of patterned scaffolds and substrates with arrays of immobilized proteins and/or cells using photopolymerizable poly(ethylene glycol) (PEG) [14–18] and photolithographic projection-based processes for the fabrication of 3D scaffolds [19–23]. Similar to the RP technology of stereolithography (SL), these techniques use a photoreactive material and an ultraviolet (UV) source, and 3D features are created in a layered fashion.

R.B. Wicker (✉)
W.M. Keck Center for 3D Innovation, University of Texas at El Paso, El Paso, TX, USA
e-mail: rwicker@utep.edu

P.J. Bártolo (ed.), *Stereolithography: Materials, Processes and Applications*,
DOI 10.1007/978-0-387-92904-0_12, © Springer Science+Business Media, LLC 2011

The use of SL in TE has not been significantly explored, perhaps because of the lack of commercially available implantable or biocompatible materials from the SL industry. Photocrosslinkable non-toxic formulations have been developed by the TE community, however, and more recently used in SL [24–29]. Poly(propylene fumarate) (PPF) and diethyl fumarate (DEF) photocrosslinkable mixtures [26, 27] and mixtures of polyfunctional methacrylic oligomers and hydroxyapatite [28, 29] have been used in commercial SL systems to create complex porous scaffolds for bone regeneration. Photocrosslinkable PPF-DEF mixtures have recently been used in novel micro-SL systems developed using a digital micromirror device for dynamic pattern generation [23]. A mixture of poly(ethylene oxide) (PEO) and poly(ethylene glycol)-dimethacrylate (PEG-dma) was used in a commercial SL system by Dhariwala et al. [30] to create hydrogels containing a high density (50×10^6 cells/mL) of Chinese hamster ovary cells (CHO-B2) for soft TE applications. Simple constructs (rings with a 5-mm diameter) were created and the high viability of this cell line in high-density constructs made in SL was demonstrated [30]. In previous work by our group, the capabilities of commercial line-scan SL to fabricate PEG-hydrogel structures of high complexity were demonstrated using solutions of a low molecular weight (MW 1 K) photoreactive PEG, showing the potential use of SL in soft TE applications [31–33]. The following sections briefly review SL and the photocrosslinkable biomaterials available to use in SL. Subsequent sections describe in detail what is necessary to build in SL using a solution of the biopolymer PEG modified with photocrosslinkable groups, demonstrating the potential use of SL in soft TE applications.

12.1.1 Stereolithography

When SL became commercially available in the mid-1980s, the manufacturing industry was profoundly impacted by improving the way in which prototypes were fabricated [34]. Since the introduction of SL, a number of new RP technologies have been developed and commercialized with widely varying applications in engineering design, manufacturing, medicine, and more. SL has potential use in TE for fabricating 3D tissue engineered scaffolds for both soft and hard TE applications. SL is an accurate and easy to use technology to fabricate complex structures in single quantities or in mass production. SL has open and easy access to the building chamber, and the building envelope can be easily adapted to create a clean, particulate-free environment, or for accurate registration between materials in multi-material fabrication [35, 36]. The SL system laser and optics are relatively simple to control with potential for high accuracy in microfabrication. Typical laser beam diameters are 75 or 250 μm, where 75 μm is considered high resolution and 250 μm is the traditional or more typical beam diameter. In newer generation machines, variable beam diameters are used so that small features can be built using the small beam diameters and large areas can be scanned quicker using larger beam diameters (up to ~750 μm in current machines). Another advantage of SL is the availability of a wide range of photocurable materials with potential use in SL, and the fact that the industry is constantly developing new materials for use in SL with different properties.

The basic components, principles of operation, and procedures for working with SL can be found elsewhere in this text and in references [37–42]. It is important, however, to mention here that commercial line-scan SL systems are designed to control the laser scan speed across the surface of the photocurable material using equations for photospeed based on the E_C (critical exposure) and D_P (penetration depth) parameters specific to each material, and other factors such as the laser power. Based on the energy that the photopolymer receives, 3D parts are created by stacking 2D layers with layer thicknesses as thin as ~50 μm (and more typically ~100 μm) with overcures of ~200 μm. The most common laser wavelength currently available for commercial SL systems is 354.7 nm, although early generation machines used Helium Cadmium (He–Cd) lasers with wavelengths of 325 nm. The majority of the TE work described here was performed with a 3D Systems Model 250/50 SL machine (3D Systems, Rock Hill, SC) equipped with a He–Cd laser (325 nm, 40 mW).

12.1.2 Photocrosslinkable Biopolymers

Different biopolymers have been modified with photoreactive and crosslinkable groups like acrylates or methacrylates to make the biopolymer functional for processability into coatings, drug-delivery vehicles, TE scaffolds, cell carriers and more. Modification of the biocompatible polymer PEG with photocrosslinkable groups has been performed extensively for a variety of TE applications including cell encapsulation [43–46], creation of synthetic extracellular matrix analogs [47–51], and for the formation of substrates with patterned arrays of immobilized proteins and/or cells [14, 16]. PEG is a highly biocompatible material [52], and the photoreactive groups in the presence of a photoinitiator (PI) and upon exposure to UV light serve to crosslink the PEG into a hydrogel. These hydrogels are permeable to oxygen, nutrients, and other water-soluble metabolites and have mechanical properties similar to many soft tissues [53]. Although PEG-based hydrogels are not bioactive, cell-specific bioactivity can be achieved by covalently attaching adhesion ligands, growth factors, and cytokines to the hydrogel, or by simply trapping the bioactive agents within the hydrogel [47–50]. Additionally, while PEG hydrogels are normally nondegradable, they can be rendered biodegradable by incorporating proteolytically degradable peptide sequences into the PEG backbone or by creating a copolymer with a degradable polymer such as poly(lactic acid) [48, 51]. Photocrosslinkable PEG has been used in SL, as a copolymer blended with PEO to create a photoreactive solution [30], and as the main reactive species in water-based photocrosslinkable solutions [31–33, 54].

Another biomaterial that has been used in SL is a mixture of PPF and DEF [23, 26, 27]. Carbon–carbon double bonds present in the repeating unit of PPF and in

DEF make the mixture photocrosslinkable. Crosslinked PPF has been shown to be biocompatible and bioresorbable and to elicit a bone healing response in cranial defects (no/low load bearing bone defects) [55]. Mixtures of polyfunctional methacrylic oligomers and hydroxyapatite have also been used in SL to create complex porous scaffolds for bone regeneration [28, 29]. Carbon–carbon double bonds in the methacrylate group make the blend photocrosslinkable, and therefore suitable for use in SL. Other biopolymers such as the natural polysaccharide hyaluronic acid have been modified with photocrosslinkable groups [56–58], and thus may find potential use in SL as well.

12.2 Stereolithography of Photocrosslinkable PEG

The work presented here investigates the use in SL of photocrosslinkable PEG with two different molecular weights (MW) for the creation of complex 3D structures with applications in TE. Rather than having a liquid mixture of photoreactive monomers, as traditionally used with SL, a water-based solution of a photoreactive polymer and a PI was used. Complex 3D PEG hydrogel structures were fabricated in a layer-by-layer fashion, and the use of SL to fabricate multi-material 3D constructs with specified spatially-controlled characteristics is described. Different photopolymerization experiments were performed to characterize PEG-based photopolymer solutions for use in SL. In the experiments, the crosslinked depth (or gel thickness) was measured at different laser energies, and patterned features were inspected under a microscope for accuracy.

12.2.1 Effect of Photoinitiator

First, the photoinitiating conditions that allow for the photopolymerization of thin layers were determined by examining two different PI at different concentrations. Thin layers are preferred in AM as the achievable layer thickness controls the Z (or vertical) resolution of the fabricated parts. The two different PIs investigated were Irgacure 2959 (I-2959, Ciba Speciality Chemicals Corp., Tarrytown, NY) and 2-hydroxy-2-methyl-1-phenyl-1-propanone (HMPP, Sartomer Co., Exton, PA). The concentration of the PIs in a PEG-based solution ranged from 0.01 to 0.5 wt%. Poly (ethylene glycol)-dimethacrylate MW 1 K (PEG-dma 1 K, Polysciences, Inc., Warrington, PA) at a concentration of 30 wt% in solution with distilled water was used for these experiments.

Figure 12.1 shows the gel thickness curves as a function of PI concentration for the two PIs investigated. HMPP produced thicker gels than I-2959 for a given concentration, indicating that the photopolymerization is dependent on PI type. The concentration of PI also affected gel thickness, with thinner gels being created as

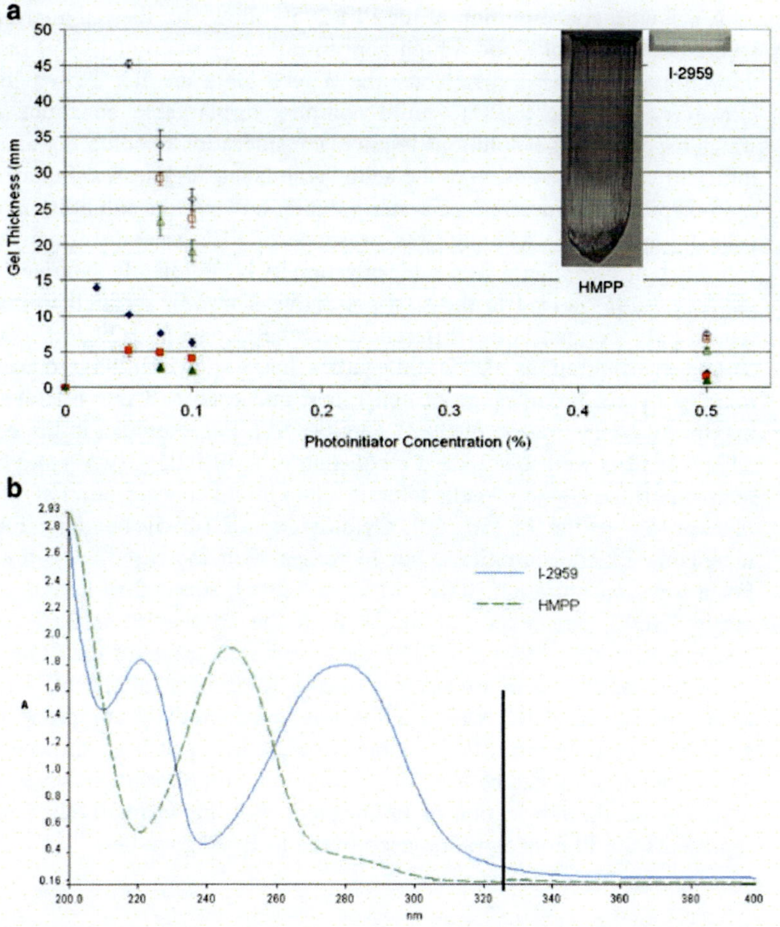

Fig. 12.1 Photoinitiator effects. (**a**) Gel thickness curves for the two types of PIs used, I-2959 (*solid markers*) and HMPP (*hollow markers*). The three different types of markers correspond to three different energy dosages (*Diamond* 1,800 mJ/cm^2 [0.258 in./s], *Square box* 820 mJ/cm^2 [0.567 in./s], *Triangle* 300 mJ/cm^2 [1.585 in./s]). Error bars not shown are contained within the symbol. PEG concentration is 30 wt% in distilled water. Insert shows representative images of gels crosslinked with the two different photoinitiators. (**b**) UV–Vis absorption curve for the two different photoinitiators (the *black line* in the graph indicates the wavelength of the laser used in the experiments)

the PI concentration increased (after a peak in concentration below ~0.01 wt%). Since each PI has a unique absorption spectrum, and radicals are formed when the initiator in solution is exposed to an initiating light of appropriate wavelength, the crosslinked thickness would be different depending on the PI type, concentration, and wavelength of the laser used. Based on these results, the PI I-2959 is preferred over HMPP for AM using SL as I-2959 allows for photopolymerization of thinner layers using a wavelength of 325 nm.

A 0.5 wt% concentration of the PI I-2959 allows for the polymerization of thin layers (~1 mm thick), and a high concentration (3 wt%) of I-2959 results in even thinner layers (~0.3 mm) (Note: the 3 wt% data are not shown in Fig. 12.1). However, selecting a PI to build complex implantable constructs using these water-based hydrogel solutions requires consideration of solubility and biocompatibility of the PI in addition to the achievable layer thicknesses. In terms of solubility, I-2959 is considered to be water-soluble; however its solubility in water is low (~0.01 g/mL) [59]. Although the presence of PEG in the solution increases the solubility of I-2959, a 3 wt% concentration of I-2959 nearly saturates the solution, and any I-2959 added to the solution above 3 wt% is difficult to dissolve. With regard to biocompatibility, different researchers consider I-2959 cytocompatible in certain concentrations. Different studies have shown I-2959 to cause minimal toxicity over a broad range of cell types, and each cell line tolerates a different maximum concentration of the PI [56, 60, 61]. For example, in the cell line NIH/3T3, I-2959 is well tolerated at a concentration of 0.05% (w/w) and below, while human aortic endothelial cells tolerate a maximum concentration of 0.01% for 48 h of exposure to the PI [56, 60]. Cytotoxicity of I-2959 not only increases with increasing PI concentrations, but increases with increased exposure times [54]. For a low concentration (0.015 wt%) of I-2959, survival of human dermal fibroblasts (HDF) cells exposed for 24 h to the PI was 99 ± 2.6%. At a higher concentration (0.5 wt%) of I-2959, the detrimental effect of the PI on cell survival is evident after 4 h of exposure as only 50 ± 5.3% of the cells remain alive. A concentration of 0.5 wt% I-2959 was considered by our group to provide a good balance between solubility, biocompatibility, and layer thickness resolution. This concentration can be used in conjunction with SL without severe cytotoxicity allowing for the fabrication of PEG-based scaffolds in relatively thin layers, and represents the PI concentration used in the following studies.

12.2.2 Effect of Energy Dosage, Photopolymer Concentration, and Type

Different photocrosslinking experiments were performed in order to understand the photocrosslinking behavior of PEG-based solutions. Cure depth curves for different PEG-based photopolymer solutions as a function of energy exposure were determined similar to previous experiments [33, 54]. The PEG-based photopolymer solutions investigated contained a 0.5 wt% concentration of the PI I-2959. Two different types of photocrosslinkable PEG (PEG dimethacrylate, MW 1 K, and PEG diacrylate, MW 3.4 K) at two different concentrations (20 and 30 wt%) were investigated.

Initially, the cure depth curves for each photoreactive formulation were determined by measuring the crosslinked thickness at different energy exposures. For the different photoreactive PEG solutions investigated, the gel thickness vs. energy dosage data

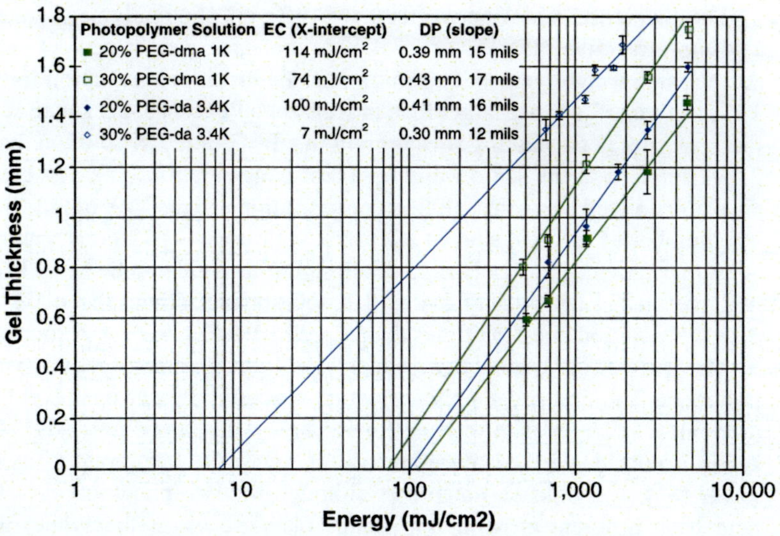

Fig. 12.2 Semi-log plot of gel thickness vs. energy dosage for PEG-dma 1 K (*square marker*) and PEG-da 3.4 K (*diamond markers*) solutions with 0.5 wt% of I-2959. *Solid markers* correspond to 20 wt%, and *hollow markers* correspond to 30 wt%

recorded on a semilogarithmic plot conformed closely to a straight line, similar to the data for commercial SL resins (see Fig. 12.2). The characteristic E_C (X-intercept) and D_P (slope) values for the PEG-based photopolymer solutions were determined from these gel thickness curves. Figure 12.2 shows the curves for the four different photopolymer solutions investigated. The slope of the curves represents the D_P value. D_P is the penetration depth of the UV energy into the photopolymer and represents the depth at which the irradiance becomes $1/e$ times that at the surface [38].

The E_C value, represents the minimum energy required to transform the photopolymer from liquid to solid [37]. For a given type of photopolymer solution, the solution with a lower concentration of polymer (20 wt%) has a higher E_C. With a lower concentration of polymer, there is a smaller number of possible crosslinking sites, and thus more energy is required to crosslink a given cure depth.

The photoreactive, water-based hydrogel solutions require further characterization prior to use in SL, as the E_C and D_P values obtained from the curves in Fig. 12.2 cannot be used directly in commercial SL systems. The characteristic photopolymerization of PEG-based solutions is quite different from commercial SL resins. This may be due to the fact that commercial SL resins are composed mostly of reactive components, while the PEG-based photopolymer solutions used here are composed mainly of water. Additionally, the crosslinking species of commercial resins is a monomer, unlike the polymer used here with reactive endgroups. The methods developed for successful building in SL using these solutions are described in the following section. The differences in concentration and reactivity of the two different photopolymers and solutions are some of the factors that affect the photopolymerization reaction.

The difference in MW between the two different photopolymers makes PEG-dma 1 K more reactive than PEG-da 3.4 K at the same weight percent concentration. At the same weight percent concentration, the molar concentration for a PEG-dma 1 K solution is 3.4 times larger than the one for a PEG-da 3.4 K solution. As a result, there are more crosslinking sites present in a PEG-dma 1 K solution than a PEG-da 3.4 K solution at the same weight percent concentration. The number of reactive sites is estimated using the solution concentration and the MW of the photopolymer. The number of reactive sites in a 30 wt% solution is ~1.72 times that of a 20 wt% solution when the photopolymer in both solutions has the same MW. The number of reactive sites in a 30 wt% PEG-dma 1 K solution is 3.4 times that of a 30 wt% PEG-da 3.4 K solution and 5.85 times that of a 20 wt% PEG-da 3.4 K solution.

The different photoreactive groups also influence the reactivity of the photopolymers. The first order polymerization reaction rate constant (k_P) for acrylated monomers is larger than the one for methacrylated monomers [62]. The acrylate group is considered more dormant than the methacrylate group as its reactivity is lower [63]. It should be noted that although the two photopolymers investigated here have different crosslinking groups (acrylate vs. methacrylate) these groups have only one double bond at each end available for crosslinking.

12.3 3D Fabrication Using PEG-Based Solutions

The characteristic photocrosslinking or photosensitivity of PEG-based solutions influences the successful fabrication of PEG-hydrogel structures using SL. The cured gel depth, specific for each photoreactive formulation, controls the layer thickness and overcuring between layers and thus the resolution in *Z* of the fabricated parts. Accuracy in the *X-Y* direction is also affected by the formulation photosensitivity and number of laser exposures. The photosensitivity of PEG-based solutions is typically higher than traditional resins used in SL, as crosslinked PEG gels continue to transmit the UV energy of the laser since these are in large part water-based solutions. As a result, each successive exposure of the laser penetrates the construct or part (in *X*, *Y*, and *Z*) further crosslinking the solution and affecting the resulting geometric accuracy of the part [33, 54].

12.3.1 Build Files

Since an early generation SL machine from 3D Systems was used in the hydrogel work described here (3D Systems SLA® 250/50 with He–Cd, 325 nm laser, 3D Systems, Rock Hill, SC), some of the specifics involved in the creation of a standard 3D part using this technology are described to provide a context for the modifications required to build using the water-based hydrogel solutions. There should be sufficient information contained in this description so these modifications can be incorporated in the latest generation of SL machines and software. Virtually any computer-aided design (CAD) software can be used to design the 3D parts where the CAD part must be saved in .stl format in order to be used with the 3D Systems preprocessing software, 3DLightyear™ (3D Systems, Rock Hill, SC). 3DLightyear™ generates 2D slices of the 3D part and creates four different files, each one with different information that is necessary to build the part. The extensions of these build files are .v, .r, .l, and .prm. The largest and most important of these build files is the vector (.v) file as it contains the information of the vectors (start and end points defined by *X*, *Y* and *Z* coordinates) that are traced by the laser to photocrosslink the polymer [64]. There are three different types of vectors that are used to define the surface boundaries and internal structures in a layer using 3DLightyear™: border, hatch, and fill vectors. Border vectors define the object boundaries and are drawn first, while hatch and fill vectors are used to fill in the space defined by the border vectors. The difference between hatch and fill vectors is that fill vectors are used for upfacing and downfacing surfaces that are not connected to additional part layers, while hatch vectors are used when the region within a layer is sandwiched between additional part layers. The main difference for the three different vectors is the speed at which they are drawn. Each vector speed is defined by the overcure parameters in 3DLightyear™. Only border and hatch vectors are considered in the discussion below since it is not necessary to differentiate between hatch and fill vectors.

Each commercial resin has its own build style conditions in 3DLightyear™ to produce the build files. In general, the default build style conditions in 3DLightyear™ produce a vector file that contains an extra border, and border and hatch vectors are traced at different speeds (and thus different energies), with the speed of the hatch vector generally being faster than the one for the border. In the default build operation, the inner border vector that defines the boundaries of the layer is drawn first, then the laser traces the hatch vectors (at a faster speed than the one for border vectors, although the hatch pattern includes traces in *X* and *Y*) that define the internal area. Once finished with hatching, the second and outermost border vector is outlined. These default conditions can be adjusted by changing the parameters in the build style. To adjust the speed of the vector files (or the speed of the laser traces), the overcure parameters for each vector type should be changed in the build file.

To improve the geometric accuracy of PEG-based gels built in SL, it was determined that the vector build files used for the fabrication of scaffolds should not contain the extra border vector and that the border and hatch vectors have equal trace speeds [33, 54]. Previous investigations on the geometric accuracy issues of the SL fabrication process of PEG-based structures showed that the successive energy exposure from the extra border, as well as the differences in speed between the border and hatch vectors affected the geometric accuracy of the desired features. Because the solutions are primarily water-based, PEG solutions behave differently to successive UV energy exposure than typical SL resins. Crosslinked PEG gels continue to transmit the UV energy of the laser, and as a result, the laser crosslinks further into the solution (in *X*, *Y*, and *Z*) affecting the geometric accuracy of the

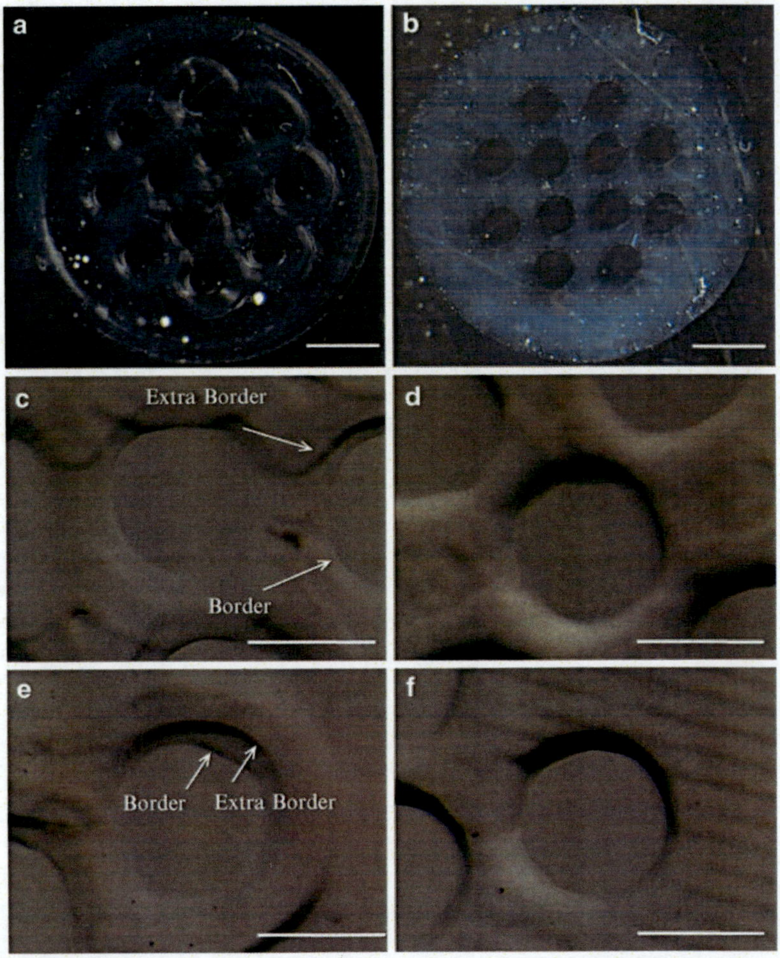

Fig. 12.3 Top view of the patterns crosslinked using two different software conditions. (**a**, **c**, and **e**) Default 3DLightyear™ parameters (extra border and different border and hatch speeds. (**b**, **d**, and **f**) Modified parameters (no extra border and same border and hatch speeds. (**c** and **d**) Inner openings. (**e** and **f**) Outer openings. Patterns crosslinked using a PEG-dma 1 K 30% solution. Marker in **a** and **b** represents 1 mm. Marker in **c**–**f** represents 0.5 mm. **a** and **b** from [54]

desired features. As can be seen in Fig. 12.3c, d, the extra border in the vector file produced a smaller diameter in the circular openings of the crosslinked pattern. The inner circular openings in the pattern crosslinked with the default conditions (extra border and different border and hatch speeds) are approximately 30 μm smaller than the outer openings, due to the successive UV energy exposure of the extra border [33, 54]. To improve the geometric accuracy, the extra border is removed by changing the build style to have zero additional borders.

It can also be seen that the pattern crosslinked with the default conditions (Fig. 12.3a) has a defined (and thicker) outline due not only to the extra border, but also because the borders are crosslinked at a slower laser speed, and thus, a higher energy than the hatch vectors. As a result, patterns crosslinked with the default conditions have different thicknesses where border vectors are thicker due to the slower laser speeds and higher exposure energies and hatch vectors are thinner due to the faster laser speeds and lower exposure energies. To crosslink patterns with more uniform thicknesses, the scanning speeds for the border and hatch vectors should be adjusted. Adjustments to the scanning speeds for the border and hatch vectors are achieved by changing the overcure parameters in the build file. To measure the exposure energy for each vector, laser paper (ZAP-48, Zap-it® Laser Alignment Products, Kentek Corporation, Pittsfield, NH) can be placed on the build platform and exposed to the border and hatch vectors until the exposures are equivalent (by adjusting the overcure parameters). Typically, setting the border and hatch overcure parameters at the same value corresponding to the maximum allowed by the hatch overcure provided the best build results. Each solution, however, may perform differently and appropriate build parameters can be determined experimentally as described in the following sections. Figure 12.3 shows the pattern of a 5-mm disk containing twelve 500-μm circular openings within a 3-mm diameter crosslinked with the default conditions (extra border and different border and hatch speeds) and crosslinked with the modified conditions (no extra border and same border and hatch speeds). While the pattern crosslinked with the default conditions has a thicker outline and smaller inner openings (Fig. 12.3a), the pattern crosslinked with the modified conditions (equivalent overcure settings between border and hatch vectors) appears flat with a uniform thickness and all the openings have the same dimensions (Fig. 12.3b).

12.3.2 Layer Thickness

As previously mentioned, the photosensitivity of PEG-based solutions is typically higher than traditional resins used in SL. Due to this effect, the overlapping of the laser in the build file (mainly in hatch vectors) affects the thickness of the cured gel. Hatch vectors define internal areas and include traces in *X* and *Y*, and default spacing between hatch vectors is 0.004 in. Figure 12.4 shows the gel thickness at a given laser speed for different hatch conditions: single hatch with no overlapping of the laser, double *X*-*Y* hatch with no overlapping of the laser (hatch spacing of 0.010 in.), and the default hatch (double *X*-*Y* with a hatch spacing of 0.004 in.). The thickness of the cured gel increases between 25 and 30% when an additional hatch (*X* and *Y*) is included to crosslink the gel (due to successive energy exposure). The gel thickness increases another ~30% using the default hatch spacing (0.004 in.) when compared to the 0.010 in. hatch spacing.

Although cure depth (or hydrogel thickness) curves as a function of energy for the PEG-based solutions provided the parameters E_C (critical energy) and D_P (penetration depth) used to build in SL, it was found that additional adjustments to the laser scan speeds (i.e. adjustments to E_C and D_P) were required to build accurate geometries

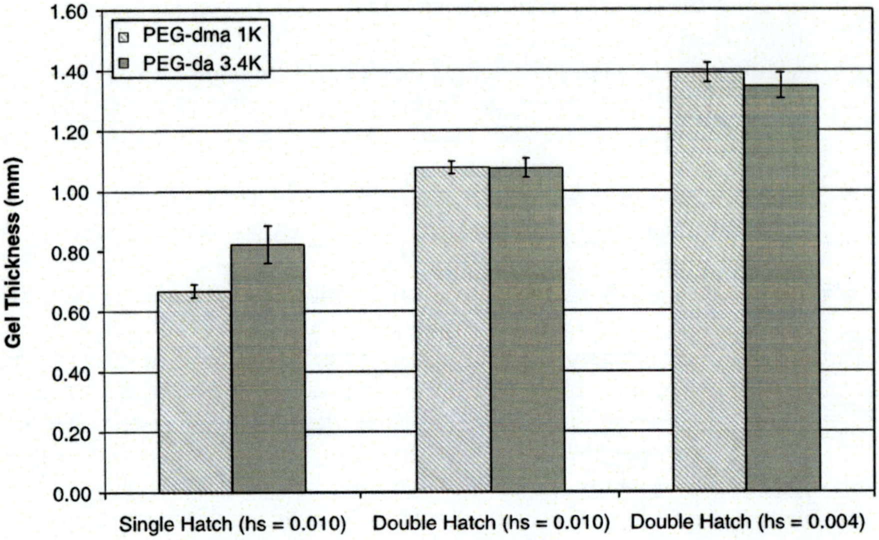

Fig. 12.4 Effect of the number of laser exposures and hatch spacing on gel thickness for a 20 wt% photocrosslinkable PEG solution with 0.5 wt% I-2959

using these hydrogel solutions. This is primarily a consequence of the water-based solutions and the effect of successive energy exposure as described above. The laser scan velocity is determined by crosslinking the desired pattern at different scan speeds and inspecting the patterns under a microscope. For example, the design pattern of a 2.94-mm disk containing seven 400-μm circular openings within a 1.72-mm diameter was crosslinked at different laser speeds, and the size of the openings was measured from optical micrographs for each crosslinked pattern. The velocity selected to fabricate the scaffold was determined as the velocity that produced (1) a crosslinked multi-lumen pattern forming a gel with sufficient mechanical strength to be handled and imaged without severe deformation (i.e. it maintained its circular shape), and (2) all of the circular openings with dimension closest to 400-μm in diameter.

Figure 12.5 shows images of the described pattern crosslinked at different speeds to determine the appropriate velocity to be used to fabricate the scaffolds. It can be seen that at a high laser speed (low energy dosages ~205 mJ/cm^2), a gel is cross-linked, although it does not have sufficient mechanical strength and deforms easily. Furthermore, the features (openings) in the pattern are not preserved (see Fig. 12.5a). When a lower laser velocity is used (high energy dosage ~570 mJ/cm^2), a gel with good mechanical strength is crosslinked (see Fig. 12.5b), although the dimensions of the features (ranging from ~195 to 265 μm) at this high energy are smaller than the dimension in the design (400 μm). Energies of ~300 mJ/cm^2 crosslink a "good" pattern with the dimensions of the features closest (ranging from ~310 to 340 μm) to the ones in the design (400 μm) (see Fig. 12.5c, d).

For each PEG-based solution used in SL, it is recommended that the layer thickness be determined experimentally taking into consideration the features in

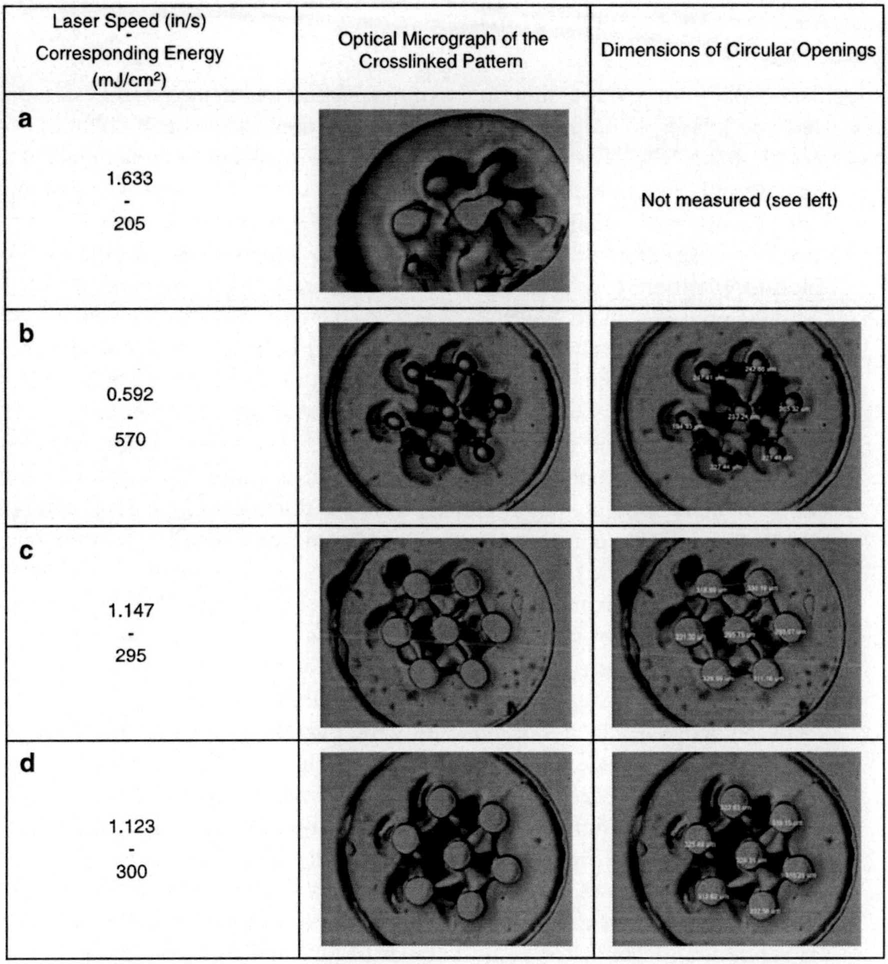

Fig. 12.5 Optical micrographs of the design pattern of a 2.94-mm disk containing seven 400-μm circular openings within a 1.72-mm diameter multi-lumen pattern crosslinked at different speeds

the design. An appropriate balance between the crosslinking density (or the mechanical strength of the crosslinked gel) and the dimensional accuracy of the features present in the *X-Y* plane should be identified first (as described above). Once the specific energy dosage required to crosslink the gel for the particular design with the appropriate mechanical strength and dimensions is found, this energy dosage produces a specific layer thickness. The specific layer thickness obtained in this procedure is then used for process planning of the layer-by-layer build process. Layered fabrication of the PEG-based 3D scaffolds using these layer thicknesses also requires taking into consideration overlapping between layers for interlayer bonding (described in the following section).

12.3.3 Interlayer Bonding

To build PEG-based structures in a commercial SL system, the system was modified in order to fabricate complex multi-material structures, using small volumes of photocrosslinkable PEG. The SL system modifications involved the removal of the original vat and replacement with a self-aligning mini-vat setup. Photocrosslinkable solution is added to and removed from the mini-vat using micro-pipettes in order to construct the part in a layer-by-layer fashion with a minimum of photocrosslinkable solution. The elevator platform, originally used to hold the part during fabrication, is now used to hold the mini-vat, and is lowered a distance equal to the amount of solution added, in order to draw the next layer and maintain the laser beam diameter at the focusing spot.

For layered fabrication, the thickness of the crosslinked gel at the specific scanning velocity (exposure energy) should be determined. A large (>500 μL) amount of volume is added to the container and the pattern is crosslinked. The crosslinked pattern should float in the solution in order to measure its thickness using calipers or optical micrographs. Once the thickness of the gel is determined, a volume of solution corresponding to a layer thickness smaller than the one crosslinked is added to the container so that the first layer is well attached to the base of the mini-vat (For example, for a measured gel thickness of 1 mm, the solution would be added to provide a first layer thickness of ~700 μm so that the ~300 μm of overcure would secure the gel to the build platform). For consecutive layer fabrication, the corresponding volume of solution that allows ~300 μm of overlapping between layers is added to provide for good interlayer bonding. Then, the platform is lowered a distance equal to the amount of solution added, and the next pattern is crosslinked. The process is repeated until the complete structure has been fabricated. Figure 12.6 shows images of two-layer multi-lumen structures with appropriate interlayer bonding (Fig. 12.6a) and inappropriate interlayer bonding (Fig. 12.6b). The overlapping between the layers in Fig. 12.6a is approximately 280 μm. Adding a large amount of volume (>300 μL) for the consecutive layers produces a weak structure that delaminates easily (similar to that in Fig. 12.6b), or a cluster of 2D patterns floating in the solution.

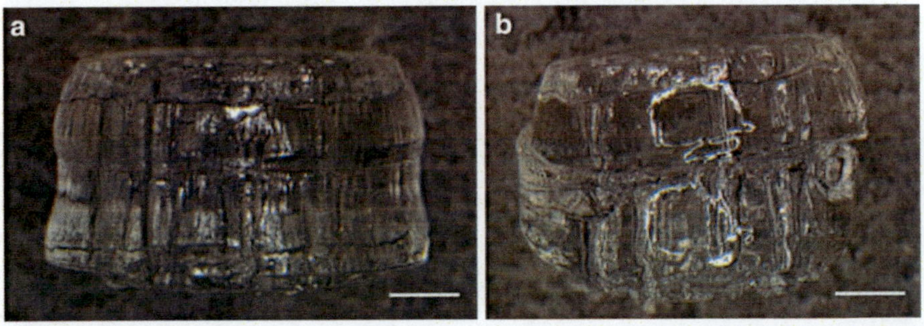

Fig. 12.6 Interlayer bonding, (**a**) appropriate interlayer bonding (overlapping ~280 μm) and (**b**) inappropriate interlayer bonding as the structure demonstrates delamination. Marker represents 1 mm

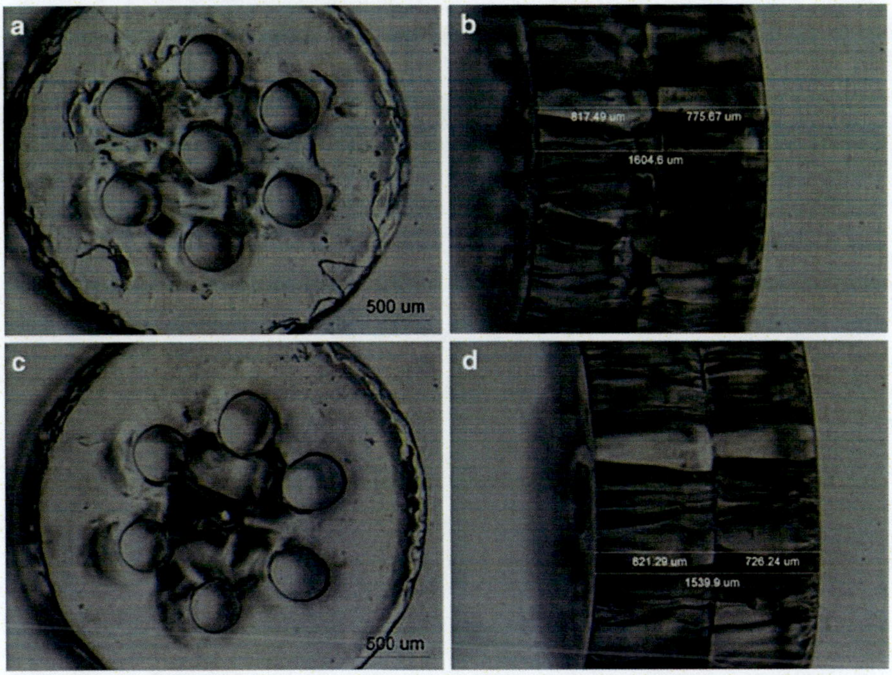

Fig. 12.7 Interlayer bonding and geometric accuracy, *Top* (**a** and **c**) and side (**b** and **d**) views of 2-layer multi-lumen structures fabricated using different volumes for the second layer (where the second layer is located on the left side of **b** and **d**). (**a** and **b**) Structure fabricated using 280 μL (~300 μm overlapping). (**c** and **d**) Structure fabricated using 260 μL (~350 μm overlapping)

As described above, the overlapping between layers also affects the accuracy of features in the *X-Y* plane. The overlapping between layers of ~300 μm was estimated by crosslinking a second pattern using different volumes, and observing the samples under a stereomicroscope. An overlapping between layers of ~300 μm produces good interlayer bonding, maintaining the geometric accuracy of the desired pattern. Figure 12.7a, b show a two-layer structure with the diameters of the lumens having similar dimensions (ranging from 349 to 422 μm). With a larger overlap, the layers of the structure are well attached; however, the geometric accuracy of the crosslinked pattern is affected as the laser penetrates into the already crosslinked layer and the features are subjected to successive UV exposure. The two-layer structure in Fig. 12.7c, d has a layer overlap of ~350 μm. It can be seen that the center lumen is smaller (~258 μm) than the rest of the lumens (339–396 μm) due to successive UV exposure that results in overcuring. The larger layer overlap produced smaller diameters in the outer lumens and an obstructed middle lumen as shown in Fig. 12.7c.

As summarized at the end of the previous section, successful layered fabrication of PEG-based 3D structures using SL requires a balance between crosslinking density and *X-Y* dimensions that will produce a specific layer thickness. As described in this section, a second balance between layer thickness overlapping (and therefore overcuring between layers that will sufficiently bond the layers together) while maintaining the *X-Y* plane feature dimensions is required for layered fabrication. The volume added for the new uncured layer should be sufficient for the laser to crosslink the uncured layer and penetrate (~300 μm) in the previously crosslinked layer so that there is sufficient overcure in *Z* without affecting the *X-Y* dimensions. Using small overlaps for consecutive layers produces weak interlayer bonding that will eventually delaminate, and if large overlaps are used, the feature accuracies in the *X-Y* plane suffer.

12.3.4 Swelling Behavior of PEG-Based Hydrogel Structures

In traditional SL fabrication, the manufactured parts are in the "green state" and are usually post-cured in a UV oven [64]. Fabricated PEG-based structures are not post-cured, but post-processed by placing them in distilled water or in phosphate buffered saline (PBS) solution (usually for more than 48 h) to allow the parts to reach equilibrium swelling. The swelling solution is changed at least twice in order to remove unreacted PI and polymer that diffuses out of the samples.

The equilibrium swelling behavior of hydrogels is usually expressed as the equilibrium volumetric swelling ratio (Q) and calculated from the equilibrium mass swelling ratio ($Q = M_S/M_D$) [44]. It is important to determine the change in dimensions of hydrogel samples at the equilibrium swollen state in order to fabricate scaffolds with the appropriate dimensions at the swollen state [65–68].

Initial experiments compared the dimensions of simple PEG-based scaffolds fabricated using SL to estimate a dimensional swelling factor (DSF). DSF is calculated from a ratio of the measured swollen dimension with the original dimension in the CAD drawing, according to the equation below.

$$\mathrm{DSF} = \frac{\text{swollen dimension}}{\text{design dimension}} \tag{12.1}$$

To determine DSFs, hydrogel samples of the four different photopolymer solutions were crosslinked in SL. A simple ring pattern (ID = 2.5 mm, OD = 8.0 mm) was drawn by the laser at a fixed laser speed to crosslink the samples under the same energy conditions. Two different energy exposures (or laser speeds) were used to crosslink the samples. The samples were allowed to swell to equilibrium by placing them in distilled water for more than 48 h, and then digital images of the samples were acquired using a stereomicroscope (MZ16, Leica Microsystems, Germany) equipped with a CCD camera (Retiga 2000R Fast 1394, QImaging Corp., Canada) to measure the samples' dimensions.

The calculated dimensional swelling factor exhibited a dependence on energy dosage, PEG type, and concentration. Overall, PEG-dma 1 K gels at equilibrium decreased in size, while PEG-da 3.4 K gels increased in size, and as the polymer

Table 12.1 Dimensional swelling factor (DSF) and swelling ratio of hydrogels fabricated at two different energy dosages for the four different PEG-based solutions investigated ($n = 6$)

	PEG-dma 1 K		PEG-da 3.4 K	
Energy	20%	30%	20%	30%
Outside diameter				
180 mJ/cm^2	1.01 ± 0.01	1.03 ± 0.01	1.24 ± 0.01	1.28 ± 0.01
300 mJ/cm^2	0.99 ± 0.01	1.02 ± 0.01	1.29 ± 0.01	1.31 ± 0.01
Inside diameter				
180 mJ/cm^2	1.12 ± 0.03	1.16 ± 0.04	1.34 ± 0.01	1.35 ± 0.02
300 mJ/cm^2	1.10 ± 0.02	1.09 ± 0.02	1.33 ± 0.01	1.36 ± 0.01
Wall thickness				
180 mJ/cm^2	0.96 ± 0.02	0.97 ± 0.03	1.19 ± 0.02	1.25 ± 0.01
300 mJ/cm^2	0.94 ± 0.02	0.99 ± 0.01	1.26 ± 0.02	1.29 ± 0.01
Swelling ratio				
180 mJ/cm^2	11.32 ± 0.90	6.72 ± 0.40	14.25 ± 0.64	8.23 ± 0.97
300 mJ/cm^2	10.88 ± 0.53	6.15 ± 0.12	15.86 ± 0.50	8.12 ± 0.08

concentration decreased, the size of the gels decreased. Table 12.1 summarizes the dimensional swelling factor of hydrogels fabricated at two different energy dosages for the four different PEG-based solutions investigated. Table 12.1 also shows the mass swelling ratio data. The data in Table 12.1 were acquired to make a direct comparison between the four different PEG-based solutions at the same cross-linking condition (same laser scan speed).

For PEG-dma 1 K gels, the swollen dimensions, both ID and OD, are slightly larger than the ones in the original CAD design while the wall thickness of the gels slightly decreased. For PEG-da 3.4 K gels, the swollen dimensions, both ID and OD, are considerably larger than the ones in the original CAD design, and the overall wall thickness of the gels also increased. It should be noted that these values are approximations, as the swollen PEG-da 3.4 K samples deformed while swelling, forming a concave and not a flat ring (making accurate measurements more difficult). Inside diameters could be measured more accurately than outside diameters due to the deformation of the samples. With regard to the measured swelling ratio, hydrogels fabricated with PEG-dma 1 K had a lower swelling ratio than hydrogels fabricated with PEG-da 3.4 K. In terms of photopolymer concentration, hydrogels fabricated with a 20 wt% concentration had a higher swelling ratio than the ones fabricated with the 30 wt% concentration (for both PEG-dma 1 K and PEG-da 3.4 K gels).

The data presented in Table 12.1 serve as a reference to compare the difference in swelling behavior of different types of PEG-based structures. However, dimensional swelling data should be determined at energies used in the actual fabrication of parts. To further investigate the changes in swollen dimensions, 3D samples were fabricated using a 20 wt% PEG-da 3.4 K solution, and the different dimensions (length, layer thickness, outside diameter, inside diameter, and center opening diameter) of the samples were measured. Dimensional swelling factor data showed that the final swollen volume of the samples depends on the specific features of the sample. For example, the sample expanded along its major axis (longitudinally) to a final dimension 25% larger than the original (the calculated DSF for both the overall length of the sample and an individual layer in the sample was 1.25). The OD of the overall structure also expanded to a final dimension 22% larger than the original, while the OD of the center lumen expanded to a final dimension 15% larger than the original.

Crosslinked hydrogel structures at the equilibrium state contain two kinds of water: the free or bulk water physically trapped in the hydrogel mesh, and the bound water that is adsorbed in the PEG molecule [65]. Due to the trapped or adsorbed water in the crosslinked hydrogel structure, both the weight and dimensions of the gels change at the equilibrium swollen state. It is important to take into consideration the dimensional changes of crosslinked PEG-based structures in the design of constructs, since these constructs will most often be used in their equilibrium swollen state.

12.3.5 *Fabrication of 3D Structures*

Using the cure depth curves for 30 wt% PEG-dma 1 K presented previously, it was possible to fabricate highly complex 3D structures in a layer-by-layer fashion. Figure 12.8 shows a complex part commonly manufactured by many SL users. The chess rook on the right in each picture was manufactured using the 3D Systems Viper™ Si^2 SL apparatus using DSM Somos® WaterShed™ 11120 resin and standard SL build parameters for the resin. The chess rook on the left in each picture was built semi-automatically in the modified SL 250/50 apparatus in 0.010-in. layers. The PEG-based chess rook includes all the details of the resin-based rook such as the staircase contained within the walls, the windows, and the bricks in the façade and measures approximately 23 mm. Figure 12.8b shows that the PEG-based chess rook is elastic and can be easily deformed without breakage. Figure 12.8c illustrates the decrease in size when the hydrogel scaffold is allowed to dry.

Complex 3D scaffolds with precisely localized materials such as growth factors and living cells and with customized properties for TE applications can be fabricated with the use of photocrosslinkable PEG and SL. The pictures in Fig. 12.9

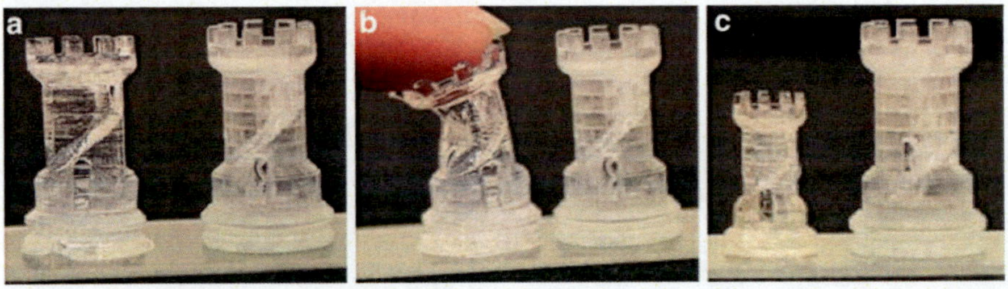

Fig. 12.8 Complex 3D structures manufactured using SL, the chess rook on the right in each picture was made with DSM Somos® WaterShed™ 11120 resin while the rook on the left is PEG-based. (**a**) PEG-based structure immediately after it was built. (**b**) Deformation of elastic PEG-based structure. (**c**) PEG-based structure after drying [54]

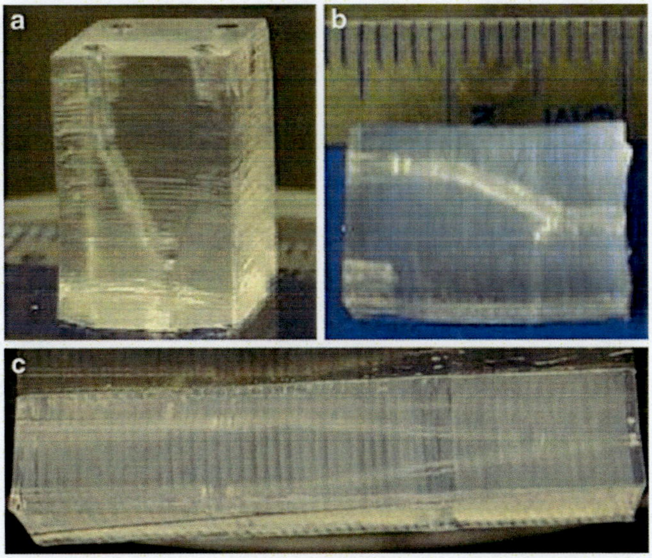

Fig. 12.9 PEG-based scaffolds with embedded channels of different sizes and orientations: (**a**) Isometric view, (**b**) Side view of a 1 × 1 × 1.6 cm block, and (**c**) Top view of a 1 × 1 × 5 cm block [54]

show PEG-based structure of a block with a single open 1-mm channel that bifurcates twice into 1-mm channels at different planes of the gel (each bifurcation is oriented 90° from one another). This geometry is representative of the arterial system and a PEG-based construct such as this one may find potential use in angiogenesis studies (along with the appropriate bioactive agents and cells). The dimensions of the PEG-based block in Fig. 12.9a, b are 1 × 1 × 1.6 cm, and scaffolds of virtually any length and geometry can be fabricated as demonstrated by the scaffold in Fig. 12.9c, which starts as a single channel and bifurcates twice with an overall length of approximately 5 cm.

Figure 12.10 show a PEG-based scaffold fabricated in SL with promising use as a nerve guidance conduit (NGC) in peripheral nerve regeneration. The design of the conduit includes single lumen end caps for suture and a multi-lumen middle section to enhance regeneration by providing a greater surface area for support cells to adhere and additional pathways for sprouting axons to be guided for regeneration. The conduit in Fig. 12.10a, b (fabricated with a 30 wt% PEG-dma 1 K solution) has an OD of 5 mm, an ID of 3 mm (these dimensions are within the dimensions of a tubular sheath offered by SaluMedica™ for peripheral nerve repair), and twelve 500-μm lumens in the middle section. The CAD design of the conduit in Fig. 12.10c–f has an OD of 2.94 mm, an ID of 1.72 mm, and seven 400-μm lumens in the middle section. The images in Fig. 12.10c, d were fabricated with a 20 wt% PEG-dma 1 K solution, while the images in Fig. 12.10e, f were fabricated with a 20% PEG-da 3.4 K solution. As PEG-da 3.4 K swells, the final dimensions of the conduits in Fig. 12.10d, f were approximately 3.67 ± 0.191 mm in OD, and 2.09 ± 0.176 mm ID.

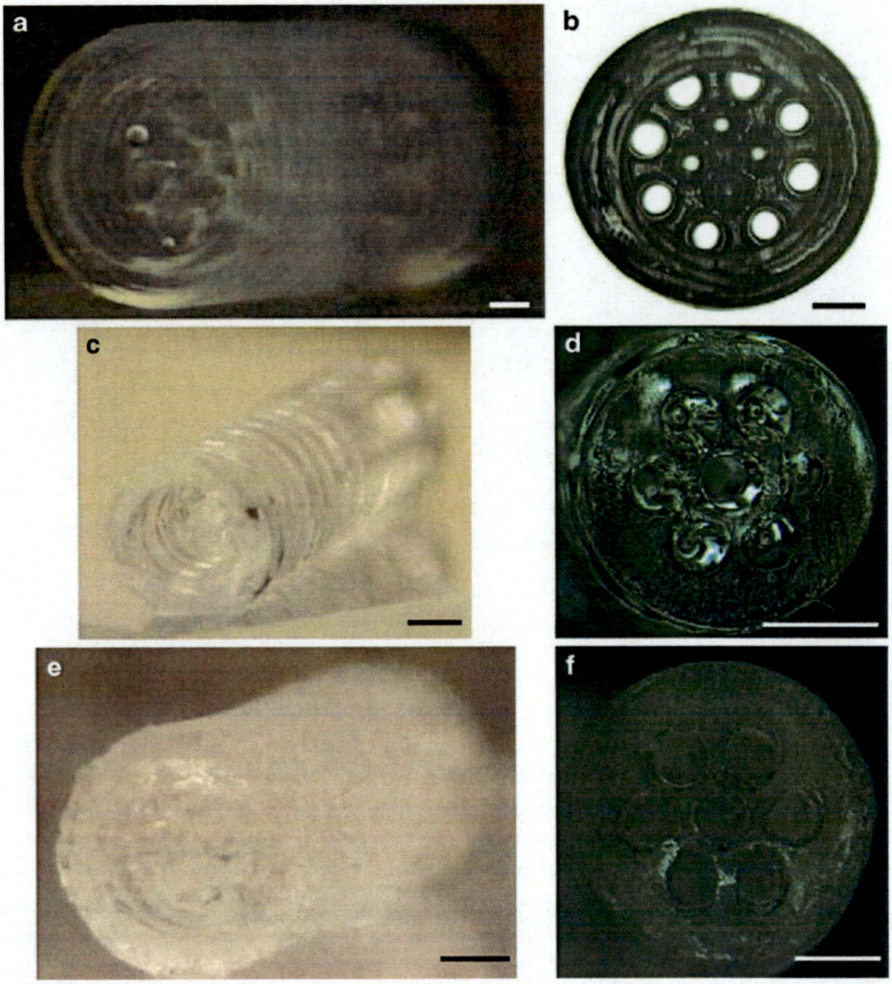

Fig. 12.10 Multi-lumen PEG-based nerve regeneration conduits. (**a** and **b**) Isometric and top view of NCG with OD = 5 mm, ID = 3 mm and twelve 500-μm lumens fabricated with a 30 wt% PEG-dma 1 K solution. (**c** and **d**) Isometric and top view of NCG with OD = 2.94 mm, ID = 1.72 mm and seven 400-μm lumens fabricated with a 20 wt% PEG-dma 1 K solution. (**e** and **f**) Isometric and top view of NCG with OD = 3.7 mm, ID = 2.09 mm and seven 400-μm lumens fabricated with a 20 wt% PEG-da 3.4 K solution. Marker represents 1 mm. **a** and **b** from [32]

12.3.6 Multi-Material Fabrication

It is hard to define the ideal tissue engineered scaffold, as each organ or tissue that needs regeneration has different and particular requirements. Proper functioning of a tissue relies on not only the overall macroproperties of the tissue, but the microenvironment surrounding the different cell types. In order to control these microenvironments

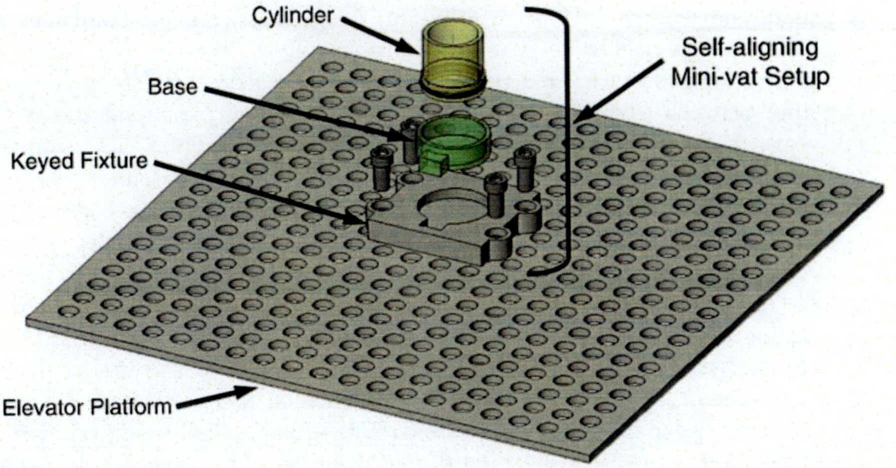

Fig. 12.11 Schematic of the self-aligning mini-vat setup [69]

in a TE scaffold, different materials and cells need to be precisely placed throughout the scaffold to elicit the specific desired cell responses. Such control will likely lead to enhanced generation of new, functional tissue. SL is a flexible technology that allows for such multi-material fabrication of 3D structures. Modifications to the system to maintain *X-Y* registration of the scaffold during fabrication, as well as management of build files are required for successful multi-material fabrication.

The machine modifications for multi-material fabrication involved the removal of the original vat and replacement with a self-aligning mini-vat setup. Figure 12.11 shows a schematic of the mini-vat setup that provides an easy method for using multiple hydrogel materials for scaffold fabrication. The setup consists of a fixture with a keyway design affixed to the elevator platform to secure the mini-vat in a specific *X-Y* orientation. The mini-vat consists of two pieces: a base and cylinder. The base includes the same keyway design as the fixture to provide a press-fit into the fixture and maintain *X-Y* registration of the scaffold during fabrication. The cylinder is designed so that it can be interchanged (from small to large) during the build as needed to hold more photopolymer solution and thus manufacture longer scaffolds. An O-ring is used on the outside of the cylinder to seal the base-cylinder assembly and provide a water-tight sealed assembly. Similar to the layered-fabrication of PEG-based structures described above, the elevator platform, originally used to hold the part during fabrication, is now used to hold the mini-vat setup, and is lowered a distance equal to the amount of solution added, in order to draw the next layer and maintain the laser beam diameter at the focusing spot.

The procedure to fabricate a multi-material scaffold started with the CAD drawing of the scaffold in stl format. Build files were obtained using 3DLightyear^{TM}. The build files did not have an extra border, and the border and hatch vectors were commanded to have equivalent speeds so improved geometric accuracy could be achieved as described previously. For multi-material fabrication, process planning plays an important role. There are as many build files as materials used, and there is a ½ beam (125-μm) overlap between the build files for the different materials to be attached forming a single structure. The multi-material fabrication process is time consuming as intermediate steps for withdrawal and rinse between materials are required. The self-aligning mini-vat allows for *X-Y* registration of the construct during fabrication so complex 3D multi-material structures could be created. The photocrosslinkable PEG-based solution is added to and removed from the mini-vat using micro-pipettes in order to construct the part in a layer-by-layer fashion. During fabrication, the base-cylinder assembly is removed so the part can be separately rinsed. Using the self-aligning feature, the base-cylinder assembly is replaced and a multiple material build continues by adding a different photocrosslinkable material to the assembly. The process is repeated until the build is completed.

To demonstrate multiple material SL fabrication, simple 2D scaffolds using two different materials (plain PEG-based solution and a PEG-based solution with a fluorescent component) were fabricated using SL. Brightfield low magnification images as wells as the corresponding high magnification fluorescent images of the multi-material scaffolds fabricated with both photopolymer solutions containing either FITC-dextran (fluorescent component physically trapped in the hydrogel, pictures A–H) or PEG-RGDS-FITC (fluorescent component covalently attached to the hydrogel, pictures I–P) are shown in Fig. 12.12. Images A–D and I–L correspond to the scaffolds fabricated using PEG-dma 1 K while E–H, and M–P correspond to the scaffolds fabricated using PEG-da 3.4 K. The fabrication of the multi-material scaffolds shown in Fig. 12.12 took approximately 10 min. During fabrication, the rinsing (using a wash bottle with distilled water) and removal of excess water (using a Kimwipe) were the most time consuming steps since care was required in order to avoid damaging the scaffold. The scan times for the scaffolds were less than 5 s for the fluorescently labeled regions and approximately 20 s for the plain PEG regions.

It can be seen in Fig. 12.12a–d that the smallest features (ø = 500 μm, *L* = 500 μm) in the fluorescent patterns were not crosslinked in the PEG-dma 1 K scaffolds. This can be seen more clearly in the corresponding high magnification fluorescent images (Fig. 12.12b, d). The empty spaces in the PEG-dma 1 K scaffolds where the smallest features should be located can be seen easily, while the smallest features for the PEG-da 3.4 K scaffolds do appear. The absence of these features is due to the SL machine building only border vectors (without hatching) and so the solution is not receiving sufficient energy to crosslink a gel in these regions. The energy delivered by the laser to crosslink the medium and large features was ~1,700 mJ/cm^2 while the energy delivered in the smallest features was ~780 mJ/cm^2. This problem was solved when creating the multi-material scaffolds using PEG-RGDS-FITC by including an additional vector file of the smallest features only, and drawing the smallest features three-times as it was determined that these features required three traces to successfully cure (with an energy of ~2,300 mJ/cm^2). The laser speed at which the features are drawn can also be adjusted so that equivalent energies could be delivered for the small and large features, but this was not performed in this case (It was simply easier to draw the small features three times using the same scanning parameters).

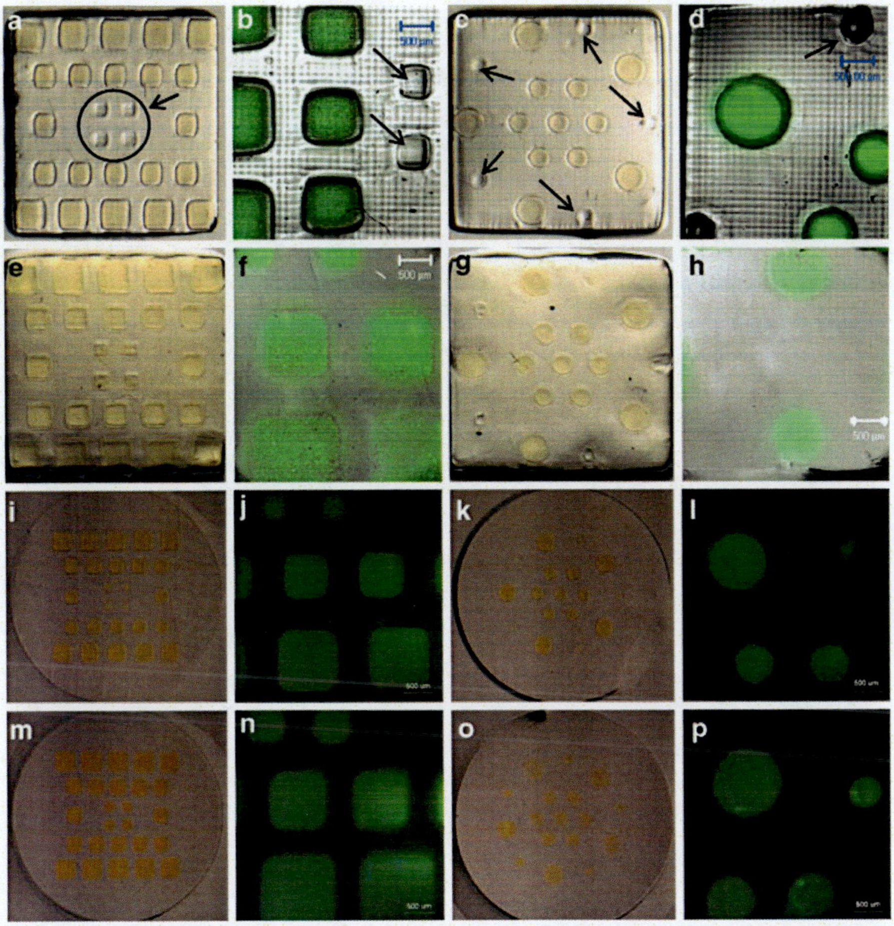

Fig. 12.12 2D multi-material scaffolds. (**a–h**) Scaffolds fabricated with fluorescently-labeled dextran (fluorescent component physically trapped within the hydrogel). (**i–p**) Scaffolds fabricated with fluorescently-labeled PEG-RGDS (fluorescent component covalently attached to the hydrogel mesh). (**a–d** and **i–l**) PEG-dma 1 K as the main material in the gels. (**e–h** and **m–p**) PEG-da 3.4 K as the main material in the gels. Arrows in (**a–d**) point to the empty spaces where the fluorescent component should be present [69]

Since the FITC-dextran was physically trapped within the crosslinked gel, it diffuses out of the hydrogel matrix once it is placed in PBS to reach equilibrium, and the FITC-dextran diffuses out faster in the scaffolds prepared with PEG-da 3.4 K. Since the MW of the PEG 3.4 K is 3.4 times that of PEG 1 K, the longer molecules of the higher MW PEG create a larger mesh size so the FITC-dextran molecules diffuse faster in the higher MW PEGs. Fluorescent images of the scaffolds taken at higher magnification corroborate these results. Also, it can be seen in Fig. 12.12f how the FITC-dextran diffuses out from the crosslinked PEG-da 3.4 K gel. The images in Fig. 12.12 were taken quickly (or immediately) after fabrication (and not at equilibrium) to avoid

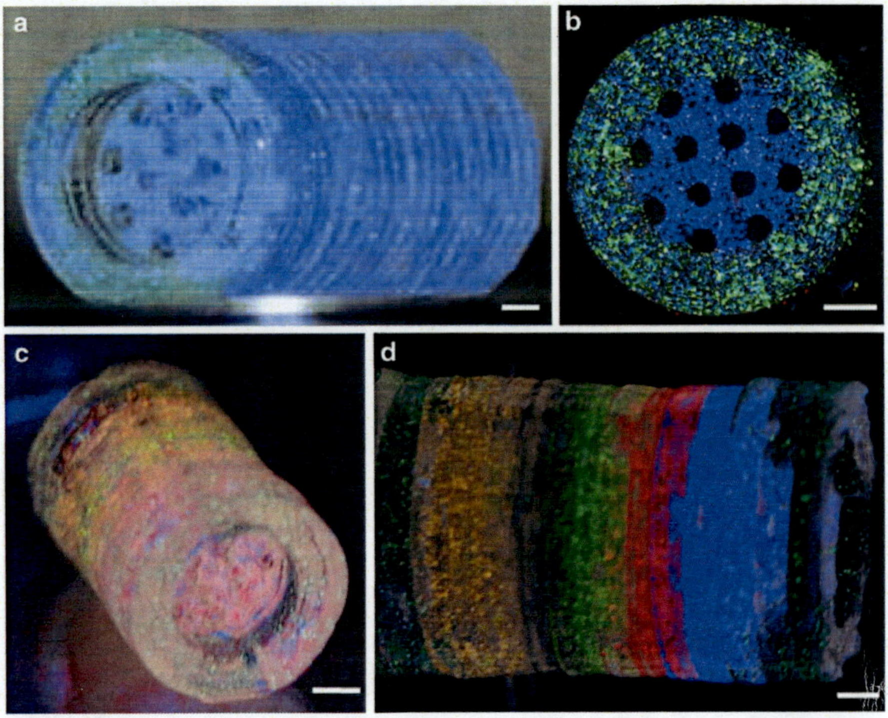

Fig. 12.13 NGC fabricated using different colors of fluorescent particles in the PEG-based solution. (**a** and **b**) Isometric and top view of scaffold showing precise placement of different materials within layers. (**c** and **d**) Isometric and side view of scaffold showing precise placement of different materials across layers. Marker represents 1 mm. **a** and **b** from [54]

diffusion of the fluorescent component. The images in Fig. 12.12i–p show the multi-material scaffolds fabricated with fluorescently labeled PEG-RGDS. All the features of the scaffolds including the 500 μm features are present. The difference in swelling between the two different hydrogel materials is also apparent in these images.

Figure 12.13 shows an NGC fabricated with different materials. The outer portion of the conduit contains 15-μm fluorescent green particles while the inner portion contains 10-μm fluorescent blue particles (Fig. 12.13a, b). The images in Fig. 12.13c, d show an NGC fabricated with different colors of fluorescent particles in each layer. These conduits illustrate the capability of SL to precisely place different agents within the hydrogel construct, allowing multi-material fabrication, both within a single layer and from layer-to-layer.

A complex, multi-material 3D structure (shown in Fig. 12.14) was fabricated using two different photopolymer solutions of PEG-dma 1 K (30 wt%), one plain and the other containing fluorescent (580/605 nm) 10-μm microspheres (FluoSpheres®, Molecular Probes, Inc., Eugene, OR) as a demonstration of our multi-material fabrication technique. The complex 3D structure was a 23-mm detailed chess rook, with a staircase contained within the walls, windows, and bricks in the façade. The staircase

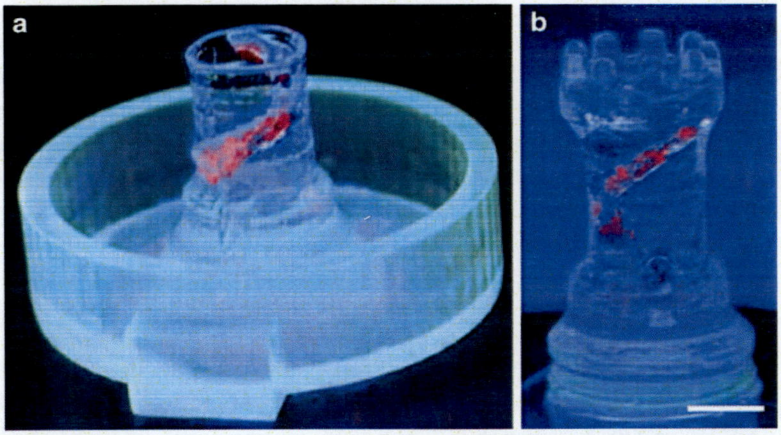

Fig. 12.14 Complex 3D structure manufactured in multi-material SL. PEG-dma 1 K at a 20 wt% concentration was used as the main material. The solution used to create the staircase contained fluorescent microspheres. (**a**) Structure shown during fabrication and attached to the base of the mini-vat setup. (**b**) Final multi-material structure. Marker represents 5 mm [69]

was fabricated with the PEG-dma 1 K solution containing fluorescent microspheres. The multi-material rook was manufactured in 0.010-in. layers, semi-automatically in three stages. In the first stage, the entire piece was fabricated except for the staircase and the top part that includes the ceiling and the parapet using a plain PEG-dma 1 K solution. Then, the staircase was built using the photopolymer solution containing fluorescent spheres. Finally, the top of the rook and the parapet were fabricated using the same photopolymer solution used in the first stage. The solutions were removed and the scaffold rinsed at each stage. The fabrication of the multi-material rook took approximately 3 h. The same rook fabricated out of plain PEG using this procedure requires less than 1 h, and a single rook fabricated out of commercial SL resin and using the automated commercial system requires 2.5 h (Fig. 12.8). Thus, the build time for the multi-material fabrication process shown here is comparable to the automated commercial system primarily because sweeping is not required between layers (the viscosity of the solution is essentially that of water), and the difference in the number of layers (~225 layers for the automated commercial system with layer thicknesses of 0.004-in, and ~90 layers for the developed fabrication procedures using PEG with layer thicknesses of 0.010-in). The multi-material build times could be reduced by automating the material fill and removal procedure, including developing an intermediate scaffold cleaning procedure.

12.4 In vitro Cell Studies of SL Parameters

Some TE applications rely on the availability of cells for the success of the application (e.g. the reduction of invasiveness in cartilage or bone repair using systems with encapsulated chondrogenic or osteoblast cells). SL parameters related to unfavorable cell growth and survival, including PI exposure [54], UV exposure, and radical formation of the PI upon exposure to UV energy were assessed *in vitro* using human dermal fibroblasts (HDFs, Cambrex BioScience, Walkersville, MD) cells to explore the capabilities of SL to create scaffolds with encapsulated cells [68]. Furthermore, HDFs were encapsulated in photocrosslinkable PEG hydrogels using SL and the viability of the cells was evaluated at 2 and 24 h following encapsulation. Two different types of PEG were used to encapsulate the cells.

Cell viability of HDFs exposed to different concentrations of two different PI was assessed using a MTT cytotoxicity assay (Sigma-Aldrich, St. Louis, MO). The two PIs used were: Irgacure 2959, or I-2959, and HMPP. HDFs were cultured in 96-well flat-bottom plates (50,000 cells/well) in the presence of decreasing concentrations of the PI, starting at 0.5 wt% with five doubling dilutions thereafter. Cells were incubated for different time periods (1, 4, or 24 h) and cytotoxicity was assessed using a MTT colorimetric assay. Cytotoxicity increased with increasing PI concentrations and increased exposure time, as expected (see Fig. 12.15). For a given PI concentration (0.5 wt%), the detrimental effect of the PIs on cell survival is evident after 4 h of exposure (see Fig. 12.15a). For prolonged exposures (24 h), cell viability was similar to the control (complete media) for HDFs treated with I-2959 only at concentrations up to 0.05 wt% (see Fig. 12.15b). These results agree with previous studies on the effect of I-2959 on cytotoxicity of cells [56, 58, 59]. It should be noted, however, that the longer exposure times used here are greater than any exposure the cells would actually experience during the SL process. Once constructs are made (30–40 min, even for complex 3D structures), the constructs would be placed in media and excess PI would begin leaching out. Thus, it is believed that 0.5 wt% I-2959 could realistically be used in conjunction with SL without severe cytotoxicity.

To assess the toxic effect of UV irradiation and the effect of PI under UV irradiation on cultured HDFs exposed to a range of UV irradiation from 0 to 20,000 mJ/cm^2 from the 325 nm laser of the SL system, a LIVE/DEAD viability/cytotoxicity kit was used. Cell viability was determined using fluorescence readings and image analysis for two different cell concentrations (5,000 and 50,000 cells/cm^2). No detrimental effect of UV irradiation or of PI under UV irradiation was observed in the experiment using a high concentration of cells (50,000 cells/cm^2). For a low concentration of cells (5,000 cells/cm^2), cell viability decreased with increasing the UV energy dosage at which the cells were exposed. Viability of cells exposed to a concentration of 0.5 wt% of the PI I-2959 and irradiated with different dosages of UV also decreased with increasing energy dosages. Figure 12.16 shows the results of cell viability of HDFs exposed to UV irradiation (solid bars) and exposed to the PI and UV irradiation (downward diagonal bars). For the UV irradiation experiment, a significant decrease (63%) in cell survival is observed at the highest energy exposure of 20,000 mJ/cm^2, while the cell survival for the other energies is comparable to the control. For the experiment including photoinitiator exposure and UV irradiation, the cell survival for the lowest energy (1,500 mJ/cm^2) is comparable to the control (cells not exposed to UV). A decrease in cell survival (not statistically significant)

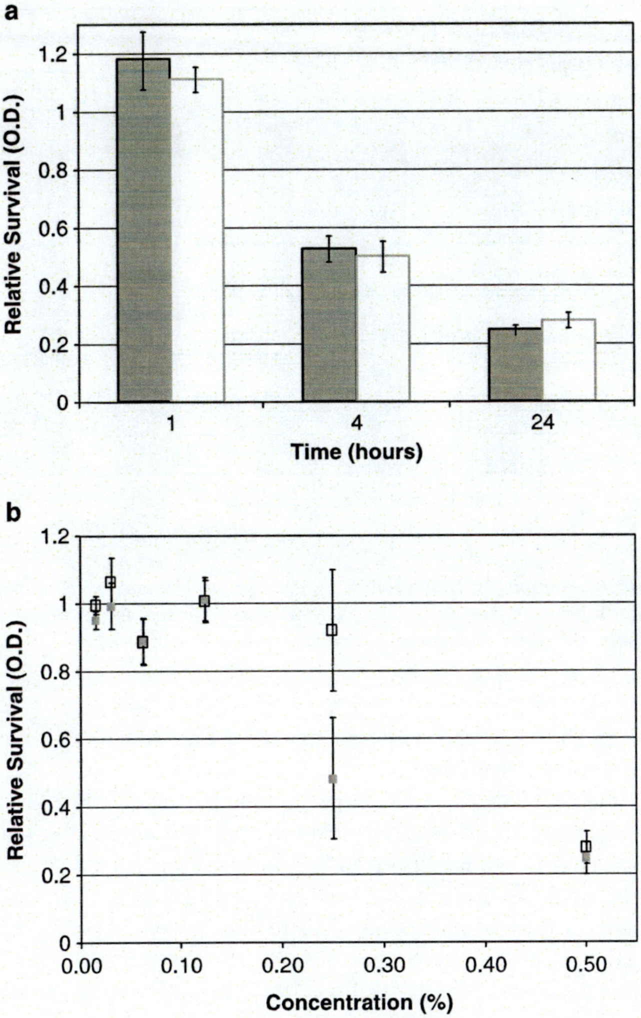

Fig. 12.15 Cytotoxicity of photoinitiators. (**a**) Relative survival as a function of time for a given photoinitiator concentration (0.05 wt%). (**b**) Relative survival as a function of photoinitiator concentration for an exposure time of 48 h. Survival is normalized to that observed in controls of cells not exposed to the PI. Solid bars and markers correspond to HMPP, hollow bars and markers correspond to I-2959. Note that the symbols in B for HMPP and I-2959 at concentrations of 0.125 wt% and 0.0625 wt% overlap [54]

is observed at energies of 2,500 and 6,000 mJ/cm^2 (77 and 82%, respectively). For the two highest energy exposures, a statistically significant decrease in cell survival is observed (52 and 63%, for 12,000 and 20,000 mJ/cm^2, respectively). The results of UV exposure and exposure to free-radicals from the PI upon UV excitation showed that SL threshold levels for unfavorable cell survival were sufficiently

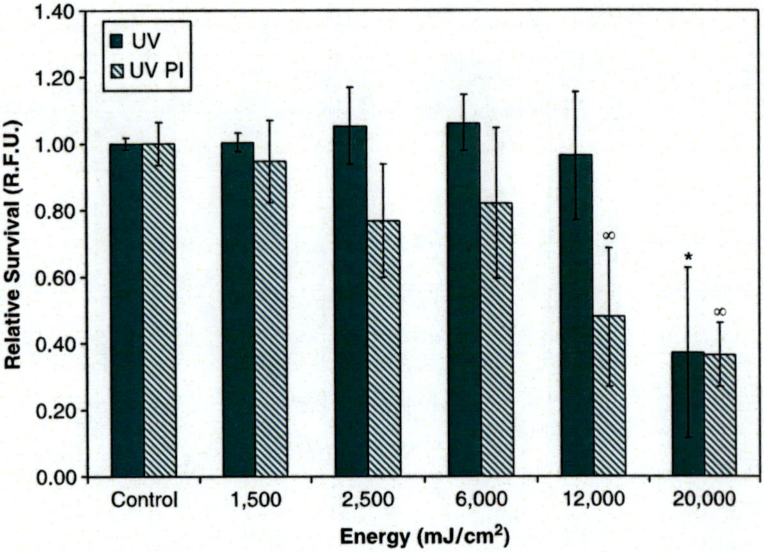

Fig. 12.16 Survival of HDFs following exposure to UV energy (*solid bars*) and to the photoinitiator I-2959 and irradiated with UV (*downward diagonal bars*), assessed using a fluorescence plate reader. * and ∞ are statistically different ($p << 0.05$) [66]

far from the exposures required to create PEG-based scaffolds in SL using the procedures described here.

In the cell encapsulation experiments, hydrogel scaffolds with encapsulated cells showed a uniform distribution of cells throughout the scaffolds, and a high percentage of viable cells at both 2 and 24 h (at least 87%, see Fig. 12.17). No significant differences in the viability of cells were observed for the two PEG types or for the two time points. The results presented here indicate that the use of SL and photocrosslinkable biomaterials, such as photocrosslinkable PEG, appears feasible for fabricating scaffolds with living cells for a variety of important TE applications.

12.5 Conclusions

In summary, SL represents a unique Additive Manufacturing technology that can create complex 3D PEG-based scaffolds with applications in TE. Flexibility and ease of use of SL allows for control over the scaffold's macro-scale design as well as precise placement of viable cells and bioactive agents within the scaffold during construction. The detailed methodology presented here to successfully fabricate complex hydrogel based structures using SL can be extended to other photocrosslinkable biomaterials and SL technologies.

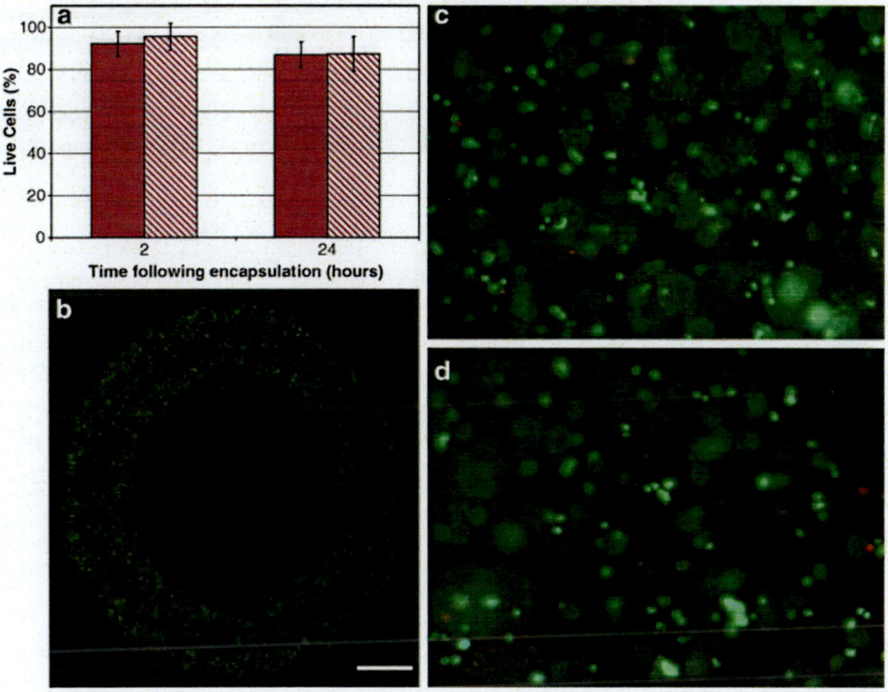

Fig. 12.17 Cell encapsulation using SL. (**a**) Viability of human dermal fibroblasts encapsulated in PEG hydrogels photocrosslinked using SL at 2 and 24 h following fabrication (*Solid bars* correspond to PEG-dma 1 K gels, downward diagonal bards correspond to PEG-da 3.4 K gels). (**b**) Low magnification image of HDFs encapsulated in PEG-based hydrogels using SL showing an even distribution of cells throughout the gel (marker represents 1 mm). (**c**) Cells encapsulated in PEG-dma 1 K imaged at 2 h following fabrication. (**d**) Cells encapsulated in PEG-da 3.4 K imaged at 2 h following fabrication. Cells were stained with a LIVE/DEAD cell viability assay in which the live cells fluoresce green and the dead cells fluoresce red

Acknowledgments The research presented here was performed at the University of Texas at El Paso (UTEP) within the W.M. Keck Center for 3D Innovation (Keck Center). Primary support for this research was provided by the National Science Foundation through Grant No. CBET-0730750. The authors are grateful to the many faculty, staff, and students within the Keck Center who assisted in various ways with this research. Equipment and facilities in the UTEP Analytical Cytology Core Facility of the Biological Sciences Department used here are maintained through NCRR Grant Number 5G12 RR008124. Any opinions, findings, and conclusions or recommendations expressed in this work are those of the authors and do not necessarily reflect the views of the National Science Foundation or any other individual or funding agency.

References

1. Leong, K.F., C.M. Cheah, and C.K. Chua. Solid freeform fabrication of three-dimensional scaffolds for engineering replacements tissues and organs. Biomaterials 24: 2363-2378, 2003.
2. Liu, V. and S.N. Bhatia. Three-dimensional tissue fabrication. Advanced Drug Delivery Reviews 56: 1635–1647, 2004.

3. Hutmacher, D.W. and M.A. Woodruff. Design, Fabrication, and Characterization of Scaffolds via Solid Free-Form Fabrication Techniques. In: Biomaterials Fabrication and Processing Handbook, edited by P.K. Chu and X. Liu. Boca Raton, FL: CRC Press, Taylor & Francis Group, 2008, pp. 45–67.
4. Vozzi, G. and A. Ahluwalia. Rapid Prototyping Methods for Tissue Engineering Applications. In: Biomaterials Fabrication and Processing Handbook, edited by P.K. Chu and X. Liu. Boca Raton, FL: CRC Press, Taylor & Francis Group, 2008, pp. 95–114
5. Ang, T.H., F.S.A. Sultana, D.W. Hutmacher, Y.S. Wong, J.Y.H. Fuh, X.M. Mo, H.T. Loh, and S.H. Teoh. Fabrication of 3D chitosan-hydroxyapatite scaffolds using a robotic dispensing system. Materials Science and Engineering C 20: 35–42, 2002.
6. Landers, R., U. Hubner, R. Schmelzeisen and R. Mulhaupt. Rapid prototyping of scaffolds derived from thermoreversible hydrogels and tailored for applications in tissue engineering. Biomaterials 23: 4437–4447, 2002
7. Vozzi, G., C. Flaim, A. Ahluwalia, and S. Bhatia. Fabrication of PLGA scaffolds using soft lithography and microsyringe deposition. Biomaterials 24: 2533–2540, 2003.
8. Vozzi, G., V. Chiono, G. Ciardelli, P. Giusti, A. Previti, C. Cristallini, N. Barbani, G. Tantussi, and A. Ahluwalia. Microfabrication of biodegradable polymeric structures for guided tissue engineering. Materials Research Society Symposium Proceedings, EXS-1: F5.22.1–3, 2004.
9. Wiria, F.E., K.F. Leong, and Y. Liu. Poly-ε-caprolactone/hydroxiapatite for tissue engineering scaffold fabrication via selective laser sintering. Acta Biomaterialia 3: 1–12, 2007.
10. Tan, K.H., C.K. Chua, K.F. Leong, C.M. Cheah, P. Cheang, M.S. Abu Bakar, and S.W. Cha. Scaffold development using selective laser sintering of polyetherketone-hydroxyapatite biocomposite blends. Biomaterials 24: 3115–3123, 2003.
11. Hutmacher, D.W., T. Schantz, I. Zein, K.W. Ng, S.H. Teoh, and K.C. Tan. Mechanical properties and cell cultural response of polycaprolactone scaffolds designed and fabricated via fused deposition modeling. Journal of Biomedical Materials Research 55(2): 203–216, 2001.
12. Zein, I., D.W. Hutmacher, K.C. Tan, and S.H. Teoh. Fused deposition modeling of novel scaffold architectures for tissue engineering applications. Biomaterials 23: 1169–1185, 2002.
13. Chim, H., D.W. Hutmacher, A.M. Chou, A.L. Oliveira, R.L. Reis, T.C. Lim, and J.T. Schantz. A comparative analysis of scaffold material modifications for load-bearing applications in tissue engineering. International Journal of Oral and Maxillofacial Surgery 35: 928–934, 2006.
14. Liu, V.A. and S.N. Bhatia. Three-dimensional photopatterning of hydrogels containing living cells. Biomedical Microdevices 4: 257–266, 2002.
15. Hahn, M.S., Miller, J.S., and J.L. West. Laser scanning lithography for surface micropatterning on hydrogels. Advanced Materials 17: 2939–2942, 2005.
16. Hahn, M.S. L.J. Taite, J.J. Moon, M.C. Rowland, K.A. Ruffino, and J.L. West. Photolithographic patterning of polyethylene glycol hydrogels. Biomaterials 27: 2519–2534, 2006.
17. Hahn, M.S., Miller J.S., and J.L. West. Three-dimensional biochemical and biomechanical patterning of hydrogels for guiding cell behavior. Advanced Materials 18: 2679–2684, 2006.
18. Luo, N., A.T. Metters, B. Hutchison, C.N. Bowman, and K.S. Anseth. A methacrylated photoiniferter as a chemical basis for microlithography: micropatterning based on photografting polymerization. Macromolecules 36: 6739–6745, 2003.
19. Starly, B., R. Chang, and W. Sun. UV-Photolithography fabrication of poly-ethylene glycol hydrogels encapsulated with hepatocytes. Proceedings of the 17th Annual Solid Freeform Fabrication Symposium, University of Texas at Austin, August 1–3, 2006, pp 102–110.
20. Han, L.H., G. Mapili, S. Chen, and K. Roy. Freeform fabrication of biological scaffolds by projection photopolymerization. Proceedings of the 18th Annual Solid Freeform Fabrication Symposium, University of Texas at Austin, August 1–3, 2007, pp 450–457.
21. Han, L.H., G. Mapili, S. Chen, and K. Roy. Projection Microfabrication of three-dimensional scaffolds for tissue engineering. Transactions of ASME: Journal of Manufacturing Science and Engineering 130: 021005-1–021005-4, 2008.

22. Choi, J.W., R.B. Wicker, S.H. Cho, C.S. Ha, S.H. Lee. Cure depth control for complex 3D microstructure fabrication in dynamic mask projection microstereolithography. Rapid Prototyping Journal 15(1): 59–70, 2009.
23. Choi, J.W., R.B. Wicker, S.H. Lee, K.H.Choi, C.S. Ha, and I. Chung. Fabrication of 3D biocompatible/biodegradable micro-scaffolds using dynamic mask projection microstereolithography. Journal of Materials Processing Technology, 209(15–16): 5494–5503, 2009.
24. Comeau, B.M., Umar, Y., Gonsalves, K.E., and Henderson, C.L. New materials and methods for hierarchically structured tissue scaffolds. Materials Research Society Symposium Proceedings, 845(A): AA4.4.1–6, 2005.
25. Bens, A.T., C. Tille, B. Leukers, G. Bermes, E. Emons, R. Sobe, A. Pansky, B. Roitzheim, M. Schulze, E. Tobiasch, and H. Seitz. Mechanical properties and bioanalytical characterization for a novel non-toxic flexible photopolymer formulation class. Proceedings of the 16th Annual Solid Freeform Fabrication Symposium, University of Texas at Austin, August 1–3, 2005, pp 162–173.
26. Cooke, M.N., J.P. Fisher, D. Dean, Rimnac, C. and A.G. Mikos. Use of stereolithography to manufacture critical-sized 3D biodegradable scaffolds for bone in growth. Materials Research Part B: Applied Biomaterials 64B: 65–69, 2002.
27. Lee, K.W., S. Wang, B.C. Fox, E.L. Ritman, M.J. Yaszemski, and L. Lu. Poly(propylene fumarate) bone tissue engineering scaffold fabrication using stereolithography: effects of resin formulations and laser parameters. Biomacromolecules 8: 1077–1084, 2007.
28. Popov, V.K., A.V. Evseev, A.L. Ivanov, V.V. Roginski, A.I. Volozhin, and S.M. Howdle. Laser stereolithography and super critical fluid processing for custom-designed implant fabrication. Journal of Materials Science: Materials in Medicine 15: 123–128, 2004.
29. Barry, J.J.A., A.V. Evseev, M.A. Markov, C.E. Upton, C.A. Scotchford, V.K. Popov, and S.M. Howdle. In vitro study of hydroxyapatite-based photocurable polymer composites prepared by laser stereolithography and supercritical fluid extraction. Acta Biomaterialia 4(6): 1603–1610, 2008.
30. Dhariwala, B., Hunt, E., and Boland, T. Rapid prototyping of tissue engineering constructs, using photopolymerizable hydrogels and stereolithography. Tissue Engineering 9(10): 1316–1322, 2004.
31. Arcaute, K., L. Ochoa, F. Medina, C. Elkins, B. Mann, and Wicker, R. Three-dimensional PEG hydrogel construct fabrication using stereolithography. Materials Research Society Symposium Proceedings, 874:L5.5.1–L5.5.7, 2005.
32. Arcaute, K., L. Ochoa, B. Mann, and R. Wicker. Hydrogels in stereolithography. Proceedings of the 16th Annual Solid Freeform Fabrication Symposium, University of Texas at Austin, August 1–3, 2005.
33. Arcaute, K., L. Ochoa, B.K. Mann, and Wicker, R.B. Stereolithography of PEG hydrogel multi-lumen nerve regeneration conduits. ASME IMECE2005-81436 American Society of Mechanical Engineers International Mechanical Engineering Congress and Exposition, November 5–11, Orlando, Florida, 2005.
34. Wohlers, T., "Wohlers Report 2004: Rapid Prototyping, Tooling and Manufacturing, State of the Industry," Wohlers Associates, Annual Worldwide Progress Report, 2004.
35. Sandoval, J.H., L. Ochoa, A. Hernandez, K.F. Soto, L.E. Murr, R.B. Wicker. Nanotailoring stereolithography resins for unique applications using carbon nanotubes. Proceedings of the 16th Annual Solid Freeform Fabrication Symposium, University of Texas at Austin, August 1–3, 2005.
36. Inamdar, A., M. Magana, F. Medina, Y. Grajeda, and R. Wicker. Development of an automated multiple material stereolithography machine. Proceedings of the 17th Annual Solid Freeform Fabrication Symposium, University of Texas at Austin, August 14–16, 2006.
37. Jacobs, P.F., Fundamental processes. In: Rapid Prototyping and Manufacturing: Fundamentals of Stereolithography, edited by P.F. Jacobs and D.T. Reid. Dearborn, Michigan: Society of Manufacturing Engineers, 1992, pp. 79–110.
38. Lee, I.H. and D.W. Cho. Micro-stereolithography photopolymer solidification patterns for various laser beam exposure conditions. International Journal of Advanced Manufacturing Technology 22: 410–416, 2003.

39. Lee, J.H., R.K. Prud'homme, and I.A. Aksay. Cure depth in photopolymerization: experiments and theory. Journal of Material Research 16(2): 3536–3544, 2001.
40. Jacobs, P.F. "Diagnostic testing." In: Rapid Prototyping and Manufacturing: Fundamentals of Stereolithography, edited by P.F. Jacobs and D.T. Reid. Dearborn, Michigan: Society of Manufacturing Engineers, 1992, pp. 249–285.
41. 3D Systems. SLA-190/250 Windowpane™ Building Procedure. In: 3D Systems Accumax™ Toolkite User Guide. Valencia, California: 3D Systems, 1993.
42. DSM Somos®. Method 2: Determination of depth of penetration of photopolymer by a laser beam scan. DSM Somos® Revision 1, pp 1–4.
43. Bryant, S.J. and K.S. Anseth. The effect of scaffold thickness on tissue engineered cartilage in photocrosslinked poly(ethylene oxide) hydrogels. Biomaterials 22: 619–626, 2001.
44. Bryant, S.J., K.S. Anseth, D.A. Lee, and D.L. Bader. Crosslinking density influences the morphology of chondrocytes photoencapsulated in PEG hydrogels during the application of compressive strain. Journal of Orthopaedic Research 22: 1143–1149, 2004.
45. Burdick, J.A. and K.S. Anseth. Photoencapsulation of osteoblasts in injectable RGD-modified PEG hydrogels for bone tissue engineering. Biomaterials 23: 4315–4323, 2002.
46. Williams, C.G., T.K. Kim, A. Taboas, A. Malik, P. Manson, and J. Elisseeff. In vitro chondrogenesis of bone marrow derived mesenchymal stem cells in a photopolymerizing hydrogel. Tissue Engineering 9(4):679–688, 2003.
47. Gunn, J.W., S.D. Turner, and B.K. Mann. Adhesive and mechanical properties of hydrogels influence neurite extension. Journal of Biomedical Materials Research, 72A (1):91–97, 2005.
48. Mann, B.K., A.S. Gobin, A.T. Tsai, R.H. Schmedlen, and J.L. West. Smooth muscle cell growth in photopolymerized hydrogels with cell adhesive and proteolytically degradable domains: synthetic ECM analogs for tissue engineering. Biomaterials 22: 3045–3051, 2001.
49. Mann, B.K., R.H. Schmedlen, and J.L. West. Tethered-TGF-β increases extracellular matrix production of vascular smooth muscle cells in peptide-modified scaffolds. Biomaterials, 22:439–44, 2001.
50. Mann, B.K. and J.L. West. Cell adhesion peptides alter smooth muscle cell adhesion, proliferation, migration, and matrix protein synthesis on modified surfaces and in polymer scaffolds. Journal of Biomedical Materials Research, 60:86–93, 2002.
51. Sawhney, A.S., C.P. Pathak, and J.A. Hubbell. Bioerodible hydrogels based on photopolymerized poly(ethylene glycol)-co-poly(alpha-hydroxy acid) diacrylate macromers. Macromolecules 26:581–587, 1993.
52. Zalispky, S. and J.M. Harris. "Introduction to chemistry and biological applications of poly (ethylene glycol)," Chapter 1. In: Poly(ethylene glycol) Chemistry and Biological Applications, edited by S. Zalispky and J.M. Harris. Washington, DC: American Chemical Society Series 680, 1997, pp. 1–13.
53. Nguyen, K.T. and J.L. West. Photopolymerizable hydrogels for tissue engineering applications. Biomaterials 23: 4307-4314, 2002.
54. Arcaute, K., B.K. Mann, and R.B. Wicker. Stereolithography of three-dimensional bioactive poly(ethylene glycol) constructs with encapsulated cells. Annals of Biomedical Engineering 34(9): 1429–1441, 2006.
55. Fisher, J.P., J.W. Vehof, D. Dean, J.P. Van der Waerden, T.A. Holland, A.G. Mikos, and J.A. Jansen. Soft and hard tissue response to photocrosslinked poly(propylene fumarate) scaffolds in a rabbit model. Journal of Biomedical Materials Research 59(3): 547–556, 2002.
56. Leach, J.B., K.A. Bivens, C.W. Patrick, and C.E. Schmidt. Photocrosslinked hyaluronic acid hydrogels: natural, biodegradable tissue engineering scaffolds. Biotechnology and Bioengineering 82(5): 578–589, 2003.
57. Burdick, J.A., C. Chung, X. Jia, M.A. Randolph and R. Langer. Controlled degradation and mechanical behavior of photopolymerized hyaluronic acid networks. Biomacromolecules 6: 386–391, 2005.
58. Masters, K.S., D.N. Shah, L.A. Leinwand, and K.S. Anseth. Crosslinked hyaluronan scaffolds as biologically active carriers for valvular interstitial cells. Biomaterials 26: 2517–2525, 2005.

59. Bryant, S.J., C.R. Nuttelman, and K.S. Anseth. Cytocompatibility of UV and visible light photoinitiating systems on cultured NIH/3T3 fibroblasts in vitro. Journal of Biomaterials Science, Polymer Edition, 11(5): 439–457, 2000.
60. Williams, C.G., A.N. Malik, T.K. Kim, P.N. Manson, and J.H Elisseeff. Variable cytocompatibility of six cell lines with photoinitiators used for polymerizing hydrogels and cell encapsulation. Biomaterials, 26: 1211–1218, 2005.
61. Ciba Specialty Chemicals, Coatings Effects Segment. Ciba® Irgacure® 2959 Technical Data Sheet. Edition 2 4 98. Ciba Specialty Chemicals.
62. McCurdy, K.G. and K.J. Laidler. Rates of polymerization of acrylates and methacrylates in emulsion systems. Canadian Journal of Chemistry 42: 825–829, 1964.
63. Klumperman, B. Pecularities in Atom Transfer Radical Copolimerization. Available online at http://academic.sun.ac.za/UNESCO/PolymerED2002/Contributions/Klumperman.pdf. Accessed on 05/2008.
64. Jacobs, P.F. "Introduction to part building." In: Rapid Prototyping and Manufacturing: Fundamentals of Stereolithography, edited by P.F. Jacobs and D.T. Reid. Dearborn, Michigan: Society of Manufacturing Engineers, 1992, pp. 171–194.
65. Gayet, J.C., and G. Fortier. "New bioatificial hydrogels: characterization and physical properties." In: Hydrogels and Biodegradable Polymers for Bioapplications, edited by R.M. Ottenbrite, S.J. Huang, and K. Park. Washington, D.C. American Chemical Society, 1996, pp. 17–24.
66. Jiankang, H, L. Dichen, L. Yaxiong, Y. Bo, L. Bingheng, and L. Qin. Fabrication and characterization of chitosan/gelatin porous scaffolds with predefined internal microstructures. Polymer, 48: 4578–4588, 2007.
67. Arcaute, K., N. Zuverza, B.K. Mann, and R.B. Wicker. Multi-material stereolithography: spatially-controlled bioactive poly(ethylene glycol) scaffolds for tissue engineering. Proceedings of the 2007 Solid Freeform Fabrication Symposium, University of Texas at Austin, August 6-8, 2007.
68. Arcaute, K. Stereolithography of Poly(Ethylene Glycol) Hydrogels with Application in Tissue Engineering as Peripheral Nerve Regeneration Scaffolds. Ph.D. Dissertation. The University of Texas at El Paso. December, 2008.
69. Arcaute, K, B.K. Mann, and R.B. Wicker. Stereolithography of spatially-controlled multi-material bioactive poly(ethylene glycol) scaffolds. Acta Biomaterialia, 6: 1047–1054, 2010.

Index

A
Active microfluidic components, 105–106
Additive manufacturing technology
classification, 3
information chain
adaptive slicing, 5–6
STL file representation, 4–5
tessellated representation, 5
Aluminium
component fabrication
characterization, 150
photocurable suspensions, 149
raw materials, 149
suspension property
dispersion reactivity, 151–153
rheological behaviour, 150–151
Amine hydrogen equivalent weight (AHEW), 66–67
Analytical irradiance model
arbitrary point
exposure term, 8–9
SL irradiance, 7–8
laser and irradiance modeling, 4
laser beam assumptions
extended model, 2
standard model, 1
refraction and absorption effects, 5–7
spherical coordinates, 5
standard cure model, 3–4
Arrhenius relation, 151

B
Beer–Lambert law, 5, 6, 83
Biological hydrogels, 163
Bioplotter, 176–177
Biopolymers
PEG (*see* Photocrosslinkable poly (ethylene glycol) (PEG))
PPF and DEF, 301–302

C
Cell encapsulation and viability, 326, 327
Ceramics
component fabrication, 145–149
curing depth, 24
particle size and light scattering effects, 24–25
suspensions, 26
turbine wheel, 24, 25
Computer-aided design (CAD)
haptic, 2
human interface design platform, 2
nD modelling, 1–2
Computer-aided engineering (CAE), 3
Computer-aided manufacturing (CAM), 3
Conventional stereolithography
definition, 12
direct or laser writing method, 12–15
mask-based writing systems, 15
resin compositions, 15–17
Cross-linking polymeric materials. *See also* Photocrosslinkable poly (ethylene glycol)
3D stability, 118
properties, 119
schematic representation, 119
translational mobility, 118
Cure depth, 13
Cure simulation model
DOC threshold model, 22–24
kinetic model
acrylate resin, photopolymerization, 18–19
boundary conditions, 20
dark reaction, 19
mass and energy balance equations, 19
reaction rates, 19
photopolymerization, 18

P.J. Bártolo (ed.), *Stereolithography: Materials, Processes and Applications*,
DOI 10.1007/978-0-387-92904-0, © Springer Science+Business Media, LLC 2011

Cure simulation model (*cont.*)
simulation domain and parameters
basic laser drawing patterns, 22
ethoxylated (4) pentaerythritol tetraacrylate, 20–21
FEMLAB software, 20
material properties, 21
single cured line
monomer conversion, 24–25
radical concentration, 24
transient response, 24
transient temperature profile, 25, 26
Curing reactions
cascade theory, 213
Flory-Stockmayer theory, 213
glass transition temperature and conversion
Couchman equation, 211
DiBenedetto equation, 210, 211
modified Couchman equation, 211
kinetic models
mechanistic models, 214
phenomenological models, 214
mechanism
exothermic polymerisation, 10
gelation, 11
generic representation, cure process, 10, 11
kinetics of, 12
vitrification, 11–12
network formation, 212–213
percolation models, 213
statistical models, 213
thermal-kinetic model (*see* Thermal-kinetic model)

D
3D case polymers
methodology, schematic representation, 127, 128
monomers and initiators, 128–130
oxetane chemistry, 131
post curing, 130
shrinkage, 130
3D fabrication, PEG-based solutions
build files
border vectors, 307
computer-aided design software, 306–307
crosslinked patterns, 308–309
3DLightyearT, 307
geometric accuracy, 307–308
hatch and fill vectors, 307
3D structures
complex 3D scaffolds, 316–317
multi-lumen PEG-based nerve regeneration conduits, 317, 318
interlayer bonding
gel thickness, 312
geometric accuracy, 313
layer thickness, 313–314
self-aligning mini-vat setup, 312
two-layer multi-lumen structures, 312
layer thickness
laser exposures and hatch spacing, 309, 310
laser scan velocity, 309–310
layer thickness, 310–311
multi-material fabrication
CAD and build files, 319–320
complex 3D structure, 322–323
2D multi-material scaffolds, 320–322
fabricated NGC, 322
microenvironment control, 318–319
self-aligning mini-vat setup, 319
swelling behavior
bound and adsorbed water, 316
dimensional swelling factor, 314–315
equilibrium, 314
free or bulk water, 316
phosphate buffered saline solution, 314
swollen dimensions, 315
Direct mold production, 245
Dispersion reactivity, aluminium suspensions
cure depth and width *vs.* ln(DE), 152, 153
density of energy, 151, 152
polymerization depth, 152
3D nonlinear processing, photonic crystals
absorption spectrum, 274
nonlinearity, 275
nonlinear optical silica sol-gel materials, 274
optical microscope image, 274
semiconductor, 273
SEM images, 274–275
3D photopatterning techniques, 299
Dynamic mask
digital micromirror device, 94–95
liquid crystal display, 93–94

E
Electrochemical fabrication (EFAB) process, 99
Epoxy/curing agent ratio
optimum
enthalpy reaction, 70, 71
kinetic effect, temperature, 70, 72
stoichiometric
AHEW, 66–67
phr (of amine) calculation, 67

F
Femtosecond laser system, 23, 290, 291
Flory–Stockmayer theory, 213
Free radical polymerization
  acrylates and methacrylates, 121
  benzoyl peroxide, photolysis, 120
  cationic systems, 122–124
Fused deposition modeling (FDM) process, 299
  advantage, 134
  crystallisation, 135
  scaffold development, 134
  schematic representation, 133
  semi-crystalline polymers, 134, 135

G
Gelation
  macroscopic, 11
  molecular, 11
Gelcasting process
  drying and debinding, 178–179
  formulation, SiC ceramics, 179
  processing steps, 177, 178
  RP-wax molds, sample parts, 179, 180
Green parts fabrication, 153–154

H
Haptic CAD, 2
Herbert's apparatus, 43–45
Hull's process, 43, 45–47
Human cartilage, 163
Hydrogels, 28–29. *See also* Polymerizable hydrogels

I
Indirect mold production, 244–245
Infrared laser stereolithography, 17–19
  laser beam characteristics
    focal spot size, 59
    laser power, 58–59
    spatial mode, 59
    Stimulated Emission, 58
    temporal mode, 59
    wavelength, 59
  laser curing chemistry, 60–61
  laser parameter and composition, 77
  physical theoretical model
    dwell time, 63
    energy flow description, 62
    finite elements method, 63
    Gaussian profile, 64
    optical characterization, 64–65
    thermal characterization, 65–77
  plastic models, 57
  system and basic products, 76
  thermo-sensitive epoxy resins
    chemical structure, 62
    curing reaction, 61–62
    kinetic characterization, 62
    silica amount, 62
Injection molding applications
  gating arrangement, 247
  glass filled materials, 250
  glass transition temperature, 246, 248
  large SL molding tool, 251
  lower settings, 248
  mold design considerations, 246–247
  molded part properties
    crystalline polymers, 252
    heat transfer, 250
    mechanical properties, 252
    non-crystalline material parts, 252
    process modifications, 252–253
    thermal conductivity, 251
  molding insert, 243, 244
  mold production
    direct production, 245
    indirect production, 244–245
    stereolithography, 247–248
  PA 66, 248, 251
  PEEK molding, 249
  polymers, 250
  prototyping, 245–246
  recommendations, 248
  SL mold production, 247–248
  thermal conductivity, 246, 249
Integral microstereolithography process
  apparatus, schematic diagram, 93
  bitmap files, 93
  dynamic mask
    Digital Micromirror Device (DMD™), 94–95
    liquid crystal display (LCD), 93–94

K
Kinetic models, curing reactions
  mechanistic models, 214
  phenomenological models
    function of conversion, 214
    maximum and onset conversion, 216
    *n*th order kinetic model, 214
    reaction rate, 214, 215
    two-step plateau kinetic model, 216

L

Laser
  beam characteristics
    focal spot size, 59
    laser power, 58–59
    spatial mode, 59
    stimulated emission, 58
    temporal mode, 59
    wavelength, 59
  curing, 60–61
Laser fabrication. *See* Infrared laser stereolithography
Layer by layer additive process, 142
Light–polymer interaction
  absorption conditions, 8
  Franck-Condon principle, 8
  initiators, 7
  Jablonski energy diagram, 8, 9
  photochemical process, 9–10
  photophysical process, 8
  radiative mechanisms, 9
  sequential bi-photonic system, 10
Liquid crystal display (LCD), 93–94
Liquid crystalline (LC) polymers, 23–24

M

Mesh size, 162
Metal component property, 155–156
Metallic stereolithography, 26–27
Microfluidic components
  active, 105–106
  passive, 103–105
Microprosthesis, 2PP technique
  bioactive materials and prostheses, 281–282
  implantation process, 283, 284
  middle ear, 280, 282
  nanoindentation tests, 283, 284
  ossiculoplasty, 280–284
  total ossicular replacement prosthesis, 282–283
Microstereolithography
  applications
    active microfluidic components, 105–106
    complex geometry objects, 110–101
    microsystems, 102–103
    passive microfluidic components, 103–105
    small mechanical components, 101–102
  collective fabrication, optical fibers, 92
  composite materials, 98–99
  EFAB process, 99
  evolution, 82
  integral process, 92–95
  lateral resolution, stereolithography, 88–89
  polymerized layer thickness
    Beer-Lambert law, 83
    liquid/solid photo-transformation, 83
    reduction, 84–87
    temporal evolution, 83
    time-resolved IR spectroscopy, 83
    transmitted photon flux (F_T), 83
    vector-by-vector process, 85
  resolution, 81
  space-resolved photopolymerization, 82
  spatial forming technique, 100
  sub-micron process
    non-linear phenomena, 95
    single photon process, 97
    two-photon microstereolithography, 95–97
  surface techniques
    constrained, 89–90
    free, 90–92
  z-overcure error, 86–88
Möbius rule, 5
Molding. *See* Injection molding applications
Monte-Carlo simulation, 25–26
Multi-material fabrication, PEG-based solutions
  CAD and build files, 319–320
  complex 3D structure, 322–323
  2D multi-material scaffolds, 320–322
  fabricated NGC, 322
  microenvironment control, 318–319
  self-aligning mini-vat setup, 319

O

Optical characterization
  absorption depth, 64
  CO_2 laser infrared radiation, semi-liquid samples, 65
  infrared spectrum analyzer, Model IR–700, 64

P

Passive microfluidic components, 103–105
Photocrosslinkable poly(ethylene glycol) (PEG), 301–302
  commercial SL resins, 305
  3D fabrication (*see* 3D Fabrication, PEG-based solutions)
  gel thickness *vs*. energy dosage data, 304–305
  photoinitiator effects

concentration, 302–304
cytotoxicity, 304
solubility and biocompatibility, 304
polymer concentration, 305
reactivity, 306
stereolithography, 300–301
types, 300
in vitro cell studies, SL parameters
cell encapsulation, 326, 327
HDF cell viability, MTT cytotoxicity assay, 324
photoinitiators, cytotoxicity, 325
unfavorable cell growth and survival, 323–324
UV irradiation, toxic effect, 324–326
Photo-curing process. *See* Thermal-kinetic model
Photolithographic systems
ascending fabrication platform, 47, 48
basic approaches, 41–42
color stereolithography, 49
2D case polymers
bisazide cross-linking, 126
Bisphenol A system, 126–127
classifications, 125
Novolac system, 125
3D model fabrication, schematics, 43, 44
graphic object reproduction, 42
Herbert's apparatus, 43–45
Hull's process, 43, 45–47
laser lithography patents, 50, 51
photochemical machining process, 43
photo-mask systems
dynamic mask projection, 52
photo-fabrication system, 50, 51
positive direct-mask exposure process, 52, 53
SGC, 50
solgel transformable photopolymer, 50
Photo-mask systems
dynamic mask projection, 52
photo-fabrication system, 50, 51
positive direct-mask exposure process, 52, 53
SGC, 50
Photopolymerisation, free radicals
acrylates and methacrylates, 121
benzoyl peroxide, photolysis, 120
cationic systems, 122–124
ketones, type II photodegradation, 120
Plastic injection molding, 243. *See also* Injection molding applications
Polyether-ether-ketone (PEEK) molding, 249
Polyethylene glycol-X-diacrylate (PEG-X-DA), 164–166
Polymeric materials
cross-linking (*see* Cross-linking polymeric materials)
polymerisation
chain-growth systems, 117, 118
Nylon 6,6 formation, 117
polymer classification, 114–116
step-growth systems, 117, 118
styrene, 117
Polymerised line width, 14
Polymerizable hydrogels
biological
human cartilage, 163
tissue engineering scaffolds, 163
chemistry
acrylates, 171
camphorquinone/ethyl dimethylamino benzoate mechanism, 170
hydroquinone monomethylether function, 171
oxygen inhibition, 169
photopolymerization principle, 169
polyacrylic acid, degradation products, 172, 173
Type I initiator Irgacure 2959®, cleavage, 170
in vitro cytotoxicity, 174
gelcasting process
drying and debinding, 178–179
formulation, SiC ceramics, 179
processing steps, 177, 178
RP-wax molds, sample parts, 179, 180
mechanical testing
cartilage samples, 167, 168
compression testing, 164, 165
elastic modulus, 164
indentation moduli E_i, 166
PEG-X-DA, compression modulus, 164, 165
relaxation time, 167, 168
structure
basic building components, 162
bioplotter, 176–177
cell incorporation, scaffolds, 174
classification, 161
mesh size, 162
microporosity, 175
RP applications, 162
Polymerized layer thickness
Beer-Lambert law, 83
liquid/solid photo-transformation, 83

Polymerized layer thickness (*cont.*)
reduction
irradiation time, 85
using highly absorbing reactive media, 85–86
using neutral absorbers, 86, 87
temporal evolution, 83
time-resolved IR spectroscopy, 83
transmitted photon flux (F_T), 83
vector-by-vector process, 85
Polymers, 143–145
Polymer scaffolds
artificial 3D matrices, 285
biocompatibility, 287–288
CAD design, 285–286
designer scaffolds, 286
gap-junctions, 288
microcapillary network model, 287
porosity levels, 285
SEM image, 286–287
tissue formation mechanisms, 287
Projection microstereolithography. *See* Integral microstereolithography process
Prototypes
additive fabrication, 37
formative methods, 41
layered manufacturing techniques (LMT), 41
manufacturing industry, changes, 38
product development process, 39
simultaneous engineering methodology, 39
subtractive methods, 40

R
Radical polymerization. *See* Free radical polymerization
Rapid manufacturing techniques
biomedical applications, 136
description, 113
fused deposition modelling process
advantage, 134
crystallisation, 135
scaffold development, 134
schematic representation, 133
semi-crystalline polymers, 134, 135
photolithography, 2D case polymers, 124–127
photopolymerisation, 119–124
polymeric materials
cross-linking, 118–119
polymerisation, 114–118
selective laser sintering, 132–133
stereolithography, 3D case polymers, 127–131
Rapid Tooling (RT), 243
Resin composition
photoinitiators
cationic, 17
radical, 16–17
prepolymer, 17
Rheology, aluminium suspensions
Arrhenius relation, 151
Sisko model, 150
viscosity *vs.* shear rates, 150, 151

S
Selective laser sintering process, 132–133
Semi-crystalline polymers, 134, 135
Simpson's 1/3 rule, 8, 9
Simulation methods
cure simulation model (*see* Cure simulation model)
error sources
laser beam fluctuations, 17
post-processing methods, 16–17
resin curing reactions, 17
Monte-Carlo, 25–26
optimized process plan, 16
simulation algorithm, 10, 11
single scan simulation, 11–12
surface finish experiments, 14–15
ten-layer flat part, 12–14
ten-layer slanted surface, 15
Sintering, 154–155
Sisko model, 150
Solid free-form fabrication. *See* Rapid manufacturing techniques
Solid ground curing (SGC) technique, 50
Spatial forming technique, 100
Spiral photonic crystals, 2PP technique
bandstop position, 272–273
dielectric filling fraction value, 271–272
Stereo-thermal-lithography (μSTLG)
advantages, 20
initiation process, 21
irradiation process, 19, 20
multimaterial functionally graded components, 19
multi-vat system, 20, 21
STL file
adjacency rule, 5
ASCII format, structure, 3, 4
facet orientation rule, 4
types, 4
STLG-FEM code
first window-screen, 225
input data organisation, 223

material properties, 228
numerical implementation scheme, 224
pre-processor unit, 226
processor and post-processor units, 227
sub-programs, 225
unsaturated polyester resins (*see also* Unsaturated polyester (UP) resins, STLG-FEM code)
composition, 228
fractional conversion, 228–230
glass transition temperature, 228, 230
light intensity variation, 231, 232
polymerisation shrinkage *vs.* irradiation time, 230
temperature variation, 231, 232
Sub-micron process
non-linear phenomena, 95
single photon process, 97
two-photon microstereolithography, 95–97
Surface techniques
constrained, 89–90
free, 90–92
Swelling behavior, PEG-based solutions
bound and adsorbed water, 316
dimensional swelling factor, 314–315
equilibrium, 314
free or bulk water, 316
phosphate buffered saline solution, 314
swollen dimensions, 315

T
Thermal characterization
DSC, 67
dynamic kinetic analysis
activation energy calculation, 75
advantage, 73
cure thermal processing, 74
isoconversional methods, 74
rate equation, 74
epoxy/curing agent ratio
optimum, 69–71
stoichiometric, 66–67
isothermal kinetic analysis, 67–69
Thermal-kinetic model
boundary conditions, 217
computer implementation
numerical implementation scheme, 223, 224
STLG-FEM code (*see* STLG-FEM code)
developmental assumptions, 216
finite element discretisation
element formulation, 222–223
implementation, flow information, 220–221
numerical integration, 223
spatial discretisation, 221–222
temporal discretisation, 222
glass transition model, 220
initial temperature and fractional conversion, 217
kinetic model, 219
light intensity values
Beer-Lambert law, 217
Fresnel diffraction theory, 218
Gaussian intensity distribution, 218
light intensity values, Gaussian intensity distribution, 218
shrinkage model, 220
two-dimensional heat conduction, 217
Thermoplastics, 114, 131, 133, 143
Thermo-sensitive epoxy resins
chemical structure, 62
curing reaction, 61–62
kinetic characterization, 62
silica amount, 62
Tissue engineering (TE)
applications (*see* Photocrosslinkable poly(ethylene glycol) (PEG))
scaffolds, 2PP technique
artificial 3D matrices, 285
biocompatibility, 287–288
CAD design, 285–286
designer scaffolds, 286
gap-junctions, 288
microcapillary network model, 287
porosity levels, 285
SEM image, 286–287
tissue formation mechanisms, 287
Total ossicular replacement prosthesis (TORP), 282–283
Transdermal drug delivery, 2PP
bioinspired microneedle, 279
DIC-fluorescence microscopy, 280, 281
fluorescein-conjugated biotin, 280
geometry, 277–278
hollow needles, 276–277
hypodermic needles, 275
load *vs.* displacement curves, 278–279
microneedle arrays, 277–278
mosquito needle, 279–280
needle openings, 277–278
off-center microneedles, 278
Two-photon-initiated polymerisation, 22–23
Two-photon microstereolithography, 95–97

Two-photon polymerization (2PP) technique
3D photonic crystals fabrication, 264–265
microoptical elements, fabrication
element design, 258–259
microprism, slicing strategies, 259–260
surface roughness, 260
microprosthesis, fabrication
bioactive materials and prostheses, 281–282
implantation process, 283, 284
middle ear, 280, 282
nanoindentation tests, 283, 284
ossiculoplasty, 280–284
total ossicular replacement prosthesis, 282–283
photonic crystals, 265–266
characterization, 265–266
colloidal self-assembly, 262
3D nonlinear processing (*see* 3D Nonlinear processing, photonic crystals)
holographic recording, 262–264
periodicity, 260
photonic bandgaps, 261
semiconductor technology, 261
spiral photonic crystals, 264–265
structural resolution, 262–263
woodpile (*see* Woodpile photonic crystals fabrication, 2PP technique)
spiral photonic crystals
bandstop position, 272–273
dielectric filling fraction value, 271–272
tissue engineering scaffolds
artificial 3D matrices, 285
biocompatibility, 287–288
CAD design, 285–286
designer scaffolds, 286
gap-junctions, 288
microcapillary network model, 287
porosity levels, 285
SEM image, 286–287
tissue formation mechanisms, 287
transdermal drug delivery, microneedles
bioinspired microneedle, 279
DIC-fluorescence microscopy, 280, 281
fluorescein-conjugated biotin, 280
geometry, 277–278
hollow needles, 276–277
hypodermic needles, 275
load *vs.* displacement curves, 278–279
microneedle arrays, 277–278
mosquito needle, 279–280
needle openings, 277–278
off-center microneedles, 278
woodpile photonic crystals fabrication
dielectric filling fraction value, bandstop quality, 269
FTIR measurements, 266
inter-layer distance, 268
woodpile layers, bandstop quality, 270–271

U

Unsaturated polyester (UP) resins, STLG-FEM code
composition, 228
fractional conversion
irradiation times, 229
lateral accuracy index *vs.* irradiation time, 235
polymerisation shrinkage *vs.* irradiation time, 230
vs. sample position, 235
glass transition temperature, 228, 230
light intensity variation, 231, 232
polymerisation shrinkage *vs.* irradiation time, 230
temperature variation, 231, 232

W

Woodpile photonic crystals fabrication, 2PP technique
bandstop quality
dielectric filling fraction value, 269
woodpile layers, 270–271
FTIR measurements, 266–268
inter-layer distance, 268
interlayer distance optimisation, 268

Z

z-overcure error, 86–88